CAMBRIDGE TRACTS IN
MATHEMATICS

General Editors
B. BOLLOBAS, P. SARNAK, C. T. C. WALL

110 **Multivalent Functions**

T0275710

MULTIVALENT FUNCTIONS

Second edition

W. K. HAYMAN
Professor Emeritus
in the University of York

CAMBRIDGE
UNIVERSITY PRESS

CAMBRIDGE UNIVERSITY PRESS
Cambridge, New York, Melbourne, Madrid, Cape Town, Singapore, São Paulo

Cambridge University Press
The Edinburgh Building, Cambridge CB2 8RU, UK

Published in the United States of America by Cambridge University Press, New York

www.cambridge.org
Information on this title: www.cambridge.org/9780521460262

First published 1958
Second edition 1994
This digitally printed version 2008

A catalogue record for this publication is available from the British Library

ISBN 978-0-521-46026-2 hardback
ISBN 978-0-521-05767-7 paperback

Contents

Preface

Suppose that we are given a function $f(z)$ regular in the unit circle, and that the equation $f(z) = w$ has there

(a) never more than one solution;
(b) never more than p solutions; or
(c) at most p solutions in some average sense,

as w moves over the open plane. Then $f(z)$ is respectively univalent, p–valent or mean p–valent in $|z| < 1$.

It is the aim of this book to study what we can say about the growth of such functions $f(z)$ and, in particular, to obtain bounds for the modulus and coefficients of $f(z)$ and related quantities. Thus our aim is entirely quantitative in character.

The univalent functions represent the classical case of this theory, and we shall study them in Chapters 1, 7 and 8. By and large the methods of these chapters do not generalize to p–valent or mean p–valent functions. The latter two are studied in Chapters 2, 3, 5 and 6. The theory of symmetrization is developed in Chapter 4, both for its applications to Chapter 5 and for its intrinsic interest. This chapter could reasonably be read by itself. Chapter 7 could be read immediately after Chapter 1 by the student interested mainly in univalent functions. Otherwise the chapters depend on preceding work.

The majority of the material here collected has not, to my knowledge, appeared in book form before, and some of it is quite new. I am, however, extremely indebted to G. M. Golusin's tract [1947] for the contents of each of Chapters 1 and 7. Montel [1933] should also be mentioned, though his approach is rather different from mine. In a tract of this size it is not possible to be exhaustive. Thus I have not been able to find space for Schiffer's variational method, nor Jenkins' theory of modules, both of which have recently scored fine successes in the general field of this book, but I have tried to give references to these results as far as possible. The variational method is developed in Schaeffer and Spencer [1950] and Jenkins has covered his theory in a tract [1958] in the Ergebnisse Series.

The book does demand certain previous knowledge of function theory. Most of this would be contained in the undergraduate course as given, for instance, in Cambridge. When something further is required I have tried to give references to Ahlfors [1979, C. A.] or Titchmarsh [1939] where the results in question can be found. Apart from such references it has been my aim to give detailed proofs of all the theorems. In several cases there is a rather difficult key theorem, from which a number of applications follow fairly simply. In such a case, the reader may omit the proof of the basic theorem on a first reading, until convinced of its value by the application.

Finally I should like to thank all those persons who have helped me with this book, and in particular, Professor Kennedy, Dr. Smithies, Dr. Kövari, Dr. Clunie and Mr. Axtell for much patient criticism in the proof stage and earlier, and Mr. Barry, who kindly prepared the index for me. I am also grateful to the editors for allowing me to publish this book in the Cambridge Tracts series, and to the Cambridge University Press for their patience and helpfulness during all the stages of the preparation of this work.

W. K. H.

LONDON
January 1958

Preface to the Second Edition

In the last 35 years the subject of multivalent and particularly univalent functions has developed rapidly so that it seemed necessary to expand this book considerably. In my choice of new material I have tried to concentrate particularly on fundamental results that are not contained in the two most important books on the subject, that by Pommerenke [1975] and Duren [1983]. Chapter 6 has been devoted to Lucas' bounds for coefficient differences of mean p-valent functions and some new results by Leung and Dawei Shen. I have included Eke's regularity theorems for the maximum, means and coefficients which make it possible to extend results to areally mean p-valent functions which had previously been obtained only for the much narrower class of circumferentially mean p-valent functions.

The most important event in the area of this book has been the proof by de Branges [1985] of Bieberbach's conjecture. Chapter 8 has been added to deal with this subject. I have also included in Chapter 3 results by Clunie, Pommerenke and Baernstein on the coefficients of univalent functions which are bounded or have restricted maximum modulus. Here we have the unusual phenomenon of a type of theorem where the results for univalent functions are significantly stronger than those for mean univalent functions.

Now that Bieberbach's conjecture is proved an analogous conjecture due to Goodman [1948] for the coefficients of p-valent functions constitutes perhaps the most interesting challenge in the area. It is mentioned at the end of Chapter 5. The conjecture is plausible but looks like being extremely difficult and has been proved only in a few very special cases.

Thus there are two completely new chapters and the other chapters all contain new material. All the chapters now contain examples to test the reader's understanding of the material.

I would like to express my debt to Professor Baernstein for his advice on new material, to Professor Pommerenke and Professor Duren for their stimulating books and to May Ghali and Detta Dickinson for their painstaking work in providing Cambridge University Press and me with a camera-ready manuscript. Any remaining mistakes are entirely my own. David Tranah of CUP has been very helpful and supportive throughout.

W. K. H.

YORK
1993

1
Elementary bounds for univalent functions

1.0 Introduction A domain is an open connected set. A function $f(z)$ regular in a domain D is said to be *univalent* in D, if $w = f(z)$ assumes different values w for different z in D. In this case the equation $f(z) = w$ has at most one root in D for any complex w. Such functions map D (1,1) conformally onto a domain in the w plane.

In this chapter we shall obtain some classical results, which give limits for the growth of functions univalent in the unit disc $|z| < 1$. Most of the rest of this tract will aim at generalizing these theorems by proving corresponding results for p-valent functions, i.e. those for which the equation $f(z) = w$ has at most p roots in D, either for every complex w, or in some average sense as w moves over the plane.

If $f(z) = \sum_0^\infty a_n z^n$ is univalent in $|z| < 1$, then so are $f(z) - a_0$ and $(f(z) - a_0)/a_1$, since $a_1 = f'(0) \neq 0$. In fact if a_1 were zero, $f(z)$ would take all values sufficiently near $w = a_0$ at least twice. We thus study the normalized class $\mathfrak{S}$ of functions

$$w = f(z) = z + a_2 z^2 + \dots$$

univalent in $|z| < 1$.

The two equivalent basic results here are due to Bieberbach [1916] and state that, if $f(z) \in \mathfrak{S},$[†] $|a_2| \leq 2$, and that $f(z)$ assumes every value w such that $|w| < \frac{1}{4}$. This latter theorem had been previously proved with a smaller absolute constant by Koebe [1910]. The results of Bieberbach are best possible. We shall first prove them and then develop some of their main consequences.

1.1 Basic results We have

[†] This symbolic statement stands for '$f(z)$ belongs to the class $\mathfrak{S}$'.

1

Theorem 1.1 *Suppose that $f(z) \in \mathfrak{S}$. Then $|a_2| \leq 2$, with equality only for the Koebe functions*

$$f_\theta(z) = \frac{z}{(1 - ze^{i\theta})^2} = z + 2z^2 e^{i\theta} + 3z^3 e^{2i\theta} + \dots \qquad (1.1)$$

We need the following preliminary result:

Lemma 1.1 *Suppose that $w = f(z) = \sum_{n=-\infty}^{+\infty} a_n z^n$ is regular in a domain containing $|z| = r$, and that the image of $|z| = r$ by $f(z)$ is a simple closed curve $J(r)$, described once. Then the area $A(r)$ enclosed by $J(r)$ is $\pi \left| \sum_{n=-\infty}^{+\infty} n|a_n|^2 r^{2n} \right|$.*

We write $w = f(re^{i\theta}) = u(\theta) + iv(\theta)$, where

$$u(\theta) = \frac{1}{2} \sum_{-\infty}^{+\infty} [a_n e^{in\theta} + \bar{a}_n e^{-in\theta}] r^n,$$

$$v(\theta) = \frac{1}{2i} \sum_{-\infty}^{+\infty} [a_n e^{in\theta} - \bar{a}_n e^{-in\theta}] r^n.$$

Thus

$$
\begin{aligned}
A(r) &= \left| \int_0^{2\pi} u \frac{dv}{d\theta} d\theta \right| \\
&= \frac{1}{4} \left| \int_0^{2\pi} \left[\sum_{m=-\infty}^{+\infty} r^m (a_m e^{im\theta} + \bar{a}_m e^{-im\theta}) \right] \right. \\
&\qquad \left. \times \left[\sum_{n=-\infty}^{+\infty} n r^n (a_n e^{in\theta} + \bar{a}_n e^{-in\theta}) \right] d\theta \right| \\
&= \left| \frac{\pi}{2} \sum_{n=-\infty}^{+\infty} [a_n(-na_{-n} + nr^{2n}\bar{a}_n) + \bar{a}_n(nr^{2n}a_n - n\bar{a}_{-n})] \right| \\
&= \pi \left| \sum_{-\infty}^{+\infty} n|a_n|^2 r^{2n} \right|,
\end{aligned}
$$

since $\sum na_n a_{-n} = \sum n\bar{a}_n\bar{a}_{-n} = 0$, as we see on replacing n by $-n$ in the summation. Thus the lemma is proved.

Suppose now that

$$w = f(z) = z + a_2 z^2 + \dots \in \mathfrak{S}.$$

Then so does $F(z) = [f(z^2)]^{\frac{1}{2}} = z + \frac{1}{2}a_2 z^3 + \dots$ In fact $f(z^2)$ does not vanish except at $z = 0$, where it has a double zero, and if $F(z_1) = F(z_2)$, then $f(z_1^2) = f(z_2^2)$, and so $z_1^2 = z_2^2$, i.e. $z_1 = \pm z_2$. But $F(z)$ is an odd function, so that $z_1 = -z_2$ gives $F(z_1) = -F(z_2)$. Hence we must have $z_1 = z_2$. Also since $f(z^2)$ has only a single zero of order two, $F(z)$ is regular. Therefore $F(z)$ is univalent.

Next write

$$g(z) = \frac{1}{F(z)} = \frac{1}{z} - \frac{1}{2}a_2 z + \ldots = \frac{1}{z} + \sum_{n=1}^{\infty} b_n z^n.$$

Then $g(z)$ is univalent in $0 < |z| < 1$, and so the image of $|z| = r$ by $g(z)$ is a simple closed curve for $0 < r < 1$. Hence by Lemma 1.1

$$-\frac{1}{r^2} + \sum_{n=1}^{\infty} n|b_n|^2 r^{2n} = \pm \frac{A(r)}{\pi}$$

does not vanish for $0 < r < 1$. The left-hand side is clearly negative for small positive r and so for $0 < r < 1$. As $r \to 1$ we deduce that

$$\sum_{n=1}^{\infty} n|b_n|^2 \le 1.$$

Thus we have $|b_1| = \frac{1}{2}|a_2| \le 1$, and equality is possible only if $b_n = 0$ $(n > 1)$, and in this case

$$g(z) = \frac{1}{z} - ze^{i\theta}, \quad F(z) = \frac{z}{1 - z^2 e^{i\theta}}, \quad f(z) = \frac{z}{(1 - ze^{i\theta})^2}.$$

This proves Theorem 1.1.

We deduce immediately

Theorem 1.2 *Suppose that $f(z) \in \mathfrak{S}$ and that $f(z) \ne w$ in $|z| < 1$. Then $|w| \ge \frac{1}{4}$. Equality is possible only if $f(z)$ is given by (1.1) and $w = -\frac{1}{4}e^{-i\theta}$.*

Since $f(z) \ne w$

$$\frac{wf(z)}{w - f(z)} = z + \left(a_2 + \frac{1}{w}\right)z^2 + \ldots \in \mathfrak{S}.$$

Thus Theorem 1.1 gives

$$\left|a_2 + \frac{1}{w}\right| \le 2, \quad \left|\frac{1}{w}\right| \le 2 + |a_2| \le 4, \quad |w| \ge \frac{1}{4},$$

as required. Equality is possible only if $a_2 = 2e^{i\theta}$, $w^{-1} = -4e^{i\theta}$, and then $f(z)$ must be given by $f_\theta(z)$ in (1.1).

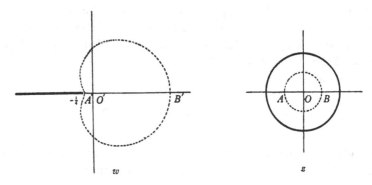

Fig. 1.

We note finally that the function

$$f_0(z) = \frac{z}{(1-z)^2} = \frac{1}{4}\left(\frac{1+z}{1-z}\right)^2 - \frac{1}{4}$$

maps $|z| < 1$ (1,1) conformally onto the w plane cut from $-\frac{1}{4}$ to $-\infty$ along the negative real axis. Thus $f_0(z) \in \mathfrak{S}$ and $f_0(z) \neq -\frac{1}{4}$ in $|z| < 1$. Hence the functions $f_\theta(z) = e^{-i\theta}f_0(ze^{i\theta})$ of (1.1) also belong to $\mathfrak{S}$ and $f_\theta(z) \neq -\frac{1}{4}e^{-i\theta}$. Thus Theorems 1.1 and 1.2 are best possible.

We shall see that these functions $f_\theta(z)$ are extreme in $\mathfrak{S}$ for a variety of other problems also.

In 1984 de Branges [1985] solved the outstanding problem in the theory, by proving the Bieberbach conjecture that $|a_n| \leq n$ holds for $f(z) \in \mathfrak{S}$ and $n > 1$ with equality only for $f(z) = f_\theta(z)$. We shall give the proof when $n = 3$, which is due to Löwner [1923] in Chapter 7. In Chapter 8 we shall prove de Branges' Theorem in full generality.

1.2 Elementary growth and distortion theorems   We can develop an interesting further group of inequalities as a direct consequence of Theorem 1.1.

Theorem 1.3   *Suppose that $f(z) \in \mathfrak{S}$. Then, for $|z| = r$ $(0 < r < 1)$ we have*[†]

$$\frac{r}{(1+r)^2} \leq |f(z)| \leq \frac{r}{(1-r)^2},\qquad(1.2)$$

[†] Bieberbach [1916], Gronwall [1916], Szegö [1928].

$$\frac{1-r}{(1+r)^3} \le |f'(z)| \le \frac{1+r}{(1-r)^3}, \qquad (1.3)$$

$$\frac{1-r}{r(1+r)} \le \left|\frac{f'(z)}{f(z)}\right| \le \frac{1+r}{r(1-r)}. \qquad (1.4)$$

Equality holds in all cases only for the functions $f_\theta(z)$ of (1.1).

We assume that $|z_0| < 1$, and set

$$\phi(z) = f\left(\frac{z_0 + z}{1 + \bar{z}_0 z}\right) = b_0 + b_1 z + b_2 z^2 + \dots \qquad (1.5)$$

Then clearly $\phi(z)$ is univalent in $|z| < 1$. Further

$$b_0 = f(z_0), \quad b_1 = \phi'(0) = (1 - |z_0|^2)f'(z_0),$$

$$b_2 = \tfrac{1}{2}\phi''(0) = \tfrac{1}{2}(1 - |z_0|^2)^2 f''(z_0) - \bar{z}_0(1 - |z_0|^2)f'(z_0).$$

We apply Theorem 1.1 to $(\phi(z) - b_0)/b_1$ in $\mathfrak{S}$ and obtain $|b_2| \le 2|b_1|$, i.e.

$$|f''(z_0)(1 - |z_0|^2)^2 - 2\bar{z}_0 f'(z_0)(1 - |z_0|^2)| \le 4(1 - |z_0|^2)|f'(z_0)|.$$

Writing $z_0 = \rho e^{i\theta}$ we deduce

$$\left|z_0 \frac{f''(z_0)}{f'(z_0)} - \frac{2\rho^2}{1 - \rho^2}\right| \le \frac{4\rho}{1 - \rho^2}. \qquad (1.6)$$

Since

$$\frac{\partial}{\partial\rho}\log|f'(\rho e^{i\theta})| = \Re e^{i\theta}\frac{f''(\rho e^{i\theta})}{f'(\rho e^{i\theta})},$$

we obtain at once

$$\frac{2\rho - 4}{1 - \rho^2} \le \frac{\partial}{\partial\rho}\log|f'(\rho e^{i\theta})| \le \frac{2\rho + 4}{1 - \rho^2}.$$

On integrating this from 0 to r with respect to ρ, we deduce (1.3).
We deduce immediately that

$$|f(re^{i\theta})| \le \int_0^r |f'(\rho e^{i\theta})|d\rho \le \int_0^r \frac{1+\rho}{(1-\rho)^3}d\rho = \frac{r}{(1-r)^2},$$

and this is the right-hand inequality in (1.2). To obtain the lower bound for $|f(re^{i\theta})|$, we assume without loss of generality, that $f(re^{i\theta}) = Re^{i\phi}$, where $R < \frac{1}{4}$, since otherwise there is nothing to prove. It then follows from Theorem 1.2 that the straight line segment λ from 0 to $Re^{i\phi}$ lies entirely in the image of $|z| < 1$ by $f(z)$. Hence λ corresponds to a path l in $|z| < 1$, which joins $z = 0$ to $re^{i\theta}$. Thus if $t = |z|$ we deduce from (1.3)

$$R = \int_\lambda |dw| = \int_l \left|\frac{dw}{dz}\right| |dz| \geq \int_l \frac{1-t}{(1+t)^3} dt = \frac{r}{(1+r)^2},$$

and this completes the proof of (1.2).

Finally, we apply (1.2) to $[\phi(z) - b_0]/b_1$, where $\phi(z)$ is defined by (1.5). This gives

$$|b_1|\frac{|z|}{(1+|z|)^2} \leq \left|f\left(\frac{z_0+z}{1+\bar{z}_0 z}\right) - f(z_0)\right| \leq |b_1|\frac{|z|}{(1-|z|)^2}.$$

Putting $z = -z_0$, $b_1 = (1 - |z_0|^2)f'(z_0)$, we deduce (1.4).

It is easily seen that the functions $f_\theta(z)$ of (1.1) yield equality in the right-hand inequalities of (1.2)–(1.4) when $z = re^{-i\theta}$, and in the left-hand inequalities if $z = -re^{-i\theta}$. On noting that by Theorem 1.1 equality is possible in (1.6) and hence in the subsequent inequalities only if $\phi(z)$ reduces to one of the functions $f_\theta(z)$, we easily see that no other functions can give equality in Theorem 1.3.

1.2.1   We develop now another proof of some of the results in Theorem 1.3, which is based on Theorem 1.2, rather than 1.1. Once Theorem 1.2 has been extended to more general classes of functions, the present proof will also generalize. To make this evident we introduce the following:

Definition   *Let $f(z) = z + a_2 z^2 + \ldots$ be regular in $|z| < 1$. We shall say that $f(z) \in \mathfrak{S}_0$ if, given any complex z_0, with $|z_0| < 1$, and any function $\omega(\zeta)$ univalent and satisfying $|\omega(\zeta)| < 1$, $\omega(\zeta) \neq z_0$ in $|\zeta| < 1$, we have for $\phi(\zeta) = f[\omega(\zeta)]$*

$$|\phi'(0)| \leq 4(|\phi(0)| + |f(z_0)|).$$

We note that $\mathfrak{S}$ is a subclass of $\mathfrak{S}_0$. For if $f(z) \in \mathfrak{S}$, $\phi(\zeta) = f[\omega(\zeta)]$ is univalent in $|\zeta| < 1$, and $\phi(\zeta) \neq f(z_0) = w_0$ say. We write $\phi(\zeta) = b_0 + b_1\zeta + \ldots$, and apply Theorem 1.2 to $(\phi(\zeta) - b_0)/b_1$ which belongs to $\mathfrak{S}$ and never takes the value $(w_0 - b_0)/b_1$. Thus

$$\left|\frac{w_0 - b_0}{b_1}\right| \geq \frac{1}{4}; \quad |b_1| = |\phi'(0)| \leq 4|w_0 - b_0| \leq 4[|f(z_0)| + |\phi(0)|],$$

and so $f(z) \in \mathfrak{S}_0$.

We shall see in Chapter 5 that the class $\mathfrak{S}_0$ is effectively a good deal larger than $\mathfrak{S}$, and so the results of Theorems 1.4 and 1.5 will apply to a significantly more general class of functions than the univalent ones.

Theorem 1.4   *Suppose that* $f(z) = z + a_2 z^2 + \ldots \in \mathfrak{S}_0$. *Then*

$$|a_2| \leq 2. \tag{1.7}$$

Further, we have for $|z| = r \,(0 < r < 1)$

$$\frac{r}{(1+r)^2} \leq |f(z)| \leq \frac{r}{(1-r)^2}, \tag{1.8}$$

$$|f'(z)| \leq \frac{1+r}{r(1-r)}|f(z)| \leq \frac{1+r}{(1-r)^3}. \tag{1.9}$$

Finally, the equation $f(z) = w$ *has exactly one root in* $|z| < 1$ *if* $|w| < \frac{1}{4}$.

In return for our greater generality we have lost only the left inequalities in (1.3) and (1.4). This is inevitable, since the derivatives of functions in $\mathfrak{S}_0$ may well vanish in $|z| < 1$.

To prove Theorem 1.4, put

$$\frac{z}{(1-z)^2} = Z = \frac{4d\zeta}{(1-\zeta)^2},$$

where $d = r/(1+r)^2$ for some fixed r satisfying $0 < r < 1$. Then $|z| < 1$ cut from $-r$ to -1 along the negative real axis is mapped (1,1) conformally onto the Z plane cut from $-d$ to $-\infty$ along the negative real axis and so onto $|\zeta| < 1$. Thus if we write $z = \omega(\zeta)$, $\phi(\zeta) = f[\omega(\zeta)]$, then $\omega(\zeta)$ is univalent, $\omega(\zeta) \neq -r$ in $|\zeta| < 1$, and so, since $f(z) \in \mathfrak{S}_0$, we have

$$|\phi'(0)| \leq 4\{|\phi(0)| + |f(-r)|\},$$

i.e.

$$4d|f'(0)| \leq 4|f(-r)|.$$

Since $f'(0) = 1$, this gives $|f(-r)| \geq d$, and on applying the argument to $e^{-i\theta} f(ze^{i\theta})$, which belongs to $\mathfrak{S}_0$ if $f(z) \in \mathfrak{S}_0$, we have the left inequality of (1.8).

It follows immediately that $f(z) \neq 0$ in $|z| < 1$ except at $z = 0$. Next it follows from Rouché's Theorem[†] that if $|w| < r(1+r)^{-2}$, $f(z)$ and $f(z) - w$ have an equal number of zeros in $|z| < r$, i.e. exactly one. Making r tend to 1, we deduce that, for $|w| < \frac{1}{4}$, $f(z) = w$ has exactly one root in $|z| < 1$.

Next choose θ so that $a_2 e^{i\theta} = -|a_2|$. Then as $r \to 0$,

$$|f(re^{i\theta})| = |r + a_2 e^{i\theta} r^2 + O(r^3)| = r - |a_2| r^2 + O(r^3),$$

[†] See e.g. Ahlfors [1979, hereafter called C.A., p. 153.]

and

$$|f(re^{i\theta})| \geq \frac{r}{(1+r)^2} = r - 2r^2 + O(r^3),$$

by the left inequality of (1.8). This gives $|a_2| \leq 2$.

It remains to prove the inequalities (1.9) and the right inequality of (1.8). We put

$$Z = \frac{z}{(1-z)^2} = k \left(\frac{1+\zeta}{1-\zeta} \right)^2, \text{ where } k = \frac{r}{(1-r)^2}.$$

Here r is a fixed positive number such that $0 < r < 1$. Then $|\zeta| < 1$ corresponds (1,1) conformally to the Z plane cut along the negative real axis and to $|z| < 1$ cut along the real axis from -1 to 0. We again write $z = \omega(\zeta)$, $\phi(\zeta) = f[\omega(\zeta)]$. Then $\omega(\zeta) \neq 0$ in $|\zeta| < 1$, and since $f(0) = 0$, $f(z) \in \mathfrak{S}_0$, we have

$$|\phi'(0)| = \frac{(1-r)^3}{1+r} 4k|f'(r)| \leq 4|\phi(0)| = 4|f(r)|.$$

Since $e^{-i\theta} f(ze^{i\theta}) \in \mathfrak{S}_0$ also, we deduce

$$|f'(re^{i\theta})| \leq \frac{1+r}{k(1-r)^3} |f(re^{i\theta})| = \frac{1+r}{r(1-r)} |f(re^{i\theta})|,$$

and this is the left inequality of (1.9). Hence

$$\frac{\partial}{\partial r} \log |f(re^{i\theta})| \leq \left| \frac{f'(re^{i\theta})}{f(re^{i\theta})} \right| \leq \frac{1+r}{r(1-r)}, \tag{1.10}$$

and integrating this from r_1 to r_2, where $0 < r_1 < r_2 < 1$, we deduce

$$\log \left| \frac{f(r_2 e^{i\theta})}{f(r_1 e^{i\theta})} \right| \leq \int_{r_1}^{r_2} \frac{(1+r)dr}{r(1-r)} = \log \left[\frac{(1-r_1)^2 r_2}{r_1(1-r_2)^2} \right],$$

or

$$\frac{(1-r_2)^2}{r_2} |f(r_2 e^{i\theta})| \leq \frac{(1-r_1)^2}{r_1} |f(r_1 e^{i\theta})|. \tag{1.11}$$

Making r_1 tend to 0 in this we deduce the right-hand inequalities of (1.8) and (1.9) with $r = r_2$. This completes the proof of Theorem 1.4.

From the point of view of later applications it is worth while to note the following consequences of (1.9):

Theorem 1.5   *Suppose that $f(z) = z + a_2 z^2 + \ldots \in \mathfrak{S}_0$ and set*

$$M(r, f) = \max_{|z|=r} |f(z)| \quad (0 < r < 1).$$

Then, unless $f(z) = f_\theta(z) = z(1 - ze^{i\theta})^{-2}$, $(1 - r)^2 r^{-1} M(r, f)$ decreases strictly with increasing r $(0 < r < 1)$, and so tends to α as $r \to 1$, where $0 \le \alpha < 1$. Hence the upper bounds for $|f(z)|, |f'(z)|$ given by (1.8) and (1.9) respectively are attained only by the functions $f_\theta(z)$.

To prove Theorem 1.5 note that equality can hold in (1.11) only if equality holds in both the inequalities of (1.10) for $r_1 < r < r_2$. This gives

$$\Re e^{i\theta} \frac{f'(re^{i\theta})}{f(re^{i\theta})} = \frac{1+r}{r(1-r)},$$

and so

$$\Im e^{i\theta} \frac{f'(re^{i\theta})}{f(re^{i\theta})} = 0 \quad (r_1 < r < r_2),$$

i.e.

$$z \frac{f'(z)}{f(z)} = \frac{1 + ze^{-i\theta}}{1 - ze^{-i\theta}}$$

for $z = re^{i\theta}$ $(r_1 < r < r_2)$ and so, by analytic continuation, throughout $|z| < 1$. In this case $f(z) = f_{-\theta}(z)$.

Otherwise strict inequality holds in (1.11) for $0 < r_1 < r_2 < 1$, and $0 \le \theta \le 2\pi$. Choose θ so that $|f(r_2 e^{i\theta})| = M(r_2, f)$. Then (1.11) gives

$$\frac{(1 - r_2)^2}{r_2} M(r_2, f) < \frac{(1 - r_1)^2}{r_1} |f(r_1 e^{i\theta})| \le \frac{(1 - r_1)^2}{r_1} M(r_1, f).$$

Hence, unless $f(z) = f_\theta(z), \psi(r) = (1 - r)^2 r^{-1} M(r, f)$ decreases strictly with increasing r $(0 < r < 1)$, and $\psi(r) \le 1$ by (1.9). Thus $\psi(r) < 1$ $(0 < r < 1)$, so that the upper bounds for $|f(z)|$ in (1.8) and for $|f'(z)|$ in (1.9) are not attained and $\lim_{r \to 1} \psi(r) = \alpha < 1$. This proves Theorem 1.5.

1.3 Means and coefficients   We referred above to de Branges' Theorem that $|a_n| \le n$ holds when $f(z) \in \mathfrak{S}$ and $n \ge 2$. We proceed to prove the simpler inequality $|a_n| < en$ due to Littlewood [1925].

Theorem 1.6   *Suppose that $f(z) = z + a_2 z^2 + \ldots \in \mathfrak{S}$. Then*

$$I_1(r, f) = \frac{1}{2\pi} \int_0^{2\pi} |f(re^{i\theta})| d\theta < \frac{r}{1 - r} \quad (0 < r < 1), \qquad (1.12)$$

and so

$$|a_n| < eI_1\left[\frac{n-1}{n}, f\right] < en \quad (n \ge 2). \qquad (1.13)$$

As we saw when proving Theorem 1.1,

$$\phi(z) = [f(z^2)]^{\frac{1}{2}} = z + b_3 z^3 + b_5 z^5 + \ldots \in \mathfrak{S},$$

and by Theorem 1.3, applied to $f(z)$, we have $|\phi(z)| \le r/(1 - r^2)$ for $|z| \le r$. We note that

$$
\frac{1}{2\pi} \int_0^{2\pi} |\phi'(re^{i\theta})|^2 d\theta
$$

$$
= \frac{1}{2\pi} \int_0^{2\pi} \phi'(re^{i\theta})\overline{\phi'(re^{i\theta})} d\theta
$$

$$
= \frac{1}{2\pi} \int_0^{2\pi} \left(\sum_{n=1}^{\infty} n b_n r^{n-1} e^{i(n-1)\theta} \right) \left(\sum_{m=1}^{\infty} m\bar{b}_m r^{m-1} e^{-i(m-1)\theta} \right) d\theta
$$

$$
= \sum_{n=1}^{\infty} n^2 |b_n|^2 r^{2n-2}.
$$

Thus

$$
\pi \sum_{n=1}^{\infty} n|b_n|^2 r^{2n} = \int_0^r \rho d\rho \int_0^{2\pi} |\phi'(\rho e^{i\theta})|^2 d\theta
$$

$$
= \{\text{area of transform of } |z| < r \text{ by } w = \phi(z)\}
$$

$$
< \pi \left(\frac{r}{1 - r^2} \right)^2. \tag{1.14}
$$

For since $\phi(z)$ is univalent, the area of the transform is at most πR^2, where R is the greatest distance of the transform from $w = 0$.[†]

Integrating term by term from 0 to r after division by r we obtain

$$
\sum_{n=1}^{\infty} |b_n|^2 r^{2n} < \frac{r^2}{1 - r^2}.
$$

But

$$
I_1(r^2, f) = \frac{1}{2\pi} \int_0^{2\pi} |f(r^2 e^{i\theta})| d\theta = \frac{1}{2\pi} \int_0^{2\pi} |\phi(re^{i\theta})|^2 d\theta
$$

$$
= \frac{1}{2\pi} \int_0^{2\pi} |\phi(re^{i\theta})\overline{\phi(re^{i\theta})}| d\theta
$$

$$
= \sum_{n=1}^{\infty} |b_n|^2 r^{2n}.
$$

[†] Equality is clearly excluded here.

Thus

$$I_1(r^2, f) < \frac{r^2}{1 - r^2}.$$

Writing r for r^2 we have (1.12). Again writing $r = 1 - 1/n$ we obtain

$$|a_n| = \frac{1}{2\pi} \left| \int_{|z|=r} \frac{f(z)dz}{z^{n+1}} \right| = \frac{1}{2\pi r^n} \left| \int_0^{2\pi} f(re^{i\theta}) e^{-in\theta} d\theta \right|$$

$$\leq \frac{1}{r^n} I_1(r, f) \leq \frac{1}{r^{n-1}(1-r)} = \left(1 + \frac{1}{n-1}\right)^{n-1} \quad n < en.$$

This gives (1.13) and completes the proof of Theorem 1.6.

Baernstein [1975] has proved that $I_1(r, f)$ and more general means of $f(z)$ attain their maximum value when $f(z) = f_\theta(z)$. In particular

$$I_1(r, f) \leq \frac{r}{1 - r^2}, \quad 0 < r < 1.$$

Baernstein's Theorem is beyond the scope of this book. (For an account see e.g. Hayman[1989, Theorem 9.8, p. 674.])

1.4 Convex univalent functions[†] It is quite simple to obtain the exact bounds for the coefficients of functions belonging to certain subclasses of $\mathfrak{S}$. In this section we consider functions mapping $|z| < 1$ onto a convex domain D. Such functions we shall call *convex univalent*. A domain D is said to be convex if given w_1 and w_2 in D the straight line segment joining w_1 to w_2 also lies in D. It is then easy to prove by induction that the centre of gravity $\frac{1}{n}(w_1 + w_2 + \ldots + w_n)$ of n points w_1 to w_n in D also lies in D.

Theorem 1.7 *Suppose that* $g(z) = \sum_{n=1}^{\infty} g_n z^n$ *is convex univalent and maps* $|z| < 1$ *onto* D. *Let* $w = h(z) = \sum_{n=1}^{\infty} h_n z^n$ *be regular in* $|z| < 1$ *and assume there only values* w *which lie in* D. *Then* $|h_n| \leq |g_1|$ *and in particular* $|g_n| \leq |g_1|$ *for* $n \geq 1$.

Consider

$$\psi(z) = g^{-1}[h(z)] = \frac{h_1}{g_1} z + \ldots$$

Then $\psi(z)$ is regular in $|z| < 1$, satisfies $|\psi(z)| < 1$ there and $\psi(0) = 0$. Hence Schwarz's Lemma[‡] yields $|h_1/g_1| \leq 1$, i.e. $|h_1| \leq |g_1|$.

[†] The results in this section are due to Löwner [1917].
[‡] See e.g. C.A. p. 135, Theorem 1.3.

Next let η_k $(1 \leq k \leq m)$ be the mth roots of unity and consider

$$H(z) = \frac{1}{m} \sum_{k=1}^{m} h(\eta_k z^{1/m}) = h_m z + h_{2m} z^2 + \ldots$$

instead of $h(z)$. Since D is convex and $h(z)$ assumes only values inside D so does $H(z)$, and we deduce $|h_m| \leq |g_1|$ $(m = 2, 3, \ldots)$. This proves Theorem 1.7.

We deduce that if $g(z) = z + g_2 z^2 + \ldots$ is convex univalent then $|g_n| \leq 1$ for $n \geq 1$. These inequalities are sharp for every n, as is shown by

$$w = g(z) = \frac{z}{1-z} = z + z^2 + \ldots,$$

which maps $|z| < 1$ onto the half-plane $\Re w > -\frac{1}{2}$.

We can also sharpen Theorems 1.2 and 1.3 for the functions $g(z)$. In each case the function $z/(1-z)$ is extremal.

Theorem 1.8 *Suppose that $w = f(z) = z + \ldots$ is convex univalent. Then $f(z)$ assumes every value in the disc $|w| < \frac{1}{2}$.*

Let D be the image of $|z| < 1$ by $f(z)$ and let $w_0 = re^{i\theta}$ be a point of smallest modulus lying outside D, By considering $-e^{-i\theta} f(-ze^{i\theta})$ instead of $f(z)$ if necessary, we may suppose that $w_0 = -r$. Then $\Re w > -r$ in D. For suppose, contrary to this, that D contains a point w_1, such that $\Re w_1 < -r$. By the convexity of D that segment, say L, of the line through w_0, w_1, which lies on the side of w_0 opposite to w_1, lies entirely outside D. But L contains points of modulus less than r, and this gives a contradiction.

Now

$$w = g(z) = \frac{2rz}{1-z} = 2rz + \ldots$$

maps $|z| < 1$ onto $D_0 = \{w : \Re w > -r\}$. Since $f(z) = z + \ldots$ assumes values in D_0 only, Theorem 1.7 applied with $f(z)$ instead of $h(z)$ gives $2r \geq 1$. This proves Theorem 1.8.

We have finally

Theorem 1.9   *Suppose that* $w = f(z) = z + \ldots$ *is convex univalent. Then we have for* $|z| = r \ (0 < r < 1)$

$$\frac{r}{1+r} \leq |f(z)| \leq \frac{r}{1-r},$$

$$\frac{1}{(1+r)^2} \leq |f'(z)| \leq \frac{1}{(1-r)^2},$$

$$\frac{1}{r(1+r)} \leq \left| \frac{f'(z)}{f(z)} \right| \leq \frac{1}{r(1-r)}.$$

All these inequalities are sharp with equality holding when $f(z) = z/(1-z)$ and $z = \pm r$. We omit the proofs which are exactly analogous to that of Theorem 1.3, starting with the inequality $|a_2| \leq 1$, which follows from Theorem 1.7, instead of $|a_2| \leq 2$.

1.5 Typically real functions[†]   Following Rogosinski we shall call $f(z)$ *typically real* if $f(z)$ is regular in $|z| < 1$ and $f(z)$ is real there if and only if z is real. We have

Theorem 1.10   *Suppose that* $f(z) = z + a_2 z^2 + \ldots$ *is typically real. In that case, and in particular if* $f(z) \in \mathfrak{S}$ *and has real coefficients, we have* $|a_n| \leq n \ (n = 2, 3, \ldots)$.

We write $f(z) = u + iv$ and suppose that $f(z)$ is typically real. Then $f(z)$ is real on the real axis and so has real coefficients. Thus

$$v(re^{i\theta}) = \sum_{1}^{\infty} a_n r^n \sin n\theta.$$

Also $v(re^{i\theta})$ has constant sign for $0 < \theta < \pi$ and so

$$
\begin{aligned}
|a_n r^n| &= \left| \frac{2}{\pi} \int_0^\pi v(re^{i\theta}) \sin n\theta d\theta \right| \\
&\leq \frac{2n}{\pi} \int_0^\pi |v(re^{i\theta}) \sin \theta| d\theta = nr \quad (0 < r < 1),
\end{aligned}
$$

since $|\sin n\theta| \leq n \sin \theta$. Letting r tend to 1, we deduce $|a_n| \leq n$.

If $f(z)$ has real coefficients then $f(\bar{z}) = \overline{f(z)}$. Thus if $f(z)$ has real coefficients and is real for some complex z_0, we have $f(z_0) = f(\bar{z}_0)$, which is impossible for a univalent $f(z)$. Thus if $f(z) \in \mathfrak{S}$ and has real coefficients, $f(z)$ is typically real and the above argument applies.

[†] Rogosinski [1931]. See also Dieudonné [1931] and Szász [1932] for the present proof.

1.6 Starlike univalent functions[†]   A domain D in the w plane is said to be starlike (or star-shaped) with respect to a fixed point O in D, if for any point P in D the straight line segment OP also lies in D. If $f(z) \in \mathfrak{S}$ and maps $|z| < 1$ onto a starlike domain with respect to $w = 0$, we shall call $f(z)$ starlike univalent. We have

Theorem 1.11   *If* $f(z) = z + a_2 z^2 + \dots$ *is starlike univalent then*

$$|a_n| \le n \quad (n = 2, 3, \dots).$$

Let G be the map of $|z| < 1, G_r$ the map of $|z| < r$, by $f(z)$. We show first that G_r is starlike with respect to $w = 0$ for $0 < r < 1$.

In fact if $w \in G$, then $tw \in G$ for $0 < t < 1$, and so the function

$$\psi(z) = f^{-1}[tf(z)]$$

is regular in $|z| < 1$ and satisfies $|\psi(z)| < 1$ there and $\psi(0) = 0$. Thus we have from Schwarz's Lemma $|\psi(z)| \le |z| < 1$.

Suppose now that $w_1 \in G_r$. Then $w_1 = f(z_1)$, where $|z_1| < r$. Hence

$$|f^{-1}(tw_1)| = |\psi(z_1)| \le |z_1| < r.$$

Thus $tw_1 = f(z_2)$ with $|z_2| < r$ and so $tw_1 \in G_r$ for $0 < t < 1$. Hence G_r is starlike.

The boundary of G_r consists of the curve $w = f(re^{i\theta})$ ($0 \le \theta \le 2\pi$). Since G_r is starlike the radius vector from $w = 0$ to $f(re^{i\theta})$ lies in G_r and so arg $f(re^{i\theta})$ increases with θ. Thus

$$\frac{\partial}{\partial \theta} \arg f(re^{i\theta}) = \Re \left[\frac{re^{i\theta} f'(re^{i\theta})}{f(re^{i\theta})} \right] \ge 0 \ (0 < r < 1, \ 0 \le \theta \le 2\pi). \quad (1.15)$$

Write now

$$z \frac{f'(z)}{f(z)} = 1 + \sum_{m=1}^{\infty} \alpha_m z^m.$$

Then it follows from (1.15) that

$$w = h(z) = z \frac{f'(z)}{f(z)} - 1 = \sum_{m=1}^{\infty} \alpha_m z^m$$

takes values lying entirely in the convex domain

$$D_0 = \{ w : \Re w > -1 \}.$$

[†] Nevanlinna [1921]. For a generalization of Theorem 1.11 to functions mapping $|z| < 1$ onto a more general class of domains introduced by Kaplan [1953] see Reade [1955].

Also

$$w = g(z) = \frac{2z}{1-z} = 2z + 2z^2 + \ldots$$

maps $|z| < 1$ onto D_0. Hence by Theorem 1.7 $|\alpha_m| \leq 2$ $(m = 1, 2, \ldots)$. Again we have

$$\left(\sum_{n=1}^{\infty} a_n z^n \right) \left(1 + \sum_{m=1}^{\infty} a_m z^m \right) = \sum_{1}^{\infty} n a_n z^n,$$

and equating coefficients we deduce

$$n a_n = a_n + \alpha_1 a_{n-1} + \ldots + \alpha_{n-1}.$$

Thus

$$(n-1)|a_n| = |\alpha_1 a_{n-1} + \ldots + \alpha_{n-1}| \leq 2(1 + |a_2| + \ldots + |a_{n-1}|).$$

If we assume that $|a_k| \leq k$ $(k = 1, 2, \ldots, n-1)$, we deduce

$$(n-1)|a_n| \leq n(n-1),$$

and the proof of Theorem 1.11 by induction is complete.

We remark finally that the function

$$\frac{z}{(1-z)^2} = z + 2z^2 + 3z^2 + \ldots$$

is both starlike univalent and typically real, so that the inequalities of Theorems 1.10 and 1.11 are sharp.

1.7 Asymptotic behaviour of the coefficients   While we defer the full proof of de Branges' Theorem to Chapter 8, we can obtain an asymptotic form of the result at this stage [Hayman 1955].

Theorem 1.12   *Suppose that $f(z) = z + a_2 z^2 + \ldots \mathfrak{S}$. Then*

$$\frac{|a_n|}{n} \to \alpha \quad as \quad n \to \infty, \tag{1.16}$$

where α is the constant of Theorem 1.5. In particular $|a_n| \leq n$ for $n > n_0(f)$.

Our proof of Theorem 1.12 is elementary but rather lengthy. Extensions covering more general functions are given in Chapters 2 and 5. They use similar arguments. For another proof of Theorem 1.12 see Milin [1970] and Duren [1983, p. 165]. An argument of Pommerenke [1975, p. 76] based on an inequality of Fitzgerald [1972] is shorter but only yields

$$\varlimsup_{n\to\infty} \frac{|a_n|}{n} < 1,$$

unless $f(z)$ is a Koebe function.

Our argument is a refinement of Littlewood's proof of Theorem 1.6. We first deal with the case $\alpha = 0$, which is relatively simple. If $\alpha > 0$ we use the Cauchy formula:

$$a_n = \frac{1}{2\pi\rho^n} \int_0^{2\pi} f(\rho e^{i\theta}) e^{-in\theta} d\theta. \tag{1.17}$$

We first prove the existence of a (unique) radius of greatest growth $\arg z = \theta_0$ such that

$$(1-r)^2 |f(re^{i\theta_0})| \to \alpha, \text{ as } r \to 1$$

while $\arg f(re^{i\theta_0})$ varies slowly. This allows us to establish a corresponding asymptotic behaviour on the major arc

$$\gamma' = |\arg z - \theta_0| \le K(1-r),$$

where K is a fixed constant. This argument works for the class $\mathfrak{S}_0$ of Theorem 1.5. We then use the univalence to show that the corresponding minor arc γ contributes relatively little to the integral in (1.17). In this way we obtain the required asymptotic formula for a_n. Theorem 1.12 shows that if, in the class $\mathfrak{S}$, $|a_n|$ has its maximal growth for large n, then $|a_n|$ behaves in a rather regular manner. Results of this nature occur frequently in function theory and are called *regularity theorems*.

1.7.1 The case $\alpha = 0$ We use the terminology and notation of Section 1.3 and suppose that $\alpha = 0$. Thus if $\varepsilon > 0$ there exists r_0, such that $r_0 < 1$, and

$$M(r, f) < \frac{\varepsilon^2 r}{(1-r)^2}, \quad r_0^2 < r < 1.$$

Recalling that $\phi(z) = \{f(z^2)\}^{\frac{1}{2}}$, we deduce that

$$M(r, \phi) < \frac{\varepsilon r}{1 - r^2}, \quad r_0 < r < 1.$$

Hence the inequality (1.14) can be sharpened to

$$\sum_{n=1}^{\infty} n|b_n|r^{2n} < \left(\frac{\varepsilon r}{1-r^2}\right)^2, \quad r_0 < r < 1.$$

Continuing as in the proof of Theorem 1.6 we obtain, as $r \to 1$

$$
\begin{aligned}
I_1(r^2, f) = \sum_1^\infty |b_n|^2 r^{2n} &= \int_0^r \left(\sum_1^\infty 2n|b_n|^2 \rho^{2n-1} d\rho \right) \\
&= \int_0^{r_0} + \int_{r_0}^r < \varepsilon^2 \int_0^r \frac{2\rho d\rho}{(1-\rho^2)^2} + O(1) \\
&= \frac{\varepsilon^2 r^2}{(1-r^2)} + O(1).
\end{aligned}
$$

Choosing $r = (1 - 1/n)$ we obtain for large n

$$
|a_n| \leq \frac{1}{r^{n-1}} I_1(r, f) < 2\varepsilon^2 en.
$$

Thus

$$
\frac{|a_n|}{n} \to 0,
$$

and Theorem 1.12 is proved in this case.

1.7.2 The radius of greatest growth We suppose from now on that

$$
\alpha > 0.
$$

We proceed in a series of steps. In this section we prove the existence of a radius $\arg z = \theta_0$, on which $|f(re^{i\theta_0})|$ has maximal growth and $\arg f(re^{i\theta_0})$ varies rather slowly. We call $\arg z = \theta_0$ a *radius of greatest growth* (r.g.g.).

Lemma 1.2 *If $\alpha > 0$ there exists a real θ_0 such that*

$$
f(re^{i\theta_0}) = R(r)e^{i\lambda(r)},
$$

where

$$
R(r) \sim \alpha(1-r)^{-2} \quad as \quad r \to 1 \tag{1.18}
$$

and

$$
\int_0^1 (1-r)\lambda'(r)^2 dr < \infty. \tag{1.19}
$$

In particular

$$
\lambda(r_2) - \lambda(r_1) \to 0, \tag{1.20}
$$

uniformly while $r_1 \to 1$ and $r_1 \leq r_2 \leq r_1 + k(1-r_1)$, where k is a constant, $k < 1$.

We recall that by (1.11) $(1 - r)^2 r^{-1} |f(re^{i\theta})|$ decreases with increasing r for fixed θ and so tends to a limit $\alpha(\theta)$ as $r \to 1$. To prove (1.18) we show that there exists θ_0 such that $\alpha(\theta_0) = \alpha$. We choose $r_n = 1 - 1/n$, and θ_n, $0 \le \theta_n \le 2\pi$, such that

$$\alpha_n = \frac{(1 - r_n)^2}{r_n} |f(r_n e^{i\theta_n})| = \frac{(1 - r_n)^2}{r_n} M(r_n) \to \alpha, \quad \text{as } n \to \infty.$$

We use (1.11) again and deduce that

$$\frac{(1 - r)^2}{r} |f(re^{i\theta_n})| \ge \alpha_n, \quad 0 < r < r_n.$$

Let θ_0 be a limit point of the sequence θ_n, and suppose that

$$\theta_{n_k} \to \theta_0 \quad \text{as } k \to \infty.$$

We fix r such that $0 < r < 1$. Then as $\theta \to \theta_0$ through the sequence θ_{n_k}

$$\frac{(1 - r)^2}{r} |f(re^{i\theta_0})| = \lim_{k \to \infty} \frac{(1 - r)^2}{r} |f(re^{i\theta_{n_k}})| \ge \lim \alpha_n = \alpha.$$

Thus

$$R(r) = |f(re^{i\theta_0})| \ge \frac{\alpha r}{(1 - r)^2}, \quad 0 < r < 1$$

and by Theorem 1.5

$$R(r) \le M(r, f) < \frac{\alpha + o(1)}{(1 - r)^2} \quad \text{as } r \to 1.$$

This yields (1.18).

Next we prove (1.19). For this we use (1.9). We write

$$\log f(re^{i\theta_0}) = \log R(r) + i\lambda(r) = \sigma(r) + i\lambda(r)$$

say, where

$$\sigma'(r)^2 + \lambda'(r)^2 \le \left\{ \frac{1 + r}{r(1 - r)} \right\}^2$$

by (1.9). We write $a = \sigma'(r)$, $b = \lambda'(r)$ and $c = (1 + r)/\{r(1 - r)\}$. Then

$$b^2 \le (c - a)(c + a) \le 2c(c - a).$$

Thus

$$\frac{b^2}{2c} \le c - a,$$

and so

$$\frac{r(1 - r)}{2(1 + r)} \lambda'(r)^2 \le \frac{1 + r}{r(1 - r)} - \sigma'(r).$$

We integrate both sides from r_1 to r_2 and deduce that

$$\int_{r_1}^{r_2} (1-r)\lambda'(r)^2 dr \leq \frac{(1+r_1)}{r_1} \int_{r_1}^{r_2} \frac{r(1-r)}{1+r}\lambda'(r)^2 dr$$

$$\leq 2\frac{(1+r_1)}{r_1} \left\{ \int_{r_1}^{r_2} \frac{1+r}{r(1-r)} dr - \int_{r_1}^{r_2} \sigma'(r)dr \right\}$$

$$= \frac{2(1+r_1)}{r_1} \left[\log \frac{r}{(1-r)^2 R(r)} \right]_{r_1}^{r_2}.$$

Letting r_2 tend to 1, we deduce that

$$\int_{r_1}^{1} (1-r)\lambda'(r)^2 dr \leq \frac{4}{r_1} \log \left\{ \frac{|f(r_1 e^{i\theta_0})|(1-r_1)^2}{r_1 \alpha} \right\} < \infty.$$

We choose $r_1 = \frac{1}{2}$ and note that $\lambda'(r)$ is continuous in the interval $\left[0, \frac{1}{2}\right]$. This yields (1.19).

Finally we suppose that $\varepsilon > 0$, that $0 < r < 1$ and choose r_0 so close to 1 that

$$\int_{r_0}^{1} (1-r)\lambda'(r)^2 dr < \frac{\varepsilon^2}{\log \frac{1}{k}}.$$

Then if $r_0 \leq r_1 < r_2 < 1$, and $(1-r_2) \geq k(1-r_1)$ we deduce that

$$|\lambda(r_2) - \lambda(r_1)| = \left| \int_{r_1}^{r_2} \lambda'(r)dr \right| \leq \left\{ \int_{r_1}^{r_2} (1-r)\lambda'(r)^2 dr \int_{r_1}^{r_2} \frac{dr}{1-r} \right\}^{\frac{1}{2}}$$

$$< \left\{ \frac{\varepsilon^2}{\log \frac{1}{k}} \log \frac{1}{k} \right\}^{\frac{1}{2}} = \varepsilon.$$

This proves (1.20).

1.7.3 Behaviour on the major arc Our next aim is to find asymptotic formulae for $f(z)$ and $f'(z)$, valid on an arc $|\theta - \theta_0| < K(1-\rho)$, where $z = \rho e^{i\theta}$. For this purpose we define

$$r_n = 1 - \frac{1}{n}, \quad z_n = r_n e^{i\theta_0}, \quad n \geq 1$$

where θ_0 is an r.g.g. It will turn out that the r.g.g. is unique for univalent f, but we do not assume this for the time being. Given a fixed positive ε we denote by $\Delta_n = \Delta_n(\varepsilon)$ the domain given by

$$\Delta_n = \left\{ z : \frac{\varepsilon}{n} < |1 - ze^{-i\theta_0}| < \frac{1}{\varepsilon n}, \ \arg (1 - ze^{-i\theta_0}) < \frac{1}{2}\pi - \varepsilon \right\}.$$

We also define

$$\alpha_n = n^{-2}f(z_n),$$

$$f_n(z) = \frac{\alpha_n}{(1 - ze^{-i\theta_0})^2}.$$

Thus α_n, $f_n(z)$ are defined for $n \geq 1$ and by (1.18)

$$|\alpha_n| \to \alpha, \text{ as } n \to \infty.$$

Lemma 1.3   *We have as $z \to e^{i\theta_0}$ through the regions $\Delta_n(\varepsilon)$*

$$f(z) \sim f_n(z) \quad and \quad f'(z) \sim f'_n(z). \tag{1.21}$$

We shall suppose without loss of generality that $\theta_0 = 0$, since otherwise we may consider $f(ze^{i\theta_0})$, $f_n(ze^{i\theta_0})$ instead of $f(z)$, $f_n(z)$. Then if

$$z = l_n(Z) = r_n + \frac{1}{n}Z, \quad (1 - z) = \frac{1}{n}(1 - Z), \tag{1.22}$$

the domain $\Delta_1(\varepsilon)$ corresponds in the Z plane to

$$\Delta_1(\varepsilon) = \left\{ Z : \varepsilon < |1 - Z| < \frac{1}{\varepsilon}, \text{ arg } (1 - Z) < \frac{1}{2}\pi - \varepsilon \right\}.$$

We write

$$g_n(Z) = (1 - Z)^2 \frac{f[l_n(Z)]}{f(z_n)}.$$

Then, for a fixed ε, $\Delta_n(\varepsilon)$ lies in $|z| < 1$ when n is sufficiently large and so $g_n(Z)$ is defined in $\Delta_1(\varepsilon)$ for all large n. We now define

$$\Delta'_n = \Delta_n(\tfrac{1}{2}\varepsilon), \quad \Delta''_n = \Delta_n(\tfrac{1}{4}\varepsilon).$$

Then the corresponding domains in the Z plane are

$$\Delta' = \Delta_1(\tfrac{1}{2}\varepsilon) \text{ and } \Delta'' = \Delta_1(\tfrac{1}{4}\varepsilon).$$

Let d be the distance of Δ' from the complement of Δ''. Then the distance of Δ'_n from the complement of Δ''_n is d/n, and hence the distance of Δ'_n from $|z| = 1$ is at least d/n for large n. Thus by Theorem 1.4 there exist constants n_0 and C_1 such that

$$|f(z)| < C_1 n^2, \text{ if } n \geq n_0 \text{ and } z \in \Delta'_n.$$

Hence if $Z \in \Delta'$ and $n \geq n_1$ we deduce that

$$|f\{l_n(Z)\}| < C_1 n^2, \quad |f(z_n)| > \tfrac{1}{2}\alpha n^2,$$

and so there is another positive constant C_2, such that

$$|g_n(Z)| < C_2, \text{ if } n \geq n_1, \text{ and } Z \in \Delta'.$$

Suppose now that Z is real, $-1 \leq Z \leq 0$, and let $z = \rho$ correspond to Z by (1.22). Then $\rho \leq r_n \leq (1 + \rho)/2$ and so by (1.18) and (1.20)

$$\frac{f(\rho)}{f(r_n)} \sim \frac{(1 - r_n)^2}{(1 - \rho)^2}, \text{ i.e. } f(\rho) \sim \frac{\alpha_n}{(1 - \rho)^2},$$

so that

$$g_n(Z) = (1 - Z)^2 \frac{f(\rho)}{n^2 \alpha_n} = \frac{(1 - \rho)^2 f(\rho)}{\alpha_n} \to 1, \quad -1 \leq Z \leq 0.$$

Thus $g_n(z)$ is uniformly bounded in Δ' for $n \geq n_0$, and $g_n(Z) \to 1$ as $n \to \infty$ for $-1 \leq Z \leq 0$. It now follows from Vitali's convergence theorem[†] that

$$g_n(Z) \to 1, \quad g_n'(Z) \to 0 \quad (n \to \infty)$$

uniformly for $Z \in \Delta_1(\varepsilon)$. Translating back into the z plane we deduce

$$\frac{f(z)}{f_n(z)} \to 1, \quad \frac{1}{n} \frac{d}{dz} \frac{n^2 (1 - z)^2 f(z)}{f(r_n)} \to 0.$$

The first of these limiting relations is the first relation in (1.21). The second gives

$$(1 - z)^2 f'(z) - 2(1 - z) f(z) = o(n)$$

i.e.

$$f'(z) = \frac{2}{1 - z} f(z) + o(1 - z)^{-3},$$

since $z \in \Delta_n(\varepsilon)$. Using the first relation in (1.21), we obtain the second one and Lemma 1.3 is proved.

1.7.4 Behaviour on the minor arc   In order to complete the proof of Theorem 1.12 we show next that $f(z)$ grows relatively slowly away from the r.g.g. For this we need f to be univalent, at least in an average sense.

Lemma 1.4   *Suppose that $f(z) \in \mathfrak{S}$, $\alpha > 0$ and that θ_0 is as in Lemma 1.3. Then given a positive η, there exist positive constants, C_0, C_1 and r_0, where $r_0 < 1$, such that if $z_0 = r e^{i\theta}$, where*

$$r_0 \leq r < 1, \quad C_0(1 - r) \leq |\theta - \theta_0| \leq \pi, \tag{1.23}$$

[†] (Titchmarsh [1939], p. 168)

we have

$$|f(z_0)| < C_1, \quad \text{or else} \quad \left|\frac{f'(z_0)}{f(z_0)}\right| < \frac{\eta}{1-r}. \tag{1.24}$$

We note that (1.21) and (1.24) are incompatible for the same z, so that the r.g.g. is unique.

We suppose that ε is as in Lemma 1.3. It follows from that lemma that, as $z \to 1$ through the union of the sectors $\Delta_n(\varepsilon)$,

$$|f(z)| \sim \alpha|1-z|^{-2}, \quad |f'(z)| \sim 2\alpha|1-z|^{-3}. \tag{1.25}$$

Thus (1.25) holds uniformly as $z \to 1$ in the Stolz angle

$$S : |\arg(1-z)| < \frac{\pi}{2} - \varepsilon, \quad |1-z| < \sin\varepsilon. \tag{1.26}$$

If $R > R_0(\varepsilon)$ we consider the image $D(R)$ by $w = f(z)$ of the sector

$$S(R) : \left\{\frac{\alpha}{R(1-\varepsilon)}\right\}^{\frac{1}{2}} < |1-z| < \left\{\frac{\alpha}{\varepsilon R(1+\varepsilon)}\right\}^{\frac{1}{2}}, \quad |\arg z| < \frac{\pi}{2} - \varepsilon.$$

We suppose that R_0 is so large that $S(R)$ lies in S. It follows from the first relation in (1.25) that, if R_0 is sufficiently large, $D(R)$ lies in

$$\varepsilon R < |w| < R.$$

Next the area $A(R)$ of $D(R)$ is given by

$$A(R) = \int_{S(R)} |f'(z)|^2 |dz|^2.$$

We set $z = 1 - \rho e^{i\phi}$, so that $|dz|^2 = \rho \, d\rho \, d\phi$, and deduce from the second relation in (1.25) that, for large R,

$$\begin{aligned}
A(R) &> 4\alpha^2(1-\varepsilon) \int_{\{\alpha/R(1-\varepsilon)\}^{\frac{1}{2}}}^{\{\alpha/\varepsilon R(1+\varepsilon)\}^{\frac{1}{2}}} \rho \, d\rho \int_{-\frac{\pi}{2}+\varepsilon}^{\frac{\pi}{2}-\varepsilon} \rho^{-6} d\phi \\
&= (\pi - 2\varepsilon)(1-\varepsilon)R^2 \left[(1-\varepsilon)^2 - \varepsilon^2(1+\varepsilon)^2\right] \\
&= (\pi - \varepsilon_1)R^2,
\end{aligned}$$

where ε_1 tends to zero with ε.

Suppose next that $\frac{3}{4} \leq |z_0| < 1$, that the disc $|z - z_0| < \frac{1}{4}(1 - |z_0|)$ does not meet $S(R)$ and that $|f(z_0)| = R/e$. Then by (1.9) we have, if $|z| = \rho$,

$$\left|\frac{f'(z)}{f(z)}\right| \leq \frac{1+\rho}{\rho(1-\rho)}.$$

In the disc $|z - z_0| \leq \frac{1}{4}(1 - |z_0|)$, we have $\frac{11}{16} \leq \rho \leq \frac{1}{4}(3|z_0| + 1)$ so that

$$\left| \frac{f'(z)}{f(z)} \right| \leq \left(\frac{1}{\rho} + 1 \right) \frac{1}{1 - \rho} \leq \frac{27}{11} \frac{4}{3(1 - |z_0|)} < \frac{4}{1 - |z_0|}.$$

Thus if $|z - z_0| < \frac{1}{4}(1 - |z_0|)$, we deduce, integrating along the segment from z_0 to z_1, that

$$\log \left| \frac{f(z_1)}{f(z_0)} \right| < \int_{z_0}^{z_1} \left| \frac{f'(z)}{f(z)} \right| |dz| \leq \frac{4}{1 - |z_0|} \frac{1 - |z_0|}{4} = 1.$$

Hence $|f(z_1)| < e|f(z_0)|$, and so the image $d(R)$ of $|z - z_0| < \frac{1}{4}(1 - |z_0|)$ lies in $|w| < R$. Since $f(z)$ is univalent in $|z| < 1$, $d(R)$ is disjoint from $D(R)$ so that the area $a(r)$ of $d(R)$ satisfies

$$a(R) < \pi R^2 - A(R) < \varepsilon_1 R^2.$$

Thus

$$\int_{(z-z_0)<\frac{1}{4}(1-|z_0|)} |f'(z)|^2 |dz|^2 < \varepsilon_1 R^2. \qquad (1.27)$$

If we expand $f'(z)$ in a power series

$$f'(z) = \sum_{0}^{\infty} b_n (z - z_0)^n$$

we deduce that

$$\frac{1}{2\pi} \int_0^{2\pi} |f'(z_0 + \rho e^{i\theta})|^2 d\theta = \sum_{0}^{\infty} |b_n|^2 \rho^{2n} \geq |b_0|^2 = |f'(z_0)|^2.$$

Thus

$$\int_{|z-z_0|<\frac{1}{4}(1-|z_0|)} |f'(z)|^2 |dz|^2 \geq \pi \left[\frac{1}{4}(1 - |z_0|) \right]^2 |f'(z_0)|^2.$$

Now (1.27) yields

$$|f'(z_0)|^2 < \frac{16}{\pi(1 - |z_0|)^2} \varepsilon_1 R^2 = \frac{16 e^2 \varepsilon_1}{\pi(1 - |z_0|)^2} |f(z_0)|^2 < \frac{\eta^2 |f(z_0)|^2}{(1 - |z_0|)^2},$$

if $\varepsilon_1 < \pi \eta^2 / (16 e^2)$. We suppose that ε is so small that ε_1 satisfies this inequality and deduce that $z = z_0$ satisfies (1.24) provided that $R = e|f(z_0)| > R_0(\varepsilon)$ and that the disc $|z - z_0| < \frac{1}{4}(1 - |z_0|)$ is disjoint from the Stolz angle S given by (1.26).

Suppose contrary to this that $z = \rho e^{i\theta}$ lies in S and also that $|z - z_0| < \frac{1}{4}(1 - |z_0|)$. Thus

$$\frac{3}{4}(1 - |z_0|) < 1 - \rho < \frac{5}{4}(1 - |z_0|).$$

Suppose further that $1 - |z_0|$ and consequently $1 - \rho$ is so small that

$$|\theta| < C(1 - \rho),$$

where C is constant depending on ε. In fact this will be the case if $C > \cot \varepsilon$ for ρ close to 1. We also have, for ρ close to 1

$$|\theta - \arg z_0| < C'(1 - \rho).$$

Thus

$$|\arg z_0| < (C + C')(1 - \rho) < \frac{5}{4}(C + C')(1 - |z_0|).$$

Hence if (1.23) holds, where r_0 is sufficiently close to 1 and C_0 is sufficiently large, the disc $|z - z_0| < \frac{1}{4}(1 - z_0)$ is disjoint from S and (1.24) holds. This proves Lemma 1.4.

1.7.5 Completion of the proof of Theorem 1.12

We now suppose that $\alpha > 0$ and $\theta_0 = 0$ in Theorem 1.12. We write $r_n = 1 - 1/n$, $f_n(z) = \alpha_n(1 - z)^{-2}$, where $\alpha_n = n^2 f(r_n)$. Then

$$a_n = \frac{1}{2\pi r^n} \int_{-\pi}^{\pi} f(r_n e^{i\theta}) e^{-in\theta} d\theta,$$

$$\alpha_n(n + 1) = \frac{1}{2\pi r_n^n} \int_{-\pi}^{\pi} f_n(r_n e^{i\theta}) e^{-in\theta} d\theta.$$

Thus

$$a_n - \alpha_n(n + 1) = \frac{1}{2\pi} \int_{-\pi}^{\pi} \left\{ f(r_n e^{i\theta}) - f_n(r_n e^{i\theta}) \right\} e^{-in\theta} d\theta. \tag{1.28}$$

We divide the right-hand integral into the ranges

$$I_1 : |\theta| < \frac{K}{n} \text{ and } I_2 : \frac{K}{n} < |\theta| \le \pi,$$

where K is a suitably large positive constant. In I_1 we have by Lemma 1.3, (1.21)

$$f(r_n e^{i\theta}) - f_n(r_n e^{i\theta}) = o(n^2).$$

Thus

$$\int_{I_1} \left\{ f(r_n e^{i\theta}) - f_n(r_n e^{i\theta}) \right\} e^{-in\theta} d\theta = o(n), \text{ as } n \to \infty. \tag{1.29}$$

Next we fix $\eta = \frac{1}{2}$ and suppose that r_0, C_0, C_1 are the corresponding constants of Lemma 1.4. We suppose also that $C_0 \geq 1$, that

$$C_0(1 - r) \leq |\theta| \leq \pi,$$

and that $|f(re^{i\theta})| > C_1$. Let r_1 be the smallest number such that $|f(te^{i\theta})| \geq C_1$ for $r_1 \leq t \leq r$, further $C_0(1 - r_1) \leq |\theta|$ and $r_0 \leq r_1$.

Then we deduce from Lemma 1.4 that

$$\frac{\partial}{\partial t} \log|f(te^{i\theta})| = \Re \frac{f'(te^{i\theta})}{f(te^{i\theta})} \leq \frac{1}{2(1 - t)}, \; r_1 \leq t \leq r.$$

Integrating this inequality we obtain

$$|f(re^{i\theta})| \leq |f(r_1 e^{i\theta})| \left(\frac{1 - r_1}{1 - r} \right)^{\frac{1}{2}}.$$

Our conditions for r_1 ensure that

$$r_1 = r_0 \; \text{or} \; |f(r_1 e^{i\theta})| = C_1 \; \text{or} \; C_0(1 - r_1) = |\theta|,$$

since otherwise we could replace r_1 by a smaller quantity. In the first case we obtain

$$|f(re^{i\theta})| \leq M(r_0, f) \left(\frac{1 - r_0}{1 - r} \right)^{\frac{1}{2}}.$$

In the second case we have

$$|f(re^{i\theta})| \leq C_1 \left(\frac{1 - r_1}{1 - r} \right)^{\frac{1}{2}} \leq C_1 \left(\frac{1 - r_0}{1 - r} \right)^{\frac{1}{2}}.$$

The third case yields

$$\begin{aligned}
|f(re^{i\theta})| &\leq |f(r_1 e^{i\theta})| \left(\frac{1 - r_1}{1 - r} \right)^{\frac{1}{2}} \\
&\leq (1 - r_1)^{-2} \left(\frac{1 - r_1}{1 - r} \right)^{\frac{1}{2}} \\
&= (1 - r)^{-\frac{1}{2}} (1 - r_1)^{-\frac{3}{2}} \\
&= C_0^{\frac{3}{2}} |\theta|^{-\frac{3}{2}} (1 - r)^{-\frac{1}{2}}.
\end{aligned}$$

Thus we obtain in all cases

$$|f(re^{i\theta})| \leq C_2 (1 - r)^{-\frac{1}{2}} |\theta|^{-\frac{3}{2}}, \; \text{if} \; C_0(1 - r) \leq |\theta| \leq \pi, \quad r_0 \leq r \leq 1,$$

where the constants C_0, C_2 depend only on the function f. We also have

$$|f_n(re^{i\theta})| \leq |1 - re^{i\theta}|^{-2} \leq C_3 |\theta|^{-2}, \; C_0(1 - r) \leq |\theta| \leq \pi.$$

We now choose $r = r_n = 1 - 1/n$ and $K \geq C_0 \geq 1$. Then

$$\left| \int_{K/n \leq |\theta| \leq \pi} \left\{ f(r_n e^{i\theta}) - f_n(r_n e^{i\theta}) \right\} e^{-in\theta} d\theta \right|$$

$$\leq C_4 \int_{K/n}^{\pi} n^{\frac{1}{2}} \theta^{-\frac{3}{2}} d\theta \leq 2 C_4 n^{\frac{1}{2}} \left(\frac{n}{K} \right)^{\frac{1}{2}} = \frac{2 C_4 n}{K^{\frac{1}{2}}}. \tag{1.30}$$

Given a small positive δ we choose K so large that $2 C_4 / K^{\frac{1}{2}} < \delta$. Combining (1.28) to (1.30) we obtain for sufficiently large n

$$\left| 2\pi \left(1 - \frac{1}{n} \right)^n \left\{ a_n - \alpha_n(n+1) \right\} \right| < 2\delta n.$$

Thus

$$a_n \sim n \alpha_n \tag{1.31}$$

and (1.16) is proved. In fact when $\alpha > 0$ (1.31) is stronger than (1.16), since (1.31) gives information about the argument of a_n as well as $|a_n|$. We note that either $\alpha = 1$, and so by Theorem 1.5 $|a_n| = n$ for all n, or else $0 \leq \alpha < 1$ in which case $|a_n| < n$ for $n > n_0$. This completes the proof of Theorem 1.12.

Examples

1.1   If $\lambda > 0$ and

$$b_\lambda(z) = (1 - z)^{-\lambda} = \sum_0^\infty d_n(\lambda) z^n,$$

prove that

$$d_n(\lambda) = \frac{\Gamma(\lambda + n)}{\Gamma(\lambda)\Gamma(n+1)} \sim \frac{n^{\lambda - 1}}{\Gamma(\lambda)} \quad \text{as} \quad n \to \infty,$$

where $\Gamma(x)$ denotes the Gamma function (cf. Titchmarsh [1939, p. 58]).

1.2   Deduce that, if $\lambda > \frac{1}{2}$,

$$\frac{1}{2\pi} \int_{-\pi}^{\pi} |b_\lambda(re^{i\theta})|^2 d\theta = \sum_0^\infty d_n(\lambda)^2 r^{2n}$$

$$\sim \frac{1}{2\sqrt{\pi}} \frac{\Gamma(\lambda - \frac{1}{2})}{\Gamma(\lambda)} (1 - r)^{1 - 2\lambda} \quad \text{as} \quad r \to 1.$$

(Use the duplication formula for $\Gamma(x)$, Titchmarsh [1939, p. 57])

1.3   With the hypotheses of Theorem 1.12 we define

$$I_\lambda(r,f) = \frac{1}{2\pi} \int_{-\pi}^{\pi} |f(re^{i\theta})|^\lambda d\theta.$$

Prove that, if $\lambda > \frac{1}{2}$,

$$(1-r)^{2\lambda-1} I_\lambda(r,f) \to \alpha^\lambda \frac{\Gamma(\lambda - \frac{1}{2})}{2\sqrt{\pi}\Gamma(\lambda)} \quad \text{as} \quad r \to 1.$$

1.4   Obtain corresponding asymptotic formulae for

$$I_2(r,f') \text{ and } A(r,f) = \int_0^{2\pi} d\theta \int_0^r |f'(te^{i\theta})|^2 t dt d\theta.$$

1.5   Prove that, if $\lambda > \frac{1}{2}$, there exists r_0 such that $0 < r_0 < 1$ and

$$I_\lambda(r,f) \le I_\lambda(r,f_\theta(z)), \quad r_0 < r < 1 \qquad (1.32)$$

where $f_\theta(z) = z(1 - ze^{i\theta})^{-2}$. Obtain corresponding results for $I_2(r,f')$ and $A(r,f)$. (It follows from Baernstein's Theorem, referred to in Section 1.3, that (1.32) holds for $0 < r < 1$.)

2
The growth of finitely mean valent functions

2.0 Introduction We saw in the previous chapter that the assumption that $f(z)$ is univalent in $|z| < 1$ imposes a number of restrictions on the growth of $f(z)$, such as were proved in Theorems 1.3, 1.5, 1.6 and 1.12. In this and the next chapter we see what results we can obtain under the more general assumption that the equation $f(z) = w$ has at most p roots in some average sense, as w moves over the plane.

In the present chapter we confine ourselves to obtaining bounds for $|f(z)|$. We set

$$M(r,f) = \max_{|z|=r} |f(z)| \quad (0 < r < 1), \tag{2.1}$$

and show that under suitable averaging assumptions

$$M(r,f) = O(1-r)^{2p} \quad (r \to 1). \tag{2.2}$$

The first result of this type is due to Cartwright [1935], who proved (2.2) when p is a positive integer and the equation $f(z) = w$ never has more than p roots in $|z| < 1$ for any w. Such functions are called p-valent . Her method, based on a distortion theorem of Ahlfors [1930], was extended by Spencer [1940b] to the more general case.

If $f(z) = a_0 + a_1 z + ...$ has q zeros in $|z| < 1$, bounds for $M(r, f)$ can be obtained depending on

$$\mu_q = \sum_{v=0}^{q} |a_v|. \tag{2.3}$$

This dependence is essential. In fact any polynomial of degree p is p-valent and has at most p zeros in $|z| < 1$, but bounds for $M(r)$ must clearly depend on all the coefficients.

Our method also leads to a number of theorems, which show that $f(z)$

28

cannot grow too rapidly near several points of $|z| = 1$ simultaneously, and in particular that, if $M(r)$ attains the growth $(1-r)^{-2p}$, then $|f(re^{i\theta})|$ attains this magnitude for a single fixed value of θ, and is quite small for other constant θ as $r \to 1$. Results of this type appear to need the full strength of the methods of this chapter even for univalent functions, and were first proved by Spencer [1940b]. We complete the chapter with Eke's [1967b] regularity theorem for the maximum modulus.

2.1 A length-area principle   Let $f(z)$ be regular in an open set Δ and let $n(w)$ be the number of roots in Δ of the equation $f(z) = w$. We write

$$p(R) = p(R, \Delta, f) = \frac{1}{2\pi} \int_0^{2\pi} n(Re^{i\phi})d\phi. \tag{2.4}$$

The integral exists as a finite or infinite Lebesgue integral, since $n(w) \geq 0$.[†]

The function $p(R)$ will play a dominant role in the sequel. It is clear that $p(R)$, like $n(w)$, increases with expanding domain Δ. Further if $z = t(\zeta)$ maps Δ_1 (1,1) conformally onto Δ, then if $p(R, \Delta_1)$ refers to $f[t(\zeta)]$, we have $p(R, \Delta) = p(R, \Delta_1)$.

We might make the averaging assumption

$$p(R) \leq p \quad (0 < R < \infty).$$

This is certainly satisfied if p is a positive integer and $f(z)$ is p-valent in $|z| < 1$. We shall use this assumption in Chapter 5; at present weaker hypotheses will be sufficient.

We shall base our results on a length–area inequality. A result of this type was first proved by Ahlfors [1930], and its use in this context is due to Cartwright [1935]. Such results play a key role in conformal and quasi-conformal mapping (see Lelong-Ferrand [1955], Ahlfors [1973, Chapter 4]).

Theorem 2.1   *Suppose that $f(z)$ is regular in an open set Δ and that $p(R) = p(R, \Delta)$ is defined by (2.3). Let $l(R)$ be the total length of the curves in Δ on which $|f(z)| = R$, and suppose that the area A of Δ is finite. Then we have*

$$\int_0^\infty \frac{l(R)^2 dR}{Rp(R)} \leq 2\pi A,$$

[†] It will be shown in the course of Lemma 5.2 that the sets $n(w) \geq K$ are open in the finite plane. Thus $n(Re^{i\phi})$ is certainly measurable.

where the integrand is taken be zero if $l(R) = 0$ or $p(R) = +\infty$. In particular, $l(R) < +\infty$ for almost all R for which $p(R) < +\infty$.

2.1.1   We first prove Theorem 2.1 when Δ is an open rectangle and $f(z)$ is univalent and not zero in a domain containing Δ and its sides. We say that $f(z)$ is univalent and not zero *on* Δ. Then any fixed branch of

$$s(z) = \log f(z) = \sigma + i\tau$$

is also univalent on Δ and maps Δ onto a domain Ω in the s plane. The boundary of Ω is the image of the boundary of Δ by $s(z)$ and is a sectionally analytic Jordan curve. Let θ_σ be the intersection of Ω with the line $\sigma = $ constant. Then θ_σ consists of a finite number of straight-line segments

$$\tau_1 < \tau < \tau_1', \quad \tau_2 < \tau < \tau_2', \quad \ldots$$

As $s = \sigma + i\tau$ describes θ_σ, $z = x + iy$ describes the set γ_σ in Δ on which $|f(z)| = e^\sigma$.

Since $f(z) = e^{s(z)}$ is univalent, $\tau_\nu' - \tau_\nu \le 2\pi$. On the arc of γ_σ corresponding to the segment $\tau_\nu < \tau < \tau_\nu'$, the equation

$$f(z) = e^{\sigma + i\tau_0}$$

has no roots if $\tau_\nu' < \tau_0 < \tau_\nu + 2\pi$ and one root if $\tau_\nu < \tau < \tau_\nu'$. Thus the corresponding contribution to $p(e^\sigma, \Delta)$ is exactly $(\tau_\nu' - \tau_\nu)/(2\pi)$. Writing $\theta(\sigma)$ for the total measure of θ_σ we see by addition that

$$p(e^\sigma, \Delta) = \frac{1}{2\pi} \sum (\tau_\nu' - \tau_\nu) = \frac{\theta(\sigma)}{2\pi}.$$

We now have, using Schwarz's inequality,[†]

$$
\begin{aligned}
l(e^\sigma, \Delta)^2 &= \left[\int_{\theta_\sigma} \left|\frac{dz}{ds}\right| d\tau\right]^2 \le \int_{\theta_\sigma} d\tau \int_{\theta_\sigma} \left|\frac{dz}{ds}\right|^2 d\tau \\
&= \theta(\sigma) \int_{\theta_\sigma} \left|\frac{dz}{ds}\right|^2 d\tau = 2\pi p(e^\sigma, \Delta) \int_{\theta_\sigma} \left|\frac{dz}{ds}\right|^2 d\tau.
\end{aligned}
$$

If σ_1, σ_2 are the lower and upper bounds of σ on Ω this gives

$$\int_{-\infty}^{+\infty} \frac{l(e^\sigma, \Delta)^2}{p(e^\sigma, \Delta)} d\sigma = \int_{\sigma_1}^{\sigma_2} \frac{l(e^\sigma, \Delta)^2}{p(e^\sigma, \Delta)} d\sigma \le 2\pi \int_{\sigma_1}^{\sigma_2} d\sigma \int_{\theta_\sigma} \left|\frac{dz}{ds}\right|^2 d\tau = 2\pi A,$$

[†] $\left|\sum ab\right|^2 \le \sum |a|^2 \sum |b|^2$ or $\left|\int f g dx\right|^2 \le \int |f|^2 dx \int |g|^2 dx$. For a proof see, for example, Titchmarsh [1939, p. 381].

where A is the area of Δ. Writing $e^\sigma = R$ we have

$$\int_0^\infty \frac{l(R,\Delta)^2}{p(R,\Delta)} \frac{dR}{R} \leq 2\pi A, \qquad (2.5)$$

as required.

2.1.2   To extend our result to the general case we need the theory of the Lebesgue integral. We may assume without loss of generality that $f(z)$ is regular and $f(z) \neq 0$, $f'(z) \neq 0$ in Δ. For otherwise we may consider $f(z)$ in the set Δ_0, consisting of Δ except for zeros of $f(z)$ and $f'(z)$, without affecting any of the quantities $p(R), l(R)$ and A in Theorem 2.1. Next we express Δ as a countable union of non-overlapping rectangles Δ_v ($v = 1$ to ∞).[†] At each point z_0 of such a rectangle $f'(z_0) \neq 0$, and so there is a neighbourhood of z_0 in which $f(z)$ is univalent. Thus we may subdivide each Δ_v into a finite number of smaller rectangles on each of which $f(z)$ is univalent. We therefore assume without loss of generality that $f(z)$ is univalent, $f(z) \neq 0$ on each Δ_v.

The set of curves γ_R, on which $|f(z)| = R$, meets a side of one of the rectangles Δ_v in only a finite number of points, unless $|f(z)| \equiv R$ on the side. Thus if we omit the finite or countable set of values of R for which $|f(z)| \equiv R$ on some side of a rectangle Δ_v, we have

$$p(R,\Delta) = \sum_{v=1}^\infty p(R,\Delta_v), \qquad (2.6)[‡]$$

$$l(R,\Delta) = \sum_{v=1}^\infty l(R,\Delta_v), \qquad (2.7)[§]$$

since the sides of the Δ_v do not then contribute to $p(R,\Delta)$ or $l(R,\Delta)$. Also if A_v, A are the areas of Δ_v, Δ respectively, we have[¶]

$$A = \sum_{v=1}^\infty A_v.$$

We now set

$$a_v^2 = \frac{l(R,\Delta_v)^2}{p(R,\Delta_v)}, \ b_v^2 = p(R,\Delta_v)$$

[†] Burkill [1951], B, p. 8.
[‡] This follows from (2.3) and [B], p. 41.
[§] [B], p. 35.
[¶] [B], p. 21.

in Schwarz's inequality $(\sum a_v b_v)^2 \le \sum a_v^2 \sum b_v^2$ and sum over those values of v for which $p(R, \Delta_v), l(R, \Delta_v)$ are not zero. Using (2.6) and (2.7) we obtain

$$\frac{l(R, \Delta)^2}{p(R, \Delta)} \le \sum_{v=1}^{\infty} \frac{l(R, \Delta_v)^2}{p(R, \Delta_v)},$$

where indeterminate terms are taken to be zero. We now integrate from 0 to ∞ and obtain, using (2.5) applied to each Δ_v,

$$\int_0^\infty \frac{l(R, \Delta)^2}{p(R, \Delta)} \frac{dR}{R} \le \int_0^\infty \left[\sum_{v=1}^{\infty} \frac{l(R, \Delta_v)^2}{p(R, \Delta_v)} \right] \frac{dR}{R}$$

$$= \sum_{v=1}^{\infty} \int_0^\infty \frac{l(R, \Delta_v)^2}{p(R, \Delta_v)} \frac{dR}{R} \le 2\pi \sum_{v=1}^{\infty} A_v = 2\pi A.$$

The inversion of integration and summation is justified since the terms are non-negative.[†] This completes the proof of Theorem 2.1.

Examples

2.1   If Δ is the rectangle $a < x < b$, $c < y < d$, and $f(z) = e^{\alpha z}$, where $\alpha > 0$, show that

$$l(R) = d - c, \; p(R) = \frac{\alpha(d-c)}{2\pi}, \; \text{if } e^{\alpha a} < R < e^{\alpha b},$$

and $l(R) = p(R) = 0$ for other values of R . Deduce that in this case equality holds in Theorem 2.1.

2.2   If Δ is the disc $|z| < r$, and $f(z) = az^n$ where a is a complex constant, $a \ne 0$, and n a positive integer, show that

$$l(R) = 2\pi(R/|a|)^{\frac{1}{n}} \text{ and } p(R) = n, \text{ for } 0 < R < |a|r^n$$

and $l(R) = p(R) = 0$ otherwise. Deduce that again equality holds in Theorem 2.1.

2.2 The growth of multivalent functions   We now prove a useful result relating the change in $|f(z)|$ from one point of Δ to another to the function $p(R)$. We suppose that

$$f(z) = \sum_{0}^{\infty} a_n z^n$$

[†] [B], p. 41.

has zeros $z_1, z_2, \ldots, z_q$ in $\Delta = \{|z| < 1\}$ and define

$$\mu = \inf \left\{ \sum_{v=0}^{q} |a_v|, |a_0| / \prod_{v=1}^{q} |z_v| \right\} \tag{2.8}$$

where an empty product is taken to be 1. Then we shall obtain a bound for the maximum modulus $M(r, f)$ in terms of μ and $p(R)$. The proof is adapted from a theorem of Jenkins and Oikawa [1971].

Theorem 2.2 *Suppose that $f(z)$ is regular in Δ with q zeros, counting multiplicity, that μ is defined by (2.8) and $p(R)$ by (2.4). Then if*

$$M = M(r, f) = \sup_{|z|=r} |f(z)|,$$

and $M > \mu$, we have

$$\int_{\mu}^{M} \frac{dR}{Rp(R)} \leq 2 \log \frac{1+r}{1-r} + \frac{\pi^2}{2} < 2 \log \frac{A_0}{1-r}, \quad 0 < r < 1,$$

where $A_0 = 2e^{\pi^2/4} = 23.58 \ldots$

We need

Lemma 2.1 *Suppose that $f(z)$ is regular in Δ and has zeros $z_1, \ldots, z_q$ there counting multiplicity. Let γ be a simple closed Jordan curve which surrounds the origin and lies in Δ. Then*

$$\inf_{z \in \gamma} |f(z)| \leq \mu$$

where μ is given by (2.8).

Landau [1922] has proved a sharp version of this result, when γ is a circle. The present simple argument is due to Hayman and Nicholls [1973]. We consider first the second term on the right-hand side of (2.8). We define

$$F(z) = f(z) / \prod_{v=1}^{q} \left(\frac{z - z_v}{1 - \bar{z}_v z} \right).$$

Then $F(z) \neq 0$ in Δ and so the maximum principle applied to $1/F(z)$ yields

$$\inf_{z \in \gamma} |f(z)| \leq \inf_{z \in \gamma} |F(z)| \leq |F(0)| = |a_0| / \prod_{v=1}^{q} |z_v|.$$

Suppose next that

$$|f(z)| > \sum_{v=0}^{q} |a_v|,$$

on γ. We write

$$g(z) = \sum_{v=0}^{q} a_v z^v,$$

so that

$$|g(z)| \leq \sum_{v=0}^{q} |a_v|$$

for $z \in \Delta$ and in particular for $z \in \gamma$. Thus by Rouché's Theorem [C. A. p. 153] $f(z)$ and

$$f(z) - g(z) = \sum_{v=q+1}^{\infty} a_v z^v,$$

have equally many zeros inside γ, i.e. at least $q + 1$. This contradicts our hypotheses and Lemma 2.1 is proved.

Lemma 2.2   *Suppose that $f(z)$ satisfies the hypotheses of Theorem 2.2. Let $z_0 = re^{i\theta}$ be a point on $|z| = r$, such that*

$$|f(z_0)| = M.$$

Then if $\mu < R < M$ and R is not one of a finite or countable exceptional set F of values, there exists an analytic open arc γ_R, which meets the line segment $l : [0, z_0]$ and approaches the boundary $|z| = 1$ of Δ as we move along γ_R in either direction. Also $|f(z)| = R$ on γ_R.

We note that $|f(0)| \leq \mu$ by (2.8). Thus there exists ρ, such that $|f(\rho e^{i\theta})| = R$ and $0 < \rho < r$.

Let E_1 be the set of zeros of $f'(z)$ in Δ and let E_2 be the set of points $z = \rho e^{i\theta}$ on l, such that

$$\frac{d}{dt}|f(te^{i\theta})|^2 = 0, \text{ when } t = \rho. \tag{2.9}$$

Then, unless f is constant in Δ, E_1 is finite or countable. Also unless the analytic function $\phi(\rho) = |f(\rho e^{i\theta})|^2$ is constant for $0 \leq \rho \leq r$, the set E_2 is finite. In all cases the image F of $E_1 \cup E_2$ by $|f(z)|$ is finite or countable.

We assume that $\mu < R < M$ and that R is not in F. Let ρ be the least positive number such that

$$|f(\rho e^{i\theta})| = R,$$

and let γ_R be that component of the level set $|f(z)| = R$, which contains $z = \rho e^{i\theta}$.

By hypothesis $f'(z) \neq 0$ on γ_R so that γ_R is an analytic curve. In fact $w = f(z)$ is locally univalent on γ_R and so γ_R is locally the image of $|w| = R$ by the regular function $z = f^{-1}(w)$.

We obtain the whole of γ_R by moving along the circle $w = Re^{i\phi}$ from $w_0 = f(\rho e^{i\theta}) = Re^{i\phi_0}$ say, in the direction of increasing or decreasing ϕ. If γ_R goes through no point of Δ more than once, γ_R must approach $|z| = 1$ as $\phi \to +\infty$ or $\phi \to -\infty$. Otherwise there exist ϕ_1, ϕ_2 such that $\phi_1 < \phi_0 < \phi_2$ and $f^{-1}(Re^{i\phi_1}) = f^{-1}(Re^{i\phi_2})$, while $f^{-1}(Re^{i\phi})$ is (1,1) for $\phi_1 \leq \phi < \phi_2$. In this case γ_R is the image of the interval $[\phi_1, \phi_2]$ by $f^{-1}(Re^{i\phi})$ and so γ_R is a closed Jordan curve. It remains to show that this case cannot occur.

Suppose contrary to this, that γ_R is a closed curve and let D be the interior of γ_R. The curve γ_R cannot touch the segment l at $\rho e^{i\theta}$ since $R \notin F$, so that (2.9) is false for $t = \rho$. Thus the segment l crosses γ_R at $z = \rho e^{i\theta}$. Also since ρ is the smallest number such that $|f(\rho e^{i\theta})| = R$, we must have

$$|f(te^{i\theta})| < R, \quad 0 \leq t < \rho. \tag{2.10}$$

Thus we cannot have

$$\frac{\partial}{\partial t}|f(te^{i\theta})| < 0$$

at $t = \rho$, and so

$$\frac{\partial}{\partial t}|f(te^{i\theta})| > 0, \quad t = \rho.$$

It follows that, for t slightly larger than ρ,

$$|f(te^{i\theta})| > R.$$

By the maximum principle $|f(z)| \leq R$ in D, so that points $te^{i\theta}$ lie outside D for t slightly larger than ρ. Since the segment l crosses γ_R at $\rho e^{i\theta}$, it follows that $t_1 e^{i\theta}$ must lie in D for t_1 slightly less than ρ. By (2.10) the segment $[0, t_1 e^{i\theta}]$ does not meet γ_R and so the whole of this segment lies in D. In particular $z = 0$ lies in D and this contradicts Lemma 2.1, since $|f(z)| = R > \mu$ on γ_R. This completes the proof of Lemma 2.2.

We can now complete the proof of Theorem 2.2 . We write

$$\zeta = \log \left(\frac{1 + ze^{-i\theta}}{1 - ze^{-i\theta}} \right).$$ (2.11)

Then ζ maps Δ (1,1) conformally onto the strip

$$S : \left\{ \zeta = \xi + i\eta \mid -\infty < \xi < \infty, \ |\eta| < \frac{\pi}{2} \right\},$$ (2.12)

so that l corresponds to the segment

$$L : 0 < \xi < \xi_0 = \log \frac{1 + r}{1 - r}, \quad \eta = 0.$$

Let S_0 be the subdomain of S consisting of all points distant less than $\frac{1}{2}\pi$ from l. Thus S_0 is given by

$$|\eta| < \sqrt{\left\{ \frac{\pi^2}{4} - \xi^2 \right\}}, \quad -\frac{\pi}{2} < \xi < 0$$

$$|\eta| < \frac{\pi}{2}, \quad 0 \le \xi \le \xi_0$$

$$|\eta| < \sqrt{\left\{ \frac{\pi^2}{4} - (\xi - \xi_0)^2 \right\}}, \quad \xi_0 \le \xi < \xi_0 + \frac{\pi}{2},$$

The total area of S_0 is

$$a = \pi \xi_0 + \frac{\pi^3}{4}.$$

Consider the function

$$g(\zeta) = f[z(\zeta)] = f \left\{ e^{i\theta} \left(\frac{e^{\zeta} - 1}{e^{\zeta} + 1} \right) \right\},$$

in S_0. Let $p(R, S_0, g)$ be defined by (2.4) with respect to g. Since S_0 corresponds by (2.11) to a subset of Δ, the equation $g(\zeta) = w$ cannot have more solutions in S_0 than the equation $f(z) = w$ has in Δ. Thus

$$p(R, S_0, g) \le p(R, \Delta, f) = p(R).$$

Again it follows from Lemma 2.2 that for all R, such that $\mu < R < M$, $R \notin F$, the level set $|g(\zeta)| = R$ contains an arc Γ going through a point P of L and approaching the boundary of S in both directions; Γ contains a subarc Γ_R passing through P, lying in S_0 and approaching the boundary of S_0 in both directions. The length of Γ_R must be at least π. Hence

$$l(R, S_0, g) \ge \pi, \quad \mu < R < M, \ R \notin F.$$

Now Theorem 2.1 yields

$$\int_\mu^M \frac{dR}{Rp(R)} \leq \int_\mu^M \frac{dR}{Rp(R,S_0,g)} \leq \frac{1}{\pi^2} \int_\mu^M \frac{l(R,S_0,g)^2}{p(R,S_0,g)} \frac{dR}{R}$$

$$\leq \frac{2a}{\pi} = 2\xi_0 + \frac{\pi^2}{2} = 2\log \frac{1+r}{1-r} + \frac{\pi^2}{2}.$$

This proves Theorem 2.2.

2.3 Some averaging assumptions on $p(R)$ If

$$p(R) \leq p \quad (0 < R < \infty), \tag{2.13}$$

Theorem 2.2 shows immediately that

$$\frac{1}{p} \left| \log \frac{M}{\mu} \right| < 2\log \left(e^{\pi^2/4} \frac{1+r}{1-r} \right),$$

$$M < \mu \left(e^{\pi^2/4} \frac{1+r}{1-r} \right)^{2p},$$

which is the type of result we are aiming at. Following Spencer [1940b, 1941a] we show in this section how we may weaken (2.13). We write

$$p(R) = p + h(R),$$

$$W(R) = \int_0^R p(\rho) d(\rho^2)$$

$$= pR^2 + \int_0^R h(\rho) d(\rho^2) = pR^2 + H(R),$$

say, and have the following inequality:

Lemma 2.3   *With the above notation, and if* $R_1 < R_2$ *we have*

$$\int_{R_1}^{R_2} \frac{d\rho}{\rho p(\rho)} \geq \frac{1}{p} \left\{ \log \frac{R_2}{R_1} - \frac{1}{2} - \frac{1}{2} \frac{H(R_2)}{2pR_2^2} - \frac{1}{p} \int_{R_1}^{R_2} \frac{H(\rho) d\rho}{\rho^3} \right\}.$$

We have

$$\int_{R_1}^{R_2} \frac{d\rho}{\rho p(\rho)} = \int_{R_1}^{R_2} \frac{d\rho}{\rho [p+h(\rho)]} = \int_{R_1}^{R_2} \frac{d\rho}{\rho} \left[\frac{1}{p} - \frac{h(\rho)}{p^2} + \frac{h^2(\rho)}{p^2 [p+h(\rho)]} \right]$$

$$\geq \frac{1}{p} \log \frac{R_2}{R_1} - \frac{1}{p^2} \int_{R_1}^{R_2} \frac{h(\rho) d\rho}{\rho}.$$

Again

$$\int_{R_1}^{R_2} \frac{h(\rho)d\rho}{\rho} = \int_{R_1}^{R_2} \frac{dH(\rho)}{2\rho^2} = \frac{H(R_2)}{2R_2^2} - \frac{H(R_1)}{2R_1^2} + \int_{R_1}^{R_2} \frac{H(\rho)d\rho}{\rho^3}.$$

Also $W(R)$ is necessarily positive and so $H(R_1) \geq -pR_1^2$. Now Lemma 2.1 follows.

Following Spencer we shall in this and the next chapter call a function $f(z)$ *mean p-valent* in a domain Δ, if $f(z)$ is regular in Δ, p is a positive number, and

$$W(R) \leq pR^2 \quad (0 < R < \infty).$$

Using the definition (2.4) we see that

$$W(R) = \frac{1}{\pi} \int_0^{2\pi} \int_0^R n(\rho e^{i\phi})\rho d\rho d\phi,$$

and so mean p-valency expresses the condition that the average number of roots in Δ of the equation $f(z) = w$ is not greater than p, as w ranges over any disc $|w| < R$.

We can now prove the theorem of Cartwright [1935] and Spencer [1940b].

Theorem 2.3   *Suppose that*

$$f(z) = a_0 + a_1 z + \dots$$

is mean p-valent and has q zeros in $|z| < 1$. Then $0 \leq q \leq p$ and

$$M(r, f) < A_0^{2p} e^{\frac{1}{2}} \mu (1 - r)^{-2p}, \quad 0 < r < 1,$$

where $A_0 = 2 \exp \{\pi^2/4\} < 47.2$, and μ is defined by (2.8).

The result was first proved by Cartwright [1935] for p-valent and by Spencer [1940b] for mean p-valent functions but with $e^{\frac{1}{2}} A_0^{2p}$ replaced by an unspecified constant depending on p. Jenkins and Oikawa [1971] first obtained the above form of the theorem, with a somewhat larger value of the constant A_0.

To prove it we note that if $f(z)$ has q zeros in Δ then $f(z)$ assumes all sufficiently small values at least q times in Δ. This contradicts mean p-valency if $q > p$. Thus $q \leq p$. By Lemma 2.3

$$\int_\mu^M \frac{dR}{Rp(R)} \geq \frac{1}{p} \left\{ \log \frac{M}{\mu} - \frac{1}{2} \right\},$$

since f is mean p-valent. Now Theorem 2.2 yields

$$\frac{1}{p}\left\{\log\frac{M}{\mu} - \frac{1}{2}\right\} \le \int_\mu^M \frac{dR}{Rp(R)} \le 2\log\frac{A_0}{1-r}.$$

Thus

$$\frac{M}{\mu} < \exp\left\{\frac{1}{2} + 2p\log\frac{A_0}{1-r}\right\}$$

$$= e^{\frac{1}{2}}A_0^{2p}(1-r)^{-2p}.$$

This proves Theorem 2.3.

The constant A_0 cannot be replaced by any number less than 2 in Theorem 2.3. This is shown by the function

$$w = f(z) = a_0\left(\frac{1+z}{1-z}\right)^{2p},$$

which maps $|z| < 1$ onto the sector (possibly self-overlapping) in the w plane given by

$$\left\{w : \left|\arg\frac{w}{a_0}\right| < p\pi \quad (0 < |w| < \infty)\right\},$$

and for which

$$p(R) = p, \quad W(R) = pR^2 \quad (0 < R < \infty).$$

Also $q = 0$, $\mu = |a_0|$ and

$$M(r, f) = \mu(1+r)^{2p}(1-r)^{-2p} \sim 2^{2p}\mu(1-r)^{-2p}, \quad r \to 1.$$

Examples

2.3 With the notation of the previous section, suppose that $f(z)$ is regular in Δ, $f(z) \ne 0$ and

$$\int_1^\infty \frac{H^+(R)dR}{R^3} < \infty, \text{ where } H^+(R) = \max(H(R), 0). \quad (2.14)$$

Prove that in this case

$$\limsup_{r\to 1}(1-r)^{2p}M(r, f) < \infty,$$

where $M(r, f) = \max_{|z|=r}|f(z)|$.

2.4   If $f(z)$ is regular in Δ, $f(z) \neq 0$ and

$$\lim_{R \to \infty} \sup \frac{H(R)}{R^2} = 0, \qquad (2.15)$$

prove that, for every positive ε,

$$\lim_{r \to 1}(1 - r)^{2p+\varepsilon} M(r, f) = 0.$$

2.5   Suppose that $f_1(z) = f(z) + C$ where C is a constant and that $W(R)$, $W_1(R)$ refer to $f(z)$, $f_1(z)$ respectively. Prove that

$$W(R - |c|) \leq W_1(R) \leq W(R + |c|).$$

Deduce that if $f(z)$ satisfies (2.14) or (2.15) then so does $f_1(z)$. Show also that $f_1(z)$ may be mean p-valent or satisfy (2.13) without $f(z)$ doing so.

2.4 Simultaneous growth near different boundary points   We have seen that a function $f(z)$ mean p-valent in $|z| < 1$ satisfies

$$|f(z)| = O(1 - r)^{-2p} \quad (|z| = r).$$

However a function can be as large as this only on a single rather small arc of $|z| = r$. We shall close this chapter by investigating in what way $f(z)$ can become large near several distant points of $|z| = r$. Our basic result is

Theorem 2.4   *Suppose that $f(z)$ is mean p-valent in a domain Δ containing k non-overlapping circles $|z - z_n| < r_n$ $(1 \leq n \leq k)$. Suppose further that $2^p|f(z_n)| \leq R_1$, $|f(z_n')| \geq R_2 > eR_1$, where*

$$\delta_n = \frac{r_n - |z_n' - z_n|}{r_n} > 0,$$

and that $f(z) \neq 0$ for $|z - z_n| < \frac{1}{2}r_n$ $(1 \leq n \leq k)$. Then

$$\sum_{n=1}^{k} \left[\log \frac{A_0}{\delta_n} \right]^{-1} < \frac{2p}{\log(R_2/R_1) - 1}.$$

Let Δ_n be the circle $|z - z_n| < r_n$ and let

$$p_n(R) = p(R, \Delta_n, f)$$

be defined as in (2.4). Consider

$$\phi(\zeta) = f(z_n + r_n\zeta).$$

Then $p_n(R)$ corresponds to $\phi(\zeta)$ and $|\zeta| < 1$. We choose ζ_n so that

$$z_n + r_n \zeta_n = z'_n, \quad \zeta_n = \frac{z'_n - z_n}{r_n}.$$

Then $2^p |\phi(0)| \leq R_1$, $|\phi(\zeta_n)| \geq R_2$, $\phi(\zeta) \neq 0$ for $|\zeta| < \frac{1}{2}$ and $\phi(\zeta)$ has at most p zeros in $|\zeta| < 1$. Thus $\mu \leq 2^p |\phi(0)| \leq R_1$ and Theorem 2.2 gives

$$\int_{R_1}^{R_2} \frac{dR}{R p_n(R)} < 2 \log \frac{A_0}{1 - |\zeta_n|} = 2 \log \frac{A_0}{\delta_n}.$$

Now we have from Schwarz's inequality

$$\left(\log \frac{R_2}{R_1} \right)^2 = \left(\int_{R_1}^{R_2} \frac{dR}{R} \right)^2 \leq \int_{R_1}^{R_2} \frac{p_n(R) dR}{R} \int_{R_1}^{R_2} \frac{dR}{R p_n(R)},$$

and so

$$\frac{1}{\log \left(\frac{A_0}{\delta_n} \right)} \leq \frac{2}{\int_{R_1}^{R_2} \frac{dR}{R p_n(R)}} \leq \frac{2}{\left(\log \frac{R_2}{R_1} \right)^2} \int_{R_1}^{R_2} \frac{p_n(R) dR}{R}.$$

Adding we deduce

$$\sum_{n=1}^{k} \left[\log \frac{A_0}{\delta_n} \right]^{-1} \leq \frac{2}{[\log(R_2/R_1)]^2} \int_{R_1}^{R_2} \left[\sum_{n=1}^{k} p_n(R) \right] \frac{dR}{R}. \tag{2.16}$$

Now

$$\sum_{n=1}^{k} p_n(R) = \sum_{n=1}^{k} p(R, \Delta_n) \leq p(R, \Delta) = p(R). \tag{2.17}$$

Using the notation of §2.3 we have, since $f(z)$ is mean p-valent in Δ,

$$
\begin{aligned}
\int_{R_1}^{R_2} p(R) \frac{dR}{R} &= p \log \frac{R_2}{R_1} + \int_{R_1}^{R_2} h(R) \frac{dR}{R} \\
&= p \log \frac{R_2}{R_1} + \frac{1}{2} \int_{R_1}^{R_2} \frac{dH(R)}{R^2} \\
&= p \log \frac{R_2}{R_1} + \frac{H(R_2)}{2 R_2^2} - \frac{H(R_1)}{2 R_1^2} + \int_{R_1}^{R_2} \frac{H(R) dR}{R^3} \\
&\leq p \left[\log \left(\frac{R_2}{R_1} \right) + \frac{1}{2} \right],
\end{aligned}
$$

since $-pR^2 \leq H(R) \leq 0$. Using (2.16) and (2.17), we deduce

$$\sum_{n=1}^{k} \left[\log \left(\frac{A_0}{\delta_n} \right) \right]^{-1} \leq \frac{2p}{[\log(R_2/R_1)]} + \frac{p}{[\log(R_2/R_1)]^2}$$

$$< \frac{2p}{\log(R_2/R_1) - 1},$$

and this proves Theorem 2.4.

2.5 Applications   We shall make two applications of Theorem 2.4. We first apply it to the concept of order of a mean p-valent function at a boundary point.

Suppose then that $f(z)$ is mean p-valent in $|z| < 1$ and that for $\zeta = e^{i\theta}$ there is a path $\gamma(\theta)$, lying except for its end-point ζ in $|z| < 1$, and also a positive δ such that

$$\underline{\lim}(1 - |z|)^\delta |f(z)| > 0,$$

as $z \to \zeta$ along $\gamma(\theta)$. Then we define the *order* $\alpha(\zeta)$ of $f(z)$ at ζ as the upper bound of all such δ. If no path $\gamma(\theta)$ and positive δ exist, we put $\alpha(\zeta) = 0$. We can then prove the following result of Spencer [1940b].

Theorem 2.5   *If $f(z)$ is mean p-valent in $|z| < 1$, then the set E of distinct ζ on $|\zeta| = 1$, such that $\alpha(\zeta) > 0$, is countable and satisfies $\sum_E \alpha(\zeta) \leq 2p$.*

It suffices to show that if $\zeta_1, \zeta_2, ..., \zeta_k$ are distinct points of $|\zeta| = 1$, then

$$\sum_{n=1}^{k} \alpha(\zeta_n) \leq 2p.$$

For then the set E_N of ζ, such that $|\zeta| = 1$ and $\alpha(\zeta) > N^{-1}$, is finite for $N = 1, 2, 3, ...$, and so the set E consisting of all the E_N is countable. Letting k tend to ∞ in the above inequality we have $\sum \alpha(\zeta) \leq 2p$ as required.

Suppose then that Theorem 2.5 is false. Then by the above remark we can find a finite set of points $\zeta_1, \zeta_2, ..., \zeta_k$ and a positive ϵ such that

$$\sum_{n=1}^{k} \alpha(\zeta_n) = 2(p + k\epsilon).$$

For each ζ_n there exists a path γ_n, approaching ζ_n from $|z| < 1$, on which

$$(1 - |z|)^{\eta_n} |f(z)| > 1,$$

where $\eta_n = \alpha(\zeta_n) - \epsilon$, and so

$$\sum_{n=1}^{k} \eta_n > 2p. \tag{2.18}$$

Hence there exists a positive R_0, such that for $R_2 > R_0$ we can find $z'_n = r_n e^{i\theta_n}$ on γ_n, such that

$$|f(z'_n)| = R_2 > \left(\frac{1}{1-r_n}\right)^{\eta_n} \quad (1 \le n \le k). \qquad (2.19)$$

Choose now δ so small that

(i) $f(z)$ has no zeros for $1 - 2\delta < |z| < 1$,
(ii) $4\delta < \min\limits_{1 \le m < n \le k} |\zeta_m - \zeta_n|$,

and set $r_0 = 1 - \delta$. Then if R_0 is sufficiently large, we have

$$|z'_m - z'_n| > 4\delta \quad (1 \le m < n \le k),$$

since $z'_n \to \zeta_n$ as $R_2 \to \infty$, and if we put $z_n = r_0 e^{i\theta_n}$ it follows that the circles $|z - z_n| < \delta$ are non-overlapping. Also

$$2^p |f(z_n)| \le R_1 = 2^p M(r_0, f) \text{ and } |f(z'_n)| = R_2,$$

where we may suppose that $R_2 > eR_1$. Thus we may apply Theorem 2.4 with

$$\delta_n = \frac{\delta - (r_n - r_0)}{\delta} = \frac{1 - r_n}{\delta},$$

and obtain

$$\sum_{n=1}^{k} \left[\log\left(\frac{A_0\delta}{1-r_n}\right)\right]^{-1} \le \frac{2p}{\log(R_2/eR_1)}.$$

In view of (2.19) this gives

$$\sum_{n=1}^{k} [\log(A_0\delta R_2^{1/\eta_n})]^{-1} \le \frac{2p}{\log(R_2/eR_1)},$$

$$\sum_{n=1}^{k} \frac{\eta_n}{\eta_n \log[A_0\delta] + \log R_2} \le \frac{2p}{\log R_2 - \log(eR_1)}.$$

As $R_2 \to \infty$ the two sides of this inequality are asymptotically equal to

$$\frac{\sum \eta_n}{\log R_2} \text{ and } \frac{2p}{\log R_2}$$

respectively, and so we obtain a contradiction from (2.18). This proves Theorem 2.5.

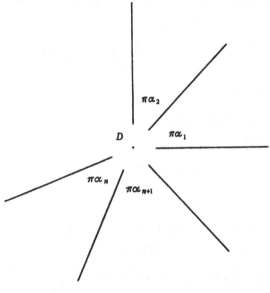

Fig. 2.

2.5.1 An example   Theorem 2.7 is sharp. Suppose in fact that α_n is any sequence of positive numbers for $1 \le n < k$, where k may be finite or infinite, such that

$$\sum \alpha_n = 2.$$

Set $S_0 = 0$, $S_N = \sum_{n=1}^{N} \alpha_n \, (1 \le N < k)$, and let D be the domain consisting of the w plane cut from $|w| = 1$ to ∞ along the lines

$$\arg w = \pi S_N \quad (0 \le N < k).$$

By Riemann's mapping theorem[†] we can find a function $z = \phi(w)$ which maps D (1,1) conformally onto $|z| < 1$.

As w moves along the boundary of that part D_N of D which lies in the angle $\pi S_{N-1} < \arg w < \pi S_N$, z moves along an arc of the unit circle. Simultaneously $W = w^{-1/\alpha_n}$ moves along a straight line segment through the origin and so by Schwarz's reflection principle[‡] can be continued as a regular function of W near $W = 0$, and conversely if $z = z_n$ corresponds

---

[†] See, for example, C.A. p. 230
[‡] See, for example, C. A., p. 172.

to $W = 0$, W becomes a regular function of z near z_n. Thus

$$W = c_1(z - z_n) + c_2(z - z_n)^2 + ..., \text{ near } z = z_n,$$

where $c_1 \neq 0$. This gives

$$w \sim [c_1(z - z_n)]^{-\alpha_n} \quad (z \to z_n).$$

Hence $w = f(z) = \phi^{-1}(z)$, which maps $|z| < 1$ onto D, is a univalent function having the orders α_n at the points z_n for $1 \leq n < k$, and the α_n are arbitrary subject to $\sum \alpha_n = 2$. Again for any positive p the function $[f(z) - 1]^p$ is mean p-valent in $|z| < 1$ and has orders $p\alpha_n$ at the points z_n, where $p\alpha_n$ is now an arbitrary sequence satisfying $\sum p\alpha_n = 2p$.

2.6 Functions of maximal growth We have seen that if $f(z)$ is mean p-valent in $|z| < 1$ then

$$M(r, f) = O(1 - r)^{2p} \quad (r \to 1).$$

We complete the chapter by investigating more closely those functions for which this order of growth is effectively attained so that

$$\alpha = \overline{\lim}_{r \to 1}(1 - r)^{2p} M(r, f) > 0. \tag{2.20}$$

We show that $f(z)$ attains this growth along a certain radius and that $|f(z)|$ is quite small except near this radius.

Theorem 2.6 *Suppose that $f(z)$ is mean p-valent in $|z| < 1$ and that (2.20) holds. Then there exists θ_0 $(0 \leq \theta_0 < 2\pi)$ such that*

$$\alpha_0 = \lim_{r \to 1}(1 - r)^{2p}|f(re^{i\theta_0})| \geq \frac{\alpha}{2e(2A_0^2)^p}.$$

Since $f(z)$ can have only a finite number of zeros in $|z| < 1$ we may suppose that $f(z) \neq 0$ in $1 - 2\delta < |z| < 1$ where $\delta > 0$. We set $r_0 = 1 - \delta$. Then there exists a sequence $\zeta_n = r_n e^{i\theta_n}$, where

$$r_0 < r_n < 1, \quad 0 \leq \theta_n < 2\pi,$$

such that $r_n \to 1$ $(n \to \infty)$ and

$$|f(\zeta_n)| > \tfrac{1}{2}\alpha(1 - r_n)^{-2p}. \tag{2.21}$$

We now apply Theorem 2.4 with $k = 1$, $z_1' = \zeta_n$, $z_1 = re^{i\theta_n}$, where

$r_0 < r < r_n$ and with $(1 - r)$ instead of r_n. This is legitimate, since $f(z)$ is mean p-valent in $|z - z_1| < 1 - r$. Then

$$\delta_1 = \frac{1 - r_n}{1 - r}.$$

We also set $R_1 = 2^p |f(z_1)|$, $R_2 = |f(z_1')|$. Thus either $R_2 \le eR_1$ or

$$R_2 \le eR_1 \left(\frac{A_0}{\delta_1}\right)^{2p}, \tag{2.22}$$

so that (2.22) holds in any case. Using (2.21) we obtain

$$R_1 = 2^p |f(re^{i\theta_n})| \ge \frac{R_2 \delta_1^{2p}}{eA_0^{2p}} > \frac{\alpha}{2eA_0^{2p}}(1 - r)^{-2p}, \quad r_0 < r < r_n.$$

If θ_0 is a limit point of the sequence θ_n, we deduce by continuity, taking r fixed, that

$$2^p |f(re^{i\theta_0})| \ge \frac{\alpha(1 - r)^{-2p}}{2eA_0^{2p}}, \quad r_0 < r < 1. \tag{2.23}$$

This proves Theorem 2.6.

2.6.1   It follows from Theorem 2.5 that $f(z)$ must have zero order at all points of $|z| = 1$ other than $e^{i\theta_0}$, so that $e^{i\theta_0}$ is necessarily unique in Theorem 2.6. However, we can prove more than this.

Theorem 2.7   *With the hypothesis of Theorem 2.6 and given ϵ such that $0 < \epsilon < 2p$, we can find a positive constant C and r_0, $0 < r_0 < 1$, such that*

$$|f(re^{i\theta})| < \frac{1}{(1 - r)^\epsilon |\theta - \theta_0|^{2p - \epsilon}} \tag{2.24}$$

for $r_0 < r < 1$, $C(1 - r) \le |\theta - \theta_0| \le \pi$. Further we have uniformly as $r \to 1$, while $\epsilon \le |\theta - \theta_0| \le \pi$,

$$\log |f(re^{i\theta})| < O\left[\log \frac{1}{1 - r}\right]^{\frac{1}{2}}. \tag{2.25}$$

In particular, we see that (2.25) holds as $re^{i\theta}$ approaches any point on $|z| = 1$ distinct from $e^{i\theta}$. This is a good deal stronger than the condition for zero order. Hayman and Kennedy [1958] have shown that O can be replaced by o in (2.25) and that the result is then sharp.

Suppose then that $f(z)$ is mean p-valent, $f(z) \neq 0$ in

$$1 - 2\delta < |z| < 1,$$

and that

$$|f(re^{i\theta_0})| \geq \frac{1}{2}\alpha_0(1 - r)^{2p} \quad (1 - \delta < r < 1). \tag{2.26}$$

These assumptions are satisfied for all sufficiently small δ. We also write

$$2^p R_1 = M(1 - \delta, f) = \max_{|z|=1-\delta} |f(z)|, \tag{2.27}$$

$$z_1 = (1 - \delta)e^{i\theta_0}, \quad z_2 = (1 - \delta)e^{i\theta},$$

and assume that $2\pi\delta \leq |\theta - \theta_0| \leq \pi$. Then the circles $|z - z_1| < \delta$, $|z - z_2| < \delta$ are disjoint.

We now suppose further that

$$|f(r_2 e^{i\theta})| = R_2 > eR_1, \tag{2.28}$$

where $1 - \delta < r_2 < 1$, and put $z_2' = r_2 e^{i\theta}$, $z_1' = r_1 e^{i\theta_0}$, where r_1 is chosen to be the smallest number such that

$$|f(z_1')| = |f(r_1 e^{i\theta_0})| = R_2.$$

Such a number exists by Theorem 2.6 and (2.27), (2.28). Then $r_1 > 1 - \delta$ by definition of R_1.

We now apply Theorem 2.4 with

$$\delta_1 = \frac{\delta - [r_1 - (1 - \delta)]}{\delta} = \frac{1 - r_1}{\delta}, \quad \delta_2 = \frac{1 - r_2}{\delta},$$

and obtain

$$\left[\log\left(\frac{A_0\delta}{1 - r_2}\right)\right]^{-1} \leq \frac{2p}{\log(R_2/R_1) - 1} - \left[\log\left(\frac{A_0\delta}{1 - r_1}\right)\right]^{-1}. \tag{2.29}$$

Now by (2.26)

$$R_2 \geq \tfrac{1}{2}\alpha_0(1 - r_1)^{-2p},$$

and by Theorem 2.3

$$R_1 \leq A_0^{2p} e^{\frac{1}{2}} \mu\delta^{-2p}.$$

Thus

$$\frac{R_2}{R_1} \geq C_1 \left(\frac{\delta}{1 - r_1}\right)^{2p}$$

where $C_1, C_2, \dots$ will denote constants depending on $f(z)$ and p only.

We deduce from this and (2.29) that

$$\left[\log \frac{A_0 \delta}{1 - r_2}\right]^{-1} \leq \frac{2p}{\log(R_2/R_1) - 1} \quad \frac{2p}{\log(R_2/R_1) + C_2},$$

and this gives

$$\log\left(\frac{A_0 \delta}{1 - r_2}\right) \geq C_3 \left\{\log\left(\frac{R_2}{R_1}\right) - 1\right\}^2.$$

Thus

$$\log \frac{R_2}{R_1} \leq 1 + C_4 \left[\log\left(\frac{C_5 \delta}{1 - r_2}\right)\right]^{\frac{1}{2}}. \qquad (2.30)$$

Taking δ and R_1 fixed, we deduce (2.25), provided that (2.28) holds, and the inequality is trivial otherwise.

It remains to prove (2.24). We may again without loss of generality suppose that (2.28) holds. Then (2.30) gives, if $2\pi\delta = |\theta - \theta_0| \geq 1 - r_2$,

$$
\begin{aligned}
R_2 \quad &< \quad R_1 \exp\left\{1 + C_4 \left[\log \frac{C_5 \delta}{(1 - r_2)}\right]^{\frac{1}{2}}\right\} \\
&< \quad C_6 |\theta - \theta_0|^{-2p} \exp\left\{C_4 \left[\log \frac{C_5 \delta}{1 - r_2}\right]^{\frac{1}{2}}\right\} \\
&< \quad \frac{1}{|\theta - \theta_0|^{2p-\epsilon}(1 - r_2)^\epsilon},
\end{aligned}
$$

provided that

$$\left(\frac{|\theta - \theta_0|}{1 - r_2}\right)^\epsilon > C_6 \exp\left\{C_4 \left(\log\left[\frac{C_5 |\theta - \theta_0|}{2\pi(1 - r_2)}\right]\right)^{\frac{1}{2}}\right\},$$

which is true for $|\theta - \theta_0| \geq C(1 - r_2)$, where C depends on p, ϵ and $f(z)$ only. This proves (2.24) and completes the proof of Theorem 2.7.

2.7 Behaviour near the radius of greatest growth It follows from Theorem 2.6 that if $f(z)$ is mean p-valent in $\Delta : |z| < 1$ and has maximal growth there, so that (2.20) holds, then $f(z)$ attains this growth along the radius $\arg z = \theta_0$. Further, by Theorem 2.7, $f(z)$ is quite small except in the immediate neighbourhood of the radius, which we shall call henceforth the *radius of greatest growth* (r.g.g.).

There is a general principle in the theory of functions, that if under certain hypotheses there is a bound on the growth of a class $\mathfrak{F}$ of functions, then those functions in $\mathfrak{F}$ which have the extremal growth display a regular behaviour. Such theorems were first found by Heins [1948] for the class $\mathfrak{F}$ of entire functions with bounded minimum modulus and are called *regularity theorems*. We had an example for univalent functions in Theorems 1.5 and 1.12.

We now proceed to prove some regularity theorems due to Eke [1967a,b] for mean p-valent functions.

It is convenient to use the transformation

$$\zeta = \xi + i\eta = \frac{1}{2} \log \left\{ \frac{z e^{-i\theta_0}}{(1 - z e^{-i\theta_0})^2} \right\}, \tag{2.31}$$

where $\arg z = \theta_0$ is the r.g.g. Then the domain D_{θ_0}, consisting of the unit disc cut along the radius $\arg z = \theta_0 + \pi$ is mapped onto the strip S, given by (2.12). If $g(\zeta) = f[z(\zeta)]$ then $g(\zeta)$ is mean p-valent in S and by (2.23) we have

$$\log |g(\zeta)| = 2p\xi + O(1), \text{ as } \zeta \to +\infty \tag{2.32}$$

along the real axis. We shall, following Eke [1967b], prove

Theorem 2.8   *Suppose that $g(\zeta)$ is mean p-valent in S and satisfies (2.32), where $\zeta = \xi + i\eta$. Then we have as $\xi \to +\infty$, uniformly for $|\eta| < \frac{\pi}{2} - \delta$, where $\delta > 0$,*

$$\log |g(\zeta)| = 2p\xi + \beta + o(1), \tag{2.33}$$

where β is a real constant and

$$g'(\zeta)/g(\zeta) \to 2p. \tag{2.34}$$

We note that $g(\zeta)$ is mean p-valent in S and so in any subdomain S_1 of S. We first consider $g(\zeta)$ in a rectangle

$$S_1 : \xi_0 - \pi < \xi < \xi_1 + \pi, \quad |\eta| < \frac{\pi}{2}.$$

We choose ξ_0 so that $\xi_0 > \pi$ and so that all the zeros of $g(\zeta)$ lie in $\xi \leq \xi_0 - \pi$.

Then it follows from (2.32) that there is also a positive constant C_0, such that

$$\log |g(\xi)| < 2p\xi_0 + C_0, \quad \xi_0 - 2\pi \leq \xi \leq \xi_0 \tag{2.35}$$

and

$$\left| \log |g(\xi)| - 2p\xi \right| < C_0, \quad \xi > \xi_0. \tag{2.36}$$

In fact (2.36) is an immediate consequence of (2.32) and $g(\xi) \neq 0$, and

(2.35) follows from the fact that $|g(\xi)|$ is continuous and so bounded on any compact subset of S. We note that if (2.35), (2.36) hold for a given ξ_0, with a constant C_0, then they also hold with the same C_0 when ξ_0 is replaced by a large number. We shall need a sequence of lemmas for our proof of Theorem 2.8.

2.7.1 Construction of some level curves

Lemma 2.4 *We define* $C_1 = \frac{13}{5}C_0 + 1 + \frac{16p\pi}{5}$, *suppose that* $p(\xi_1 - \xi_0) \geq C_0 + C_1$, *that*

$$2p\xi_0 + C_1 < \log R < 2p\xi_1 - C_1 \tag{2.37}$$

and that R does not belong to a certain exceptional set F, which is finite or countable. Then there exists a level curve γ_R on which $|g(\zeta)| = R$, and such that γ_R joins the sides $\eta = \pm\frac{\pi}{2}$ of S_1 in S_1 and so crosses the segment $[\xi_0, \xi_1]$ of the real axis an odd number of times.

Let F be the (finite or countable) set of values of R such that $R = |g(\zeta)|$ for some value of ζ for which $g'(\zeta) = 0$, or $R = |g(\xi)|$ where ξ is real and $\frac{d}{dt}|g(t)| = 0$ at $t = \xi$. We suppose first that R is not in F, that $2p(\xi_1 - \xi_0) \geq C_2 = C_0 + C_1$ and that

$$2p\xi_0 + C_0 < \log R < 2p\xi_1 - C_0. \tag{2.38}$$

Then by (2.36) there is an odd number of values of ξ such that $\xi_0 < \xi < \xi_1$ and $|g(\xi)| = R$. The level curves $|g(\xi)| = R$ through these points meet the real axis only on the segment $[\xi_0, \xi_1]$ by (2.35), (2.36) and (2.38). Hence at least one of these level curves γ_R crosses the real axis an odd number of times and so either joins the boundary segments $\eta = \pm\frac{\pi}{2}$ of S_1 in S_1 or else meets one of the vertical boundary segments $\xi = \xi_0 - \pi$, or $\xi = \xi_1 + \pi$ of S_1. In the latter case the length $l(R)$ of γ_R is at least $3\pi/2$, since γ_R contains one arc going from a point ξ on $[\xi_0, \xi_1]$ to one of the above segments, whose length is at least π, and another arc going to the boundary of S_1 whose length is at least $\frac{1}{2}\pi$. Suppose that this latter situation occurs for each R not in F and satisfying (2.38). We shall show that this leads to a contradiction.

We apply Theorem 2.1 and Lemma 2.3 to the function $g(\zeta)$ which is mean p-valent in S_1, so that $H(R) \leq 0$ in Lemma 2.3. We write $R_1 = \exp\{2p\xi_0 + C_0\}$, $R_2 = \exp\{2p\xi_1 - C_0\}$ and obtain

$$2\pi^2 \{\xi_1 - \xi_0 + 2\pi\} \geq \int_{R_1}^{R_2} \frac{(3\pi/2)^2 dR}{Rp(R)} \geq \frac{9\pi^2}{4p} \left\{ \log \frac{R_2}{R_1} - \frac{1}{2} \right\}$$

$$= \frac{9\pi^2}{2}(\xi_1 - \xi_0) - \frac{9\pi^2}{4p} \left\{ 2C_0 + \frac{1}{2} \right\},$$

i.e.

$$\frac{5\pi^2}{2}(\xi_1 - \xi_0) \leq \frac{\pi^2}{2p} \left\{ 9C_0 + \frac{9}{4} + 8p \right\}, 2p(\xi_1 - \xi_0) \leq \frac{18C_0}{5} + \frac{9}{10} + \frac{16p}{5} < C_2,$$

and this contradicts $2p(\xi_1 - \xi_0) \geq C_2$. Thus there exists R satisfying (2.38) such that γ_R joins $\eta = \pm\frac{\pi}{2}$ in S_1.

Suppose now that $p(\xi_1 - \xi_0) \geq C_2$. We define ξ_3, ξ_4 by

$$p\xi_3 = p\xi_0 + \frac{1}{2}C_2, \quad p\xi_4 = p\xi_1 - \frac{1}{2}C_2.$$

By what we have just proved, there exist R_1, R_2 such that

$$2p\xi_0 + C_0 < \log R_1 < 2p\xi_3 - C_0, \quad 2p\xi_4 + C_0 < \log R_2 < 2p\xi_1 - C_0,$$

and such that $\gamma_{R_1}, \gamma_{R_2}$ joins $\eta = \pm\frac{\pi}{2}$ in S_1. Hence if $R_1 < R < R_2$ and R does not belong to F then a level curve γ_R exists, which crosses the segment $[\xi_0, \xi_1]$ an odd number of times. This level curve separates γ_{R_1} from γ_{R_2} and so also joins $\eta = \pm\frac{\pi}{2}$ in S_1. The condition is certainly satisfied if $2p\xi_3 - C_0 < \log R < 2p\xi_4 + C_0$, i.e

$$2p\xi_0 + C_2 - C_0 < \log R < 2p\xi_1 + C_0 - C_2,$$

i.e. if (2.37) holds. This proves Lemma 2.4.

2.7.2 Basic estimates

Our next aim is to prove that when R is large the levels curves γ_R, whose existence is asserted in Lemma 2.4, are close to vertical segments. We define

$$\xi_1(R) = \inf_{\zeta \in \gamma_R} \Re\zeta, \quad \xi_2(R) = \sup_{\zeta \in \gamma_R} \Re\zeta, \qquad (2.39)$$

$$\omega(R) = \xi_2(R) - \xi_1(R), \qquad (2.40)$$

and note that the length $l(R)$ of γ_R satisfies

$$l(R)^2 \geq \pi^2 + \omega(R)^2. \qquad (2.41)$$

We leave this result as an exercise for the reader.

Examples

2.6   Show that if γ is a Jordan arc which meets all four sides of a rectangle Q, then the length of γ is not less than that of the diagonal of Q.

It is a consequence of (2.34) that $|g(\xi)|$ is finally increasing so that γ_R is unique, when R is large. For the time being we do not assume this, but choose for γ_R the first such arc which we meet on moving along the real axis from ξ_0 in the direction of increasing ξ. Given a sufficiently large R, we may define ξ_0, ξ_1 by

$$2p\xi_0 = \log R - C_1 - C_0, \quad 2p\xi_1 = \log R + C_1 + C_0,$$

since (2.35) and (2.36) continue to hold if ξ_1 is replaced by a larger quantity. We may then apply Lemma 2.4 and deduce that

$$\omega(R) \le \xi_1 - \xi_0 + 2\pi = \frac{2C_1 + 2C_0}{2p} + 2\pi = C_3, \qquad (2.42)$$

provided that $R \ge R_0$ say. Our next result is

Lemma 2.5   *If $R_2 > R_1 \ge R_0$ we have*

$$\frac{1}{2\pi^2} \int_{R_1}^{R_2} \frac{(\pi^2 + \omega(R)^2)dR}{Rp(R)} \;\le\; \{\xi_2(R_2) - \xi_1(R_1)\}$$

$$\le \; \{\xi(R_2) - \xi(R_1) + \omega(R_2) + \omega(R_1)\}$$

$$\le \; \frac{1}{2p}\log\frac{R_2}{R_1} + C_4, \qquad (2.43)$$

where $\xi(R)$ denotes the least value of ξ at which $\gamma(R)$ meets the real axis, and $C_4 = C_0/p + 2C_3$.

We apply Theorem 2.1 to that subdomain S' of the strip S which is bounded by γ_{R_1}, γ_{R_2} and the lines $\eta = \pm\frac{1}{2}\pi$. In S' we have

$$\xi_1(R_1) \le \xi \le \xi_2(R_2)$$

so that the area A of S' satisfies

$$A \le \pi(\xi_2(R_2) - \xi_1(R_1)).$$

Also if $l(R)$ denotes the length of γ_R then (2.41) holds, and if $R_1 < R < R_2$, (2.41) remains true if we replace $l(R)$ by the total length of all levels curves $|g(\xi)| = R$ in S'. Now the first inequality in (2.43) follows from Theorem 2.1.

Next if $\xi = \xi(R)$ denotes the least real value of ξ on the level curve γ_R, we have

$$\xi_1(R) \leq \xi(R) \leq \xi_2(R) \leq \xi_1(R) + \omega(R), \quad R_1 \leq R \leq R_2$$

by (2.40). Thus

$$\xi_2(R_2) - \xi_1(R_1) \leq \xi(R_2) - \xi(R_1) + \omega(R_2) + \omega(R_1).$$

This proves the second inequality in (2.43). Finally we deduce from (2.36) that if $\xi = \xi(R)$

$$|\log R - 2p\xi| < C_0, \quad R_1 \leq R \leq R_2.$$

Applying this with $R = R_1$, R_2, we obtain

$$\xi(R_2) - \xi(R_1) \leq \frac{1}{2p}\left\{\log\frac{R_2}{R_1} + 2C_0\right\}.$$

Using also (2.42) we obtain

$$\xi(R_2) - \xi(R_1) + \omega(R_2) + \omega(R_1) \leq \frac{1}{2p}\log\frac{R_2}{R_1} + \frac{C_0}{p} + 2C_3.$$

This completes the proof of Lemma 2.5.

Lemma 2.6 *We have*

$$\int_{R_0}^{R} \frac{d\rho}{\rho p(\rho)} = \frac{1}{p}\log R + \beta_1 + o(1), \quad as \quad R \to \infty \qquad (2.44)$$

where β_1 is a real constant. Also

$$\int_{R_0}^{\infty} \frac{\omega^2(\rho)d\rho}{\rho p(\rho)} < \infty. \qquad (2.45)$$

To prove (2.44) we use the argument of Lemma 2.3 and of its proof. With the terminology of Section 2.3 we have

$$\int_{R_0}^{R_2} \frac{d\rho}{\rho p(\rho)} = \frac{1}{p}\log\frac{R_2}{R_0} + \frac{1}{p^2}\left\{\frac{H(R_0)}{2R_0^2} - \frac{H(R_2)}{2R_2^2} - \int_{R_0}^{R_2}\frac{H(\rho)d\rho}{\rho^3}\right\}$$

$$+ \int_{R_0}^{R_2}\frac{h^2(\rho)d\rho}{\rho(p + h(\rho))}. \qquad (2.46)$$

Here $h^2(\rho), -H(\rho)$ are nonnegative for $\rho \geq R_0$. It follows from Lemma 2.5 that

$$\int_{R_0}^{R_2} \frac{d\rho}{\rho p(\rho)} \leq \frac{1}{p}\log R_2 + O(1), \quad as \quad R_2 \to \infty.$$

Thus

$$\int_{R_0}^{\infty} \frac{-H(\rho)d\rho}{\rho^3} = \beta_2, \quad \int_{R_0}^{\infty} \frac{h^2(\rho)d\rho}{\rho^2 \rho(p + h(\rho))} = \beta_3, \qquad (2.47)$$

where β_2, β_3 are finite nonnegative constants. Next we show that

$$H(R_2)/R_2^2 \to 0, \quad \text{as } R_2 \to \infty. \qquad (2.48)$$

Suppose that (2.48) is false. Then there exist arbitrarily large values $\rho = \rho_n$ such that

$$H(\rho_n) < -\delta\rho_n^2,$$

where δ is a positive number. Also $W(R) = pR^2 + H(R)$ is a nondecreasing function of R. Thus

$$pR^2 + H(R) < p\rho_n^2 + H(\rho_n) < (p - \delta)\rho_n^2, \quad R < \rho_n$$

i.e.

$$H(R) < (p - \delta)\rho_n^2 - pR^2 < -\frac{1}{2}\delta\rho_n^2,$$

if $R < \rho_n$ and

$$p(\rho_n^2 - R^2) < \frac{1}{2}\delta\rho_n^2, \text{ i.e. } (\rho_n - R) < \frac{\delta\rho_n^2}{2p(\rho_n + R)},$$

and so certainly if

$$\rho_n - \frac{\delta\rho_n}{4p} < R < \rho_n.$$

This yields

$$\int_{\rho_n(1-\delta/4p)}^{\rho_n} \frac{H(\rho)d\rho}{\rho^3} < -\frac{1}{2}\delta\rho_n^2 \int_{\rho_n(1-\delta/4p)}^{\rho_n} \frac{d\rho}{\rho^3}$$

$$< -\frac{\delta\rho_n^2}{4} \left\{ \left[\rho_n\left(1 - \frac{\delta}{4p}\right)\right]^{-2} - \rho_n^{-2} \right\}$$

$$= \frac{\delta}{4} \left\{ 1 - \left(1 - \frac{\delta}{4p}\right)^{-2} \right\} = -\eta,$$

say. This inequality holds for some arbitrialy large ρ_n contrary to the conclusion that the first integral in (2.47) converges. Thus (2.48) holds. Substituting (2.47) and (2.48) in (2.46) we deduce (2.44).

Next we prove (2.45). We deduce from (2.43) and (2.44) that

$$\frac{1}{2\pi^2} \int_{R_0}^{R_2} \frac{\omega^2(R)dR}{Rp(R)} \le \frac{1}{2p} \log R_2 - \frac{1}{2} \int_{R_0}^{R_2} \frac{dR}{Rp(R)} + O(1) \le O(1).$$

This proves (2.45) and completes the proof of Lemma 2.6.

Lemma 2.7   *We have as $R \to \infty$*

$$\omega(R) \to 0 \tag{2.49}$$

and

$$\xi_j(R) = \frac{1}{2p} \log R - \frac{\beta}{2p} + o(1), \quad j = 1, 2 \tag{2.50}$$

where β is a real constant.

We shall prove (2.50) and note that (2.49) is an immediate consequence of (2.50) and (2.40). We define

$$\beta_4 = \lim_{R \to \infty} \sup \left\{ \xi_2(R) - \frac{1}{2p} \log R \right\},$$

and

$$\beta_5 = \lim_{R \to \infty} \inf \left\{ \xi_1(R) - \frac{1}{2p} \log R \right\}.$$

It follows from Lemma 2.4 that β_4 and β_5 are finite. Also since $\xi_1(R) \leq \xi_2(R)$ we have

$$\lim_{R \to \infty} \sup \left\{ \xi_1(R) - \frac{1}{2p} \log R \right\} \leq \beta_4$$

and

$$\lim_{R \to \infty} \inf \left\{ \xi_2(R) - \frac{1}{2p} \log R \right\} \geq \beta_5.$$

Thus $\beta_4 \geq \beta_5$. Hence to prove (2.50) with $-\beta/2p = \beta_4 = \beta_5$ it is enough to show that $\beta_5 = \beta_4$. Suppose then contrary to this, that $\beta_4 > \beta_5$. We define ε by

$$\beta_4 - \beta_5 = 4\varepsilon \left(1 + \frac{1}{p} \right).$$

We chose R_0' so large that

$$\int_{R_0'}^{\infty} \frac{\omega(R)^2 dR}{R p(R)} < \frac{\varepsilon^3}{2p}, \tag{2.51}$$

and also that for $R > R' > R_0'$ we have

$$\left| \int_R^{R'} \frac{d\rho}{\rho p(\rho)} - \frac{1}{p} \log \frac{R}{R'} \right| < \frac{\varepsilon}{2p}. \tag{2.52}$$

We can satisfy (2.51) by using (2.45) and (2.52) by using (2.44).

Next we choose $R_1 > R_0'$, so that

$$\xi_2(R_1) > \frac{1}{2p}\log R_1 + \beta_4 - \varepsilon,$$

and then R_2, so that $R_2 > R_1$ and

$$\xi_1(R_2) < \frac{1}{2p}\log R_2 + \beta_5 + \varepsilon.$$

Subtracting we deduce that

$$
\begin{aligned}
\xi_1(R_2) - \xi_2(R_1) \ &< \ \frac{1}{2p}\log\frac{R_2}{R_1} + \beta_5 - \beta_4 + 2\varepsilon \\
&= \ \frac{1}{2p}\log\frac{R_2}{R_1} - 2\varepsilon - \frac{4\varepsilon}{p}.
\end{aligned}
\tag{2.53}
$$

We next note that by (2.51)

$$\int_{R_1}^{R_1 e^{\varepsilon}} \frac{\omega^2(\rho)d\rho}{\rho p(\rho)} < \frac{\varepsilon^3}{2p},$$

while by (2.52)

$$\int_{R_1}^{R_1 e^{\varepsilon}} \frac{d\rho}{\rho p(\rho)} > \frac{\varepsilon}{p} - \frac{\varepsilon}{2p} = \frac{\varepsilon}{2p}.$$

Thus there exists $\rho = R_1'$, such that

$$\omega(R_1') < \varepsilon, \text{ and } R_1 < R_1' < R_1 e^{\varepsilon}.$$

Similarly there exists R_2', such that

$$\omega(R_2') < \varepsilon, \text{ and } R_2 e^{-\varepsilon} < R_2' < R_2.$$

If $R_1 < R_2$, γ_{R_1} separates $-\infty$ from γ_{R_2} in the strip S and so $\xi_1(R)$, $\xi_2(R)$ are strictly increasing functions of R. We deduce that

$$
\begin{aligned}
\xi_2(R_2') - \xi_1(R_1') \ &= \ \xi_1(R_2') - \xi_2(R_1') + \omega(R_2') + \omega(R_1') \\
&< \ \xi_1(R_2') - \xi_2(R_1') + 2\varepsilon < \xi_1(R_2) - \xi_2(R_1) + 2\varepsilon.
\end{aligned}
$$

On the other hand Lemma 2.5 and (2.52) show that

$$
\begin{aligned}
\xi_2(R_2') - \xi_1(R_1') \ &\geq \ \int_{R_1'}^{R_2'} \frac{d\rho}{2\rho p(\rho)} > \frac{1}{2p}\log\frac{R_2'}{R_1'} - \frac{\varepsilon}{4p} \\
&> \ \frac{1}{2p}\log\frac{R_2}{R_1} - \frac{\varepsilon}{p} - \frac{\varepsilon}{4p}.
\end{aligned}
$$

Thus

$$\frac{1}{2p}\log\frac{R_2}{R_1} - \frac{2\varepsilon}{p} < \xi_1(R_2) - \xi_2(R_1) + 2\varepsilon.$$

This contradicts (2.53). Thus $\beta_4 = \beta_5$ and Lemma 2.7 is proved with $\beta_4 = \beta_5 = -\beta/(2p)$.

2.7.3 Proof of Theorem 2.8

We deduce from (2.50) that on the level curve γ_R we have

$$\xi = \frac{1}{2p} \log R - \frac{\beta}{2p} + o(1).$$

Since $R = |g(\zeta)|$ this gives (2.33) on the level curves γ_R. To complete the proof of (2.33) we need a final lemma.

Lemma 2.8 *If $g(\zeta)$ is mean p-valent and $g'(\zeta) \neq 0$ in the disc $|\zeta - \zeta_0| < \delta$, then*

$$\left| \frac{g'(\zeta_0)}{g(\zeta_0)} \right| < \frac{A_1(p)}{\delta}, \tag{2.54}$$

where $A_1(p) = 2e^{\frac{1}{2}}(2A_0)^{2p}$ depends only on p.

We may suppose without loss of generality that $\zeta_0 = 0$, $\delta = 1$, since otherwise we consider $g(\zeta_0 + \delta\zeta)$ instead of $g(\zeta)$. It then follows from Theorem 2.3 with $q = 0$, that

$$|g(\zeta)| < (2A_0)^{2p} e^{\frac{1}{2}} |g(0)|, \quad \text{for } |\zeta| \le \tfrac{1}{2}.$$

Now Cauchy's inequality yields

$$|g'(0)| \le 2e^{\frac{1}{2}}(2A_0)^{2p}|g(0)|,$$

and this proves Lemma 2.8.

Suppose now that $\zeta_1 = \xi_1 + i\eta_1$, where ξ_1 is large and $|\eta_1| < \frac{\pi}{2} - \delta$. We suppose that $\varepsilon > 0$ and choose R, so that

$$\left| \xi_1 - \frac{1}{2p} \log R + \frac{\beta}{2p} \right| < \varepsilon$$

where β is the quantity in Lemma 2.7, and so that the level curve γ_R exists. This is possible by Lemma 2.4. It follows that γ_R contains a point $\zeta_2 = \xi_2 + i\eta_1$, since γ_R joins the lines $\eta = \pm\pi/2$, and by Lemma 2.7 we have, if ξ_1 is sufficiently large,

$$\left| \xi_2 - \frac{1}{2p} \log R + \frac{\beta}{2p} \right| < \varepsilon,$$

so that $|\xi_2 - \xi_1| < 2\varepsilon$. Integrating (2.54) along the segment $[\zeta_1, \zeta_2]$ we deduce that

$$\left| \log|g(\zeta_2)| - \log|g(\zeta_1)| \right| = \left| \log|g(\zeta_1)| - \log R \right| < \frac{2A_1(p)\varepsilon}{\delta},$$

so that, if ξ_1 is sufficiently large, depending on ε, we have

$$\left| \xi_1 - \frac{1}{2p} \log|g(\zeta_1)| + \frac{\beta}{2p} \right| < \varepsilon \left(1 + \frac{A_1(p)}{p\delta} \right).$$

Thus

$$\log|g(\zeta)| - 2p\xi \to \beta, \text{ as } \xi \to +\infty, \text{ while } |\eta| < \frac{\pi}{2} - \delta.$$

This proves (2.33).

To prove (2.34) we again suppose that $\varepsilon > 0$ and choose ξ_0 so large that

$$\left| \log|g(\zeta)| - 2p\xi - \beta \right| < \frac{\varepsilon\delta}{2} \text{ when } |\eta| < \frac{\pi}{2} - \delta, \quad \xi > \xi_0. \qquad (2.55)$$

Suppose now that $\zeta_1 = \xi_1 + i\eta_1$, where $\xi_1 > \xi_0 + \delta$, $|\eta_1| < \frac{\pi}{2} - 2\delta$. Consider

$$F(z) = \log g(\zeta_1 + z) - 2p(\zeta_1 + z) - \beta = u + iv, \quad |z| \le \delta.$$

Then

$$F(z) = \sum_0^\infty a_n z^n, \quad |z| \le \delta.$$

We write $a_n = \alpha_n + i\beta_n$ and note that

$$u(\delta e^{i\theta}) = \sum_0^\infty \{\alpha_n \cos(n\theta) - \beta_n \sin(n\theta)\} \delta^n.$$

In particular

$$a_1 = F'(\zeta_1) = \frac{1}{\pi\delta} \int_{-\pi}^\pi u(\delta e^{i\theta})(\cos\theta - i\sin\theta)d\theta,$$

so that

$$|a_1| \le \frac{1}{\pi\delta} \int_{-\pi}^\pi |u(\delta e^{i\theta})|d\theta < \varepsilon$$

by (2.55). Hence given positive numbers δ and ε, there exists $\xi_0 = \xi_0(\delta, \varepsilon)$ such that

$$\left| \frac{g'(\zeta_1)}{g(\zeta_1)} - 2p \right| < \varepsilon, \text{ if } \xi_1 > \xi_0 + \delta, \text{ and } |\eta_1| < \frac{\pi}{2} - 2\delta.$$

This proves (2.34) and completes the proof of Theorem 2.8.

2.7.4 A bound for β We shall obtain an analogue of Theorem 1.5 for mean p-valent functions. In order to do so we proceed to prove

Theorem 2.9 *Suppose that* $g(\zeta)$ *is mean p-valent in S, and satisfies (2.33) and also that*

$$\log|g(\zeta)| = 2p\zeta + o(1), \text{ as } \zeta \to -\infty, \text{ through real values.} \quad (2.56)$$

Then $\beta \leq 0$ *in* (2.33). *Equality holds if and only if*

$$g(\zeta) = \exp(2p\zeta + i\lambda), \quad (2.57)$$

where λ *is a real constant.*

Since (2.56) holds we deduce that for all positive R we can find ξ_0, ξ_0' such that

$$|g(\xi)| < R, \quad \xi \leq \xi_0'$$

$$|g(\xi)| > R, \quad \xi \geq \xi_0.$$

Thus if R does not belong to a countable exceptional set F the level curve γ_R joining $\eta = \pm\frac{\pi}{2}$ in S exists and meets the real axis in an odd number of points in the interval $[\xi_0', \xi_0]$.

We can thus apply the inequalities of Lemma 2.5 for positive R_1, R_2. We apply Lemma 2.3 with a fixed R_2 and allow R_1 to tend to zero, and deduce that

$$\int_{R_1}^{R_2} \frac{dR}{Rp(R)} \geq \frac{1}{p}\log\frac{R_2}{R_1} + O(1).$$

Thus (2.43) becomes

$$\int_0^{R_2} \frac{\omega^2(\rho)d\rho}{\rho p(\rho)} < \infty, \text{ so that } \int_0^\infty \frac{\omega^2(\rho)d\rho}{\rho p(\rho)} < \infty.$$

Also Lemma 2.3 shows that

$$\int_0^{R_0} \frac{H(\rho)d\rho}{\rho^3} < \infty, \text{ so that } \frac{H(\rho)}{\rho^2} \to 0, \text{ as } \rho \to 0 \text{ and as } \rho \to \infty.$$

We now apply the proof of Lemma 2.3 with R_1, R_2 where $R_1 \to 0$, $R_2 \to \infty$ and deduce that

$$\begin{aligned}
\int_{R_1}^{R_2} \frac{d\rho}{\rho p(\rho)} &\geq \frac{1}{p}\log\frac{R_2}{R_1} + \frac{H(R_1)}{2p^2R_1^2} - \frac{H(R_2)}{2p^2R_2^2} \\
&= \frac{1}{p}\log\frac{R_2}{R_1} + o(1), \text{ as } R_1 \to 0, R_2 \to \infty.
\end{aligned}$$

We allow R_1 to tend to zero and R_2 to tend to ∞ through sequences of values, such that $\omega(R_1) \to 0$, $\omega(R_2) \to \infty$. Then Lemma 2.5 yields

$$\xi(R_2) - \xi(R_1) \geq \frac{1}{2} \int_{R_1}^{R_2} \frac{d\rho}{\rho p(\rho)} + \frac{1}{2\pi^2} \int_{R_1}^{R_2} \frac{\omega^2(\rho)d\rho}{\rho p(\rho)} + o(1).$$

$$\geq \frac{1}{2p} \log \frac{R_2}{R_1} + \int_{R_1}^{R_2} \frac{\omega^2(\rho)d\rho}{\rho p(\rho)} + o(1).$$

Using (2.55) and (2.33) with $R_2 = |g(\xi(R_2))|$, $R_1 = |g(\xi(R_1))|$, we deduce that

$$\xi(R_2) - \xi(R_1) = \frac{1}{2p}(\log R_2 - \log R_1) - \frac{\beta}{2p} + o(1)$$

Thus

$$\frac{\beta}{2p} + \frac{1}{2\pi^2} \int_0^\infty \frac{\omega^2(\rho)d\rho}{\rho p(\rho)} \leq 0.$$

In particular $\beta \leq 0$, with strict inequality unless $\omega(\rho) = 0$ for almost all ρ. In the latter case almost all the γ_R are vertical segments so that $\frac{\partial}{\partial \eta} \log |g(\xi + i\eta)| = 0$, on all these γ_R and so identically.

If $\phi(\zeta) = \log g(\zeta) = u + iv$, we deduce that $\phi'(\zeta)$ is purely real, so that $\phi'(\zeta)$ is a real constant q. Hence

$$\log g(\zeta) = q\zeta + C.$$

Now (2.56) shows that $q = 2p$ and $C = i\lambda$. This completes the proof of Theorem 2.9.

2.7.5 Return to the unit disc   We can now write down our regularity theorems.

Theorem 2.10   *Suppose that $f(z)$ is mean p-valent in $|z| < 1$. Then*

$$\alpha = \lim_{r \to 1}(1 - r)^{2p} \, M(r, f) \qquad (2.58)$$

exists and $0 \leq \alpha < \infty$. Further if $\alpha > 0$ there exists a (unique) radius of greatest growth $\arg z = \theta_0$ such that

$$(1 - r)^{2p}|f(re^{i\theta_0})| \to \alpha, \text{ as } r \to 1. \qquad (2.59)$$

Also if C is a positive number, and $z = re^{i\theta}$

$$\frac{f'(z)}{f(z)} \sim \frac{2p}{(e^{i\theta_0} - z)}, \text{ as } z \to e^{i\theta_0}, \text{ while } |\arg(ze^{-i\theta_0})| < C(1 - r). \qquad (2.60)$$

Theorem 2.11 *Suppose that $f(z)/z^p$ is regular[†] and equal to 1 at $z = 0$, and that $f(z)$ is mean p-valent in $|z| < 1$, cut along a radius. Then the conclusions of Theorem 2.10 still hold and $\alpha < 1$, unless*

$$f(z) = z^p(1 - ze^{-i\theta_0})^{-2p}.$$

Also we have

$$M(r,f) < C_0 r^p (1 - r)^{-2p}, \quad 0 < r < 1 \tag{2.61}$$

where $C_0 = \exp\left\{(p\pi^2 + 1)/2\right\}$.

With the hypotheses of Theorem 2.11 the proofs of Theorems 2.6 and 2.7 still go through. Thus either

$$(1 - r)^{2p} M(r,f) \to 0, \text{ as } r \to 1$$

in which case $\alpha = 0$ and there is nothing to prove, or (2.21) holds. Thus we may suppose that (2.21) holds. In this case there is a radius of greatest growth $\arg z = \theta_0$ so that Theorems 2.6 and 2.7 hold. We now set

$$\zeta = \frac{1}{2} \log \left\{ \frac{ze^{-i\theta_0}}{(1 - ze^{-i\theta_0})^2} \right\}. \tag{2.62}$$

Then the domain $D(\theta_0)$, consisting of the disc $|z| < 1$ cut along the radius $\arg z = \theta_0 + \pi$, is mapped onto the strip

$$S : \zeta = \xi + i\eta, \quad |\eta| < \frac{\pi}{2}. \tag{2.63}$$

Suppose that $ze^{-i\theta_0} = re^{i\phi}$, where $|\phi| \le C(1 - r)$ for some constant C and that $r \to 1$. Then

$$\xi = \log \frac{r^{1/2}}{|1 - re^{i\phi}|}, \quad \eta = \tan^{-1} \left\{ \frac{1 + r}{1 - r} \tan \frac{1}{2}\phi \right\},$$

so that if r is close to 1,

$$|\eta| \le 2 \tan^{-1} C = \frac{\pi}{2} - \delta, \text{ say.}$$

Thus in this region we can apply (2.33) and obtain

$$\begin{aligned} \log |f(z)| &= \log |g(\zeta)| = 2p\xi + \beta + o(1) \\ &= 2p \log \left| \frac{1}{1 - re^{i\phi}} \right| + \beta + o(1), \end{aligned}$$

[†] cf. Example 2.8.

i.e.

$$|f(z)| \sim \frac{e^\beta}{|1 - ze^{-i\theta_0}|^{2p}}, \text{ as } r \to 1, \text{ while } |\arg z - \theta_0| \le C(1 - r). \quad (2.64)$$

We write $\alpha = \exp\beta$, and deduce (2.59). Next we choose $\varepsilon = p$ in Theorem 2.7 and choose C so large that $C^{-p} < \frac{1}{2}\alpha$. Then by Theorem 2.7 we have, when $z = re^{i\theta}, C(1 - r) < |\theta - \theta_0| \le \pi$, and r is sufficiently close to 1,

$$|f(re^{i\theta})| \le \frac{1}{C^p(1 - r)^{2p}} < \frac{\frac{1}{2}\alpha}{(1 - r)^{2p}};$$

while if $|\theta - \theta_0| \le C(1 - r)$, (2.64) yields

$$|f(re^{i\theta})| < \frac{\alpha + o(1)}{(1 - r)^{2p}}. \quad (2.65)$$

Thus (2.65) holds for all θ when r is close to 1, so that

$$M(r, f) < \frac{\alpha + o(1)}{(1 - r)^{2p}}.$$

On the other hand by (2.59)

$$M(r, f) > \frac{\alpha + o(1)}{(1 - r)^{2p}}.$$

This proves (2.58). Finally if $|\arg z - \theta_0| \le C(1 - r)$, we have

$|\eta| \le \frac{\pi}{2} - \delta$ for some positive δ, and as $z \to e^{i\theta_0}$, $\xi \to +\infty$. Further

$$\frac{d\zeta}{dz} = \frac{1}{2z} + \frac{1}{e^{i\theta_0} - z} \sim \frac{1}{e^{i\theta_0} - z},$$

and by (2.34)

$$\frac{g'(\zeta)}{g(\zeta)} \to 2p.$$

Hence

$$\frac{d}{dz} \log f(z) = \frac{d\zeta}{dz} \frac{d}{d\zeta} \log g(\zeta) \sim \frac{2p}{e^{i\theta_0} - z}.$$

This proves (2.60) and completes the proof of Theorem 2.10.

Next if $f(z)$ satisfies the hypotheses of Theorem 2.11 then $g(\zeta)$ satisfies (2.56), since as $z \to 0$

$$\xi = \frac{1}{2} \log|z| + \log \frac{1}{|1 - z|} = \frac{1}{2} \log|z| + o(1),$$

so that

$$\log|g(\zeta)| = \log|f(z)| = p \log|z| + o(1) = 2p\xi + o(1).$$

Thus, by Theorem 2.9, $\beta = \log \alpha \leq 0$. Equality holds if and only if

$$f(z) = g(\zeta) = \exp(2p\zeta + i\lambda) = e^{i\lambda} \left\{ \frac{ze^{-i\theta_0}}{(1 - ze^{-i\theta_0})^2} \right\}^p.$$

Also we must have $\lambda = p\theta_0$, since $f(z)/z^p \to 1$ as $z \to 0$.

It remains to prove (2.61). We use the transformation (2.62), but without assuming that θ_0 is a radius of greatest growth. Then $g(\zeta) = f[z(\zeta)]$ is regular in the strip S given by (2.12) and is p-valent and non zero there. We suppose that $|g(0)| = R_1, |g(\xi_0)| = R_2$, where $R_1 < R_2$, and proceed as in the proof of Theorem 2.2. Then, since $g(\zeta) \neq 0$, there are no closed level curves in S and so for $R_1 < R < R_2$ there is a level curve $|g| = R$ meeting the segment

$$L : 0 < \xi < \xi_0$$

and going to the boundary of the domain S_0 consisting of all points distant at most $\pi/2$ from L. This level curve has length at least π. The area a of S_0 is $\pi\xi_0 + \pi^3/4$ and so Theorem 2.1 and Lemma 2.1 yield

$$2\pi^2 \xi_0 + \frac{\pi^4}{2} \geq \pi^2 \int_{R_1}^{R_2} \frac{dR}{Rp(R)} \geq \frac{\pi^2}{p} \left\{ \log \frac{R_2}{R_1} - \frac{1}{2} \right\},$$

since g is mean p-valent. Thus

$$\log \left| \frac{g(\xi_0)}{g(0)} \right| \leq 2p\xi_0 + \frac{p\pi^2 + 1}{2}.$$

By a simple translation in S we obtain for $-\infty < \xi_1 < \xi_2 < \infty$

$$\log |g(\xi_2)| < \log |g(\xi_1)| + 2p(\xi_2 - \xi_1) + \frac{p\pi^2 + 1}{2}.$$

We let ξ_1 tend to $-\infty$ in this. Then since $f(z)/z^p \to 1$ as $z \to 0$,

$$\log |g(\xi_1)| - 2p\xi_1 \to 0, \text{ as } \xi_1 \to -\infty.$$

Hence we obtain

$$\log |g(\xi_1)| \leq 2p\xi_2 + \frac{p\pi^2 + 1}{2}.$$

Writing $z = re^{i\theta_0}$ and using (2.62) we obtain

$$\log |f(z)| \leq p \log \frac{r}{(1 - r)^2} + \frac{p\pi^2 + 1}{2}$$

and this proves (2.61).

Examples

2.7   If $\alpha = 0$ and $p > \frac{1}{2}$ in Theorem 2.10 or Theorem 2.11 prove, using the argument for Theorem 1.6, that

$$I_1(r, f) = o(1 - r)^{1-2p}, \text{ as } r \to 1,$$

and so that $a_n/n^{2p-1} \to 0$ as $n \to \infty$.

2.8   If $f(z)$ is regular in $0 < |z| < 1$, and $f(z)/z^{\mu}$ remains regular at $z = 0$, show that in any sector $T : \theta_1 < \arg z < \theta_2$, where $\theta_1 < \theta < \theta_2$, the value of $p(R, T)$ is independent of the branch of z^{μ} chosen. Deduce that, if $\theta_2 = \theta_1 + 2\pi, p(R, T)$ is also independent of θ_1. We say that $f(z)$ is mean p-valent in $|z| < 1$, if $f(z)$ is mean p-valent in $\theta_1 < \arg z < \theta_1 + 2\pi$ for some, and so for every real θ_1. Verify that Theorems 2.6 to 2.11 and Example 2.7 remain valid.

2.9   If

$$f(z) = z^{\mu} \sum_{0}^{\infty} a_n z^n$$

is mean p-valent in $|z| < 1$, if $\alpha > 0$, and $\arg z = \theta_0$ is the r.g.g. show that

$$f(re^{i\theta}) \sim \frac{\alpha(r)}{\left(1 - re^{i(\theta-\theta_0)}\right)^{2p}}, \text{ as } r \to 1, \text{ for } |\theta - \theta_0| \le C(1 - r),$$

where C is a fixed positive number and $\alpha(r) = (1 - r)^{2p} f(re^{i\theta_0})$ (use (2.60)).

2.10   Using Cauchy's formula

$$a_n = \frac{1}{2\pi i} \int_{|z|=r} \frac{\phi(z)dz}{z^{n+1}}, \text{ where } r = 1 - \frac{1}{n},$$

applied to $\phi(z) = f(z)/z^{\mu}$, and comparing this with the corresponding formula for

$$\phi_0(z) = \alpha(r)(1 - ze^{-i\theta_0})^{-2p} = \alpha(r) \sum_{k=0}^{\infty} \frac{\Gamma(k + 2p)}{\Gamma(2p)} z^k e^{-ik\theta_0},$$

show that if $p > \frac{1}{2}$, $\alpha > 0$, and $\arg z = \theta_0$ is the r.g.g, then

$$a_n \sim \frac{\alpha \left(1 - \frac{1}{n}\right) n^{2p-1} e^{-i(n+\mu)\theta_0}}{\Gamma(2p)}, \text{ as } n \to \infty.$$

(Use Example 2.9 and Theorem 2.7. This theorem eliminates the need for the analogue of Lemma 1.4.)

2.11   If $f(z)$ is mean p-valent in $|z| < 1$, where $p > \frac{1}{2}$ show that

$$\lim_{n\to\infty} \frac{|a_n|}{n^{2p-1}} = \frac{\alpha}{\Gamma(2p)},$$

where α is given by (2.58). (Hayman [1955], Eke [1967b]).

2.12   If $f(z) \in \mathfrak{S}$, show that $|a_n| \leq n$, $n > n_0(f)$. (Distinguish the cases $\alpha = 1$, $\alpha < 1$.)

See Hayman [1955] for the above results for circumferentially mean p-valent functions and Eke [1967b] for the general case. We shall show in Chapter 5 how to extend the conclusions to the case $p > \frac{1}{4}$. They fail when $p < \frac{1}{4}$.

2.13   Prove, with the hypotheses of Theorem 2.8 that, if $q > 1$,

$$\left(\frac{d}{dz}\right)^q \log g(\zeta) \to 0$$

as $\xi \to +\infty$ uniformly in $|\eta| < \frac{\pi}{2} - \delta$. Hence prove that, if $q \geq 1$, (2.60) can be extended to

$$\frac{f^{(q)}(z)}{f(z)} \sim \frac{2p(2p+1)...(2p+q-1)}{(e^{i\theta_0} - z)^q} \quad \text{as } z \to e^{i\theta_0},$$

while $|\arg(ze^{-i\theta_0})| < C(1 - |z|)$.

Deduce that for $q \geq 0$

$$|f^{(q)}(z)| \sim \frac{2p(2p+1)...(2p+q-1)\alpha}{|e^{i\theta_0} - z|^{2p+q}} \quad \text{as } z \to e^{i\theta_0},$$

while $|\arg\ ze^{-i\theta_0}| < C(1 - |z|)$.

2.14   If $q \geq 1$ prove that, with the hypotheses of Theorem 2.10,

$$\lim_{r\to 1}(1 - r)^{2p+q} M(r, f^{(q)}) = 2p(2p+1)...(2p+q-1)\alpha.$$

(Consider separately the cases $\alpha = 0$, $\alpha > 0$.)

3

Means and coefficients

3.0 Introduction In the last chapter we investigated the growth of a function

$$f(z) = \sum_0^\infty a_n z^n,$$

mean p-valent in $|z| < 1$. We showed in Theorem 2.3 that the maximum modulus $M(r, f)$ satisfies

$$M(r, f) < A(p)\mu_p(1 - r)^{2p} \quad (0 < r < 1). \tag{3.1}$$

In this chapter we estimate the order of magnitude of the coefficients a_n and show that, if $f(z)$ is mean p-valent in $|z| < 1$ and C, β are positive constants such that $C > 0$ and $\beta > \frac{1}{2}$, then the inequality

$$M(r, f) < C(1 - r)^{-\beta} \quad (0 < r < 1), \tag{3.2}$$

implies

$$|a_n| < A(p, \beta)C(1 + n)^{\beta - 1} \quad (n = 0, 1, 2, \ldots). \tag{3.3}$$

It will follow at once that, if $f(z)$ is mean p-valent in $|z| < 1$, so that (3.1) holds, then

$$|a_n| < A(p)\mu_p n^{2p-1} \quad (n \geq 1), \tag{3.4}$$

provided that $p > \frac{1}{4}$. The functions

$$f(z) = (1 - z)^{-2p} = \sum_0^\infty b_{n,p}\, z^n, \tag{3.5}$$

which are mean p-valent in $|z| < 1$ and for which

$$b_{n,p} = \frac{\Gamma(n + 2p)}{\Gamma(2p)\Gamma(n + 1)} \sim \frac{n^{2p-1}}{\Gamma(2p)} \quad (n \to \infty), \tag{3.6}$$

66

show that the order of magnitude of the bounds in (3.4) is correct.

The method used by Littlewood [1925] for proving Theorem 1.6 is sufficient to show that (3.2) implies (3.3) if $f(z)$ is univalent and $\beta > 1$. The idea for extending this to the case $\beta > \frac{1}{2}$ occurs first in a joint paper of Littlewood and Paley [1932]. The argument was extended to p-valent functions by Biernacki [1936] and to mean p-valent functions by Spencer [1941a].

We shall show further, by means of an example of Spencer [1940b], that (3.2) does not imply (3.3) for a general mean p-valent function $f(z)$ if $\beta < \frac{1}{2}$ and that (3.4) is false in general if $p < \frac{1}{4}$. However, we shall prove in Section 3.5 a theorem of Baernstein [1986], which yields an extension for univalent functions.

In the final sections of the chapter we shall give some further applications of our main results by estimating the coefficients of certain classes of mean p-valent functions for which more restrictive bounds than (3.1) can be obtained.

3.1 The Hardy–Stein–Spencer identities We suppose now that $f(z)$ is regular in $|z| < 1$ and further that $\lambda > 0$ and $0 < r < 1$. Let $n(r, w)$ be the number of roots of the equation $f(z) = w$ in $|z| < r$ and write

$$p(r, R) = \frac{1}{2\pi} \int_0^{2\pi} n(r, R\, e^{i\psi})\, d\psi.$$

Thus $p(R) = p(r, R)$ is defined as in (2.4) when Δ is the domain $|z| < r$. We also write

$$I_\lambda(r, f) = \frac{1}{2\pi} \int_0^{2\pi} |f(r\, e^{i\theta})|^\lambda d\theta.$$

We then have the following remarkable triple identity:[†]

Theorem 3.1 *With the above notation*

$$
\begin{aligned}
r\frac{d}{dr} I_\lambda(r) &= \frac{\lambda^2}{2\pi} \int_0^r \rho\, d\rho \int_0^{2\pi} |f(\rho e^{i\theta})|^{\lambda-2} |f'(\rho e^{i\theta})|^2 d\theta \\
&= \lambda^2 \int_0^\infty p(r, R) R^{\lambda-1}\, dR.
\end{aligned}
$$

Suppose first that $f(z)$ has no zero on $|z| = r$ and write

$$f(re^{i\theta}) = Re^{i\Phi}.$$

[†] The first equality is due to Hardy [1915] and Stein [1933] and the second to Spencer [1940a].

Then near a fixed point of $|z| = r$, we have

$$\log f = \log R + i\Phi, \quad \frac{1}{r}\frac{\partial \Phi}{\partial \theta} = \frac{1}{R}\frac{\partial R}{\partial r},$$

by the Cauchy–Riemann equations. Thus

$$r\frac{d}{dr}\int_{-\pi}^{+\pi}|f(re^{i\theta})|^\lambda d\theta = \lambda\int_{|z|=r}R^{\lambda-1}r\frac{\partial R}{\partial r}d\theta = \lambda\int_{|z|=r}R^\lambda d\Phi. \qquad (3.7)$$

We make a transformation (not conformal) by writing, when $w = Re^{i\Phi}$,

$$P = R^{\frac{1}{2}\lambda}, \quad \psi = \Phi, \quad W = Pe^{i\psi}.$$

Then the right-hand side of (3.7) reduces to

$$\lambda\int_{|z|=r}P^2 d\Phi.$$

Now $\frac{1}{2}P^2 d\Phi$ is a sectorial element of area in the W plane, and so the right-hand side of (3.7) represents 2λ times the area in the W plane corresponding to $|z| < r$, multiple points being counted multiply. This is quite evident if $f(z)$ is univalent in $|z| \le r$, so that area is the interior of the simple closed Jordan curve which is the image of $|z| = r$. In the general case we can prove our result by splitting the disc into a finite number of regions in each of which $f(z)$ is univalent and noting that

$$\int P^2 d\Phi$$

taken over the boundary of the region is additive and so is the area in the W plane.

Now the area in the W plane is equal to

$$\int_0^\infty\int_0^{2\pi}v(Pe^{i\psi})P\,dP\,d\psi,$$

where $v(Pe^{i\psi})$ is the number of points in $|z| < r$, corresponding to $W = Pe^{i\psi}$. Thus

$$v(Pe^{i\psi}) = n(r, P^{2/\lambda}e^{i\psi}),$$

and we obtain

$$\begin{aligned}
\lambda\int_{|z|=r}R^\lambda d\Phi &= \lambda\int_0^\infty d(P^2)\int_0^{2\pi}v(Pe^{i\psi})d\psi \\
&= \lambda\int_{R=0}^\infty d(R^\lambda)\int_0^{2\pi}n(r, Re^{i\psi})d\psi \\
&= 2\pi\lambda^2\int_0^\infty p(r, R)R^{\lambda-1}dR,
\end{aligned}$$

and on combining this with (3.7) we see that the first term in the identity of Theorem 3.1 is equal to the third.

Again $|f'(\rho e^{i\theta})|^2 \rho\,d\rho\,d\theta$ is the area of the image of a small element of area, $\rho < |z| < \rho + d\rho$, $\theta < \arg z < \theta + d\theta$ by $w = f(z)$ and so

$$\int_0^r \rho\,d\rho \int_0^{2\pi} |f'(\rho e^{i\theta})|^2 |f(\rho e^{i\theta})|^{\lambda-2}\,d\theta = \int_0^\infty R\,dR \int_0^{2\pi} R^{\lambda-2} n(r, Re^{i\psi})\,d\psi.$$

In fact both sides represent the total mass in the w plane of a mass density $|w|^{\lambda-2}$ is spread over the image of $|z| < r$ by $w = f(z)$.

The right-hand side becomes

$$2\pi \int_0^\infty p(r,R) R^{\lambda-1}\,dR,$$

and so we have the identity of the second and third terms in Theorem 3.1, and that theorem is proved on the assumption that $f(z)$ has no zeros on $|z| = r$.

The result follows in the general case from considerations of continuity. In fact the continuous function $I_\lambda(r)$ has a continuous derivative

$$\frac{S_\lambda(r)}{r} = \frac{\lambda^2}{2\pi r} \int_0^r \rho\,d\rho \int_0^{2\pi} |f(\rho e^{i\theta})|^{\lambda-2} |f'(\rho e^{i\theta})|^2\,d\theta,$$

except possibly at certain isolated values of r. At these latter values $r_0, I_\lambda(r)$ clearly remains continuous and so we see that the equation

$$r\frac{d}{dr}I_\lambda(r) = S_\lambda(r)$$

continues to hold, by using the strong form of the mean-value theorem

$$I_\lambda(r_1) - I_\lambda(r_0) = (\log r_1 - \log r_0)S_\lambda(\rho),$$

and making r_1 tend to r_0 from below or above.

3.2 Estimates of the means $I_\lambda(r)$[†]   Suppose again that

$$f(z) = \sum_0^\infty a_n z^n$$

is regular in $|z| < 1$. Then we have for $0 < r < 1$

$$n|a_n| = \left| \frac{1}{2\pi i} \int_{|z|=r} \frac{f'(z)\,dz}{z^n} \right| \le \frac{I_1(r, f')}{r^{n-1}}.$$

[†] The results from here to Section 3.5 inclusive are mainly due to Spencer [1941a].

We choose $r = (n-1)/n$ for $n \geq 1$, so that

$$r^{-(n-1)} = \left(1 + \frac{1}{n-1}\right)^{n-1} < e,$$

and deduce

$$|a_n| < \frac{e}{n} I_1\left(\frac{n-1}{n}, f'\right) \quad (n \geq 1). \tag{3.8}$$

It is thus important to be able to estimate $I_1(r, f')$. For this purpose we use Theorem 3.1. It follows from this theorem that $I_\lambda(r, f)$ is an increasing convex function of $\log r$, when $\lambda > 0$.

We have further

Theorem 3.2 *Suppose that $f(z)$ is mean p-valent in $|z| < 1$ and set $\Lambda = \max(\lambda, \frac{1}{2}\lambda^2)$, when $\lambda > 0$. Then*

$$S_\lambda(r, f) = r\frac{d}{dr} I_\lambda(r, f) \leq p\Lambda M(r, f)^\lambda \quad (0 < r < 1), \tag{3.9}$$

and

$$I_\lambda(r, f) \leq M(r_0, f)^\lambda + p\Lambda \int_{r_0}^{r} \frac{M(t, f)^\lambda dt}{t} \quad (0 < r_0 < r < 1). \tag{3.10}$$

We have by Theorem 3.1

$$S_\lambda(r) = \lambda^2 \int_0^\infty p(r, R) R^{\lambda-1} dR = \lambda^2 \int_0^{M(r,f)} p(r, R) R^{\lambda-1} dR.$$

Also since $f(z)$ is mean p-valent in $|z| < 1$ and so *à fortiori* in $|z| < r$, we have, using the notation of §2.3,

$$W(R) = \int_0^R p(r, \rho) d(\rho^2) \leq pR^2 \quad (0 < R < \infty).$$

Hence, writing $M = M(r, f)$, we obtain

$$\begin{aligned}
\int_0^M p(r, R) R^{\lambda-1} dR &= \frac{1}{2} \int_0^M R^{\lambda-2} dW(R) \\
&= \frac{1}{2} M^{\lambda-2} W(M) - \frac{\lambda-2}{2} \int_0^M R^{\lambda-3} W(R) dR.
\end{aligned}$$

There are now two cases. If $\lambda > 2$, we deduce, since $W(R) \geq 0$,

$$\int_0^M p(r, R) R^{\lambda-1} dR \leq \frac{1}{2} M^{\lambda-2} W(M) \leq \frac{p}{2} M^\lambda = \frac{p}{2} M(r, f)^\lambda.$$

If $0 < \lambda \leq 2$ we deduce

$$
\begin{aligned}
\int_0^M p(r,R)R^{\lambda-1}dR &\leq \tfrac{1}{2}M^{\lambda-2}pM^2 + \frac{2-\lambda}{2}\int_0^M R^{\lambda-3}pR^2dR \\
&= \frac{p}{2}M^\lambda + \frac{p(2-\lambda)}{2\lambda}M^\lambda = \frac{p}{\lambda}M(r,f)^\lambda.
\end{aligned}
$$

This gives (3.9). Also

$$
I_\lambda(r,f) = I_\lambda(r_0,f) + \int_{r_0}^r S_\lambda(t,f)\frac{dt}{t} \leq M(r_0,f)^\lambda + p\Lambda\int_{r_0}^r M(t,f)^\lambda\frac{dt}{t},
$$

and this yields (3.10).

3.3 Estimates for the coefficients We now prove our basic result.

Theorem 3.3 *Suppose that $f(z) = \sum_0^\infty a_n z^n$ is mean p-valent in $|z| < 1$ and that*

$$M(r,f) \leq C(1-r)^{-\beta} \quad (0 < r < 1), \tag{3.11}$$

where $C > 0$ and $\beta > \tfrac{1}{2}$. Then we have

$$|a_n| \leq A_1(p,\beta)Cn^{\beta-1}, \quad (n \geq 1) \tag{3.12}$$

where $A_1(p,\beta)$ depends on p,β only.

We shall need the following preliminary result:

Lemma 3.1 *Suppose that $f(z)$ is mean p-valent in $|z| < 1$ and that $\tfrac{1}{2} \leq r < 1, 0 < \lambda \leq 2$. Then there exists ρ such that $2r - 1 \leq \rho \leq r$ and*

$$\frac{1}{2\pi}\int_0^{2\pi} |f'(\rho e^{i\theta})|^2|f(\rho e^{i\theta})|^{\lambda-2}d\theta \leq \frac{4pM(r,f)^\lambda}{\lambda(1-r)}. \tag{3.13}$$

We deduce from Theorems 3.1 and 3.2 that

$$\frac{1}{2\pi}\int_{2r-1}^r \rho d\rho \int_0^{2\pi} |f'(\rho e^{i\theta})|^2|f(\rho e^{i\theta})|^{\lambda-2}d\theta \leq \frac{1}{\lambda^2}S_\lambda(r) \leq \frac{p}{\lambda}M(r,f)^\lambda.$$

Hence we can choose ρ so that $2r - 1 < \rho < r$ and

$$\frac{1}{2\pi}\int_0^{2\pi} |f'(\rho e^{i\theta})|^2|f(\rho e^{i\theta})|^{\lambda-2}d\theta \leq \frac{pM(r,f)^\lambda}{\tfrac{1}{2}\lambda[r^2 - (2r-1)^2]} \leq \frac{4pM(r,f)^\lambda}{\lambda(1-r)}.$$

This proves Lemma 3.1.

We now suppose that $r \geq \frac{1}{2}$ and that (3.11) holds with $C = 1$ and $\beta > \frac{1}{2}$. We set $\lambda = (2\beta - 1)/(2\beta)$, so that $\beta(2 - \lambda) = \beta + \frac{1}{2} > 1$ and choose ρ so that (3.13) holds. Then

$$
\begin{aligned}
I_1(\rho, f') &= \frac{1}{2\pi} \int_0^{2\pi} |f'(\rho e^{i\theta})| d\theta \\
&\leq \left(\frac{1}{2\pi} \int_0^{2\pi} |f'(\rho e^{i\theta})|^2 |f(\rho e^{i\theta})|^{\lambda - 2} d\theta \right)^{\frac{1}{2}} \\
&\quad \times \left(\frac{1}{2\pi} \int_0^{2\pi} |f(\rho e^{i\theta})|^{2 - \lambda} d\theta \right)^{\frac{1}{2}},
\end{aligned}
\tag{3.14}
$$

by Schwarz's inequality. We now take $r_0 = \frac{1}{2}$ in (3.10) and, noting that $r \geq \frac{1}{2}, r \geq \rho$, we deduce

$$
\begin{aligned}
I_{2-\lambda}(\rho, f) &\leq I_{2-\lambda}(r, f) \\
&\leq (1 - r_0)^{-\beta(2-\lambda)} + \frac{p(2 - \lambda)}{r_0} \int_0^r (1 - t)^{-\beta(2-\lambda)} dt \\
&= 2^{\beta + \frac{1}{2}} + \frac{p(2\beta + 1)}{\beta} \int_0^r (1 - t)^{-\beta - \frac{1}{2}} dt \\
&< A_2 (1 - r)^{\frac{1}{2} - \beta},
\end{aligned}
$$

where

$$
A_2 = 2^{\beta + \frac{1}{2}} + \frac{2p(2\beta + 1)}{\beta(2\beta - 1)}.
$$

We write $r_1 = 2r - 1$, so that $r_1 \leq \rho < r$ and deduce from (3.13), (3.14) and the above that

$$
\begin{aligned}
I_1(r_1, f') &\leq I_1(\rho, f') \leq \left\{ \frac{4p}{\lambda} A_2 \right\}^{\frac{1}{2}} (1 - r)^{-\frac{1}{2}(1 + \beta\lambda + \beta - \frac{1}{2})} \\
&= A_3 (1 - r)^{-\beta} = 2^{\beta} A_3 (1 - r_1)^{-\beta},
\end{aligned}
$$

where

$$
A_3 = \left(\frac{4p}{\lambda} A_2 \right)^{\frac{1}{2}} = \left(\frac{8p\beta}{2\beta - 1} A_2 \right)^{\frac{1}{2}}.
$$

Here r_1 may be any number such that $0 < r_1 < 1$. Now (3.8) yields

$$
n|a_n| \leq 2^{\beta} e A_3 n^{\beta}, \quad n \geq 1.
$$

This proves Theorem 3.3 if C=1, with

$$A_1(p,\beta) = 2^\beta e \left\{ \frac{8p\beta}{2p-1} \left(2^{\beta+\frac{1}{2}} + \frac{2p(2\beta+1)}{\beta(2\beta-1)} \right) \right\}^{\frac{1}{2}}.$$

If $0 < C < \infty$, we consider $f(z)/C$ instead of $f(z)$. If $C = 0$, $a_n = 0$. This proves Theorem 3.3.

Examples

3.1   Show that if $f(z)$ is mean p-valent in $|z| < 1$ and satisfies (3.11) and $0 \leq r < 1$, then $I_\lambda(r,f) < A(p, \beta, \lambda)C^\lambda(1 - r)^{1-\beta\lambda}$ if $\beta\lambda > 1$; $I_1(r,f) < A(p,\lambda)(1 + \log\frac{1}{1-r})$ if $\beta\lambda = 1$; and $I_\lambda(r,f) < A(p, \beta, \lambda)$ if $\beta\lambda < 1$.

3.2   By considering $f(z) = (1 - z)^{-2p}$, show that orders of magnitude in Example 3.1 cannot be improved.

3.3.1   Theorem 3.3 fails if $\beta < \frac{1}{2}$. However we have

Theorem 3.4   *Suppose that* $f(z) = \sum_0^\infty a_n z^n$ *is mean p-valent in* $|z| < 1$ *and satisfies (3.11). Then if* $\beta = \frac{1}{2}$, *we have*

$$|a_n| < A_1(p)C \, n^{-\frac{1}{2}} \log(n+1), \quad n \geq 1, \tag{3.15}$$

while if $\beta < \frac{1}{2}$,

$$|a_n| < A_1(p,\beta)C \, n^{-\frac{1}{2}}, \quad n \geq 1 \tag{3.16}$$

and

$$a_n = o(n^{-\frac{1}{2}}), \text{ as } n \to \infty. \tag{3.17}$$

The inequalities (3.16) and (3.17) are due to Pommerenke [1961/2]. We shall shown by an example in the next section that they are sharp even when $f(z)$ is bounded, i.e. $\beta = 0$. The case $\beta = \frac{1}{2}$ remains open. However for univalent functions we shall prove, following Baernstein [1986] that the implication from (3.11) to (3.12) remains true for $\beta > .491$.

To prove Theorem 3.4 we need

Lemma 3.2   *Suppose that* $f(z)$ *is mean p-valent in* $|z| < 1$. *Then*

$$\int_0^1 \rho d\rho \int_0^{2\pi} \frac{|f'(\rho e^{i\theta})|^2 \rho d\rho d\theta}{1 + |f(\rho e^{i\theta})|^\lambda} < p\pi \frac{5 \times 2^\lambda - 4}{2^\lambda - 4}, \quad \lambda > 2 \tag{3.18}$$

and if $\lambda = 2$ and (3.11) holds with $C = 1, \beta = \frac{1}{2}$,

$$\int_0^r \rho d\rho \int_0^{2\pi} \frac{|f'(\rho e^{i\theta})|^2}{1 + |f(\rho e^{i\theta})|^2} \rho d\rho d\theta < p\pi \left(5 + 3\log\frac{1}{1-r}\right), \quad 0 \le r < 1.$$
(3.19)

Let E_n be the subset of $|z| < 1$, where $|f(z)| \le 1$ if $n = 0$, and where $2^{n-1} < |f(z)| \le 2^n$, if $n \ge 1$. Then, since $f(z)$ is mean p-valent

$$I_n = \int_{E_n} |f'(\rho e^{i\theta})|^2 \rho d\rho d\theta \le p\pi 4^n = 4p\pi\, 2^{2(n-1)}.$$

Thus if

$$J_n = \int_{E_n} \frac{|f'(\rho e^{i\theta})|^2 \rho d\rho d\theta}{1 + |f(\rho e^{i\theta})|^\lambda},$$

we have

$$J_0 \le I_0 \le p\pi, \quad \text{and} \quad J_n \le 2^{\lambda(1-n)} I_n \le 4p\pi 2^{(2-\lambda)(n-1)}, \quad n \ge 1.$$

Thus, if $\lambda > 2$,

$$\sum_{n=0}^\infty J_n \le p\pi \left\{1 + 4\sum_{n=1}^\infty 2^{(2-\lambda)(n-1)}\right\} = p\pi\left(1 + \frac{4}{1 - 2^{2-\lambda}}\right) = p\pi\frac{5 \times 2^\lambda - 4}{2^\lambda - 4}.$$

This proves (3.18). If $\lambda = 2$, we have $I_0 \le p\pi$, $I_n \le 4p\pi$, $n \ge 1$. Also since $M(r,f) \le (1-r)^{-\frac{1}{2}}$, E_n does not meet $|z| < r$ if $2^{n-1} > (1-r)^{-\frac{1}{2}}$, i.e. if

$$n > 1 + \left[\frac{1}{2\log 2}\,\log\frac{1}{1-r}\right] = N(r), \text{ say.}$$

Hence

$$\int_0^r \rho d\rho \int_0^{2\pi} \frac{|f'(\rho e^{i\theta})|^2 \rho d\rho d\theta}{1 + |f(\rho e^{i\theta})|^2} \le p\pi\left\{1 + \sum_{n=1}^{N(r)} 4\right\} = p\pi[1 + 4N(r)],$$

and this yields (3.19).

Suppose now first that $\beta = \frac{1}{2}$. We suppose that $r \ge \frac{3}{4}$, and write $r_1 = 2r - 1$. Then using Theorem 3.2 and (3.19) we obtain

$$\left\{\frac{1}{2\pi}\int_{r_1}^r \rho d\rho \int_0^{2\pi} |f'(\rho e^{i\theta})| d\theta\right\}^2$$

$$\le \left(\frac{1}{2\pi}\int\int \frac{|f(\rho e^{i\theta})|^2 \rho d\rho d\theta}{1 + |f(\rho e^{i\theta})|^2}\right)\left(\frac{1}{2\pi}\int\int (1 + |f(\rho e^{i\theta})|^2)\rho d\rho d\theta\right)$$

$$\le \frac{p}{2}\left(5 + 3\log\frac{1}{1-r}\right)\int_{r_1}^r (1 + I_2(\rho, f))d\rho$$

$$\le \frac{p}{2}\left(5 + 3\log\frac{1}{1-r}\right)(r - r_1)\left(2 + 4p\log\frac{1}{1-r}\right)$$

$$\le 5p(1+p)(1-r_1)\left(1 + \log\frac{1}{1-r_1}\right)^2.$$

Thus

$$\frac{1}{2}(r^2 - r_1^2)I_1(r_1, f') \le \{5p(1+p)\}^{\frac{1}{2}}(1 - r_1)^{\frac{1}{2}}\left(1 + \log\frac{1}{1-r_1}\right).$$

Setting $r_1 = \frac{n-1}{n}$ if $n \ge 2$, $r_1 = \frac{1}{2}$ if $n = 1$, and using (3.8), we deduce (3.15).

Next suppose that $\beta < \frac{1}{2}$. We choose $\lambda = (2\beta + 1)/(2\beta)$. Then

$$\frac{1}{2\pi}\int_{r_1}^r \rho d\rho \int_0^{2\pi} |f'(\rho e^{i\theta})| d\theta$$

$$\le \left\{\frac{1}{2\pi}\int\int\left(\frac{|f'|^2}{1+|f|^\lambda}\right)\rho d\rho d\theta \ \frac{1}{2\pi}\int\int(1+|f|^\lambda)\rho d\rho d\theta\right\}^{\frac{1}{2}}$$

$$\le \{A(p,\beta)(1-r)\}^{\frac{1}{2}}.$$

In fact if $\rho \ge \frac{1}{2}$ we have by (3.10) and (3.11) with $C = 1$, $r_0 = \frac{1}{2}$

$$\frac{1}{2\pi}\int_0^{2\pi} |f(\rho e^{i\theta})|^\lambda d\theta \ \le \ 2^{\beta\lambda} + 2p\lambda\int_{\frac{1}{2}}^\rho (1-t)^{-\beta\lambda} dt$$

$$\le \ 2^{\beta\lambda} + \frac{2p\lambda}{1-\beta\lambda} = 2^{\beta+\frac{1}{2}} + \frac{2p(2\beta+1)}{\beta(1-2\beta)}.$$

Using also (3.18) we find there exists ρ, such that $r_1 = (2r - 1) \le \rho \le r$, and

$$I_1(r_1, f') \le I_1(\rho, f') \le A(p,\beta)(1-\rho)^{-\frac{1}{2}} \le 2^{\frac{1}{2}}A(p,\beta)(1-r_1)^{-\frac{1}{2}}. \qquad (3.20)$$

Choosing $r_1 = (n-1)/n$, and applying (3.8) we obtain (3.16).

Next we note that by (3.18)

$$\int_{r_1}^1 \rho d\rho \int_0^{2\pi} \frac{|f'(\rho e^{i\theta})|}{1+|f(\rho e^{i\theta})|^\lambda} d\theta \to 0 \text{ as } r_1 \to 1.$$

This enables us to sharpen (3.20) to

$$I_1(r_1, f') = o(1-r_1)^{-\frac{1}{2}}.$$

Again choosing $r_1 = (n-1)/n$ and applying (3.8) we obtain (3.17). This completes the proof of Theorem 3.4.

3.4 A counter-example   The results (3.16), (3.17) of Theorem 3.4 are best possible. Even if $f(z)$ is mean p-valent with p as small as we please and continuous in $|z| \le 1$ nothing stronger than

$$|a_n| = o(n^{-\frac{1}{2}}) \qquad (3.21)$$

is true in general. To see this let n_k be a rapidly increasing sequence of integers and put

$$f(z) = 1 + \sum_{n=1}^{\infty} a_n z^n,$$

where $a_n = \epsilon 2^{-k} n^{-\frac{1}{2}}$, if $n = n_k$ $(k = 1, 2, \ldots)$, $a_n = 0$ otherwise.

If λ_n is any pre-assigned sequence of positive numbers, tending to zero as $n \to \infty$, however slowly, we suppose $n_k \to \infty$ so rapidly with k, that

$$\lambda_{n_k} < \frac{\epsilon}{2^k} \quad (k = 1, 2, \ldots).$$

Thus for the infinite sequence of values of n given by $n = n_k$ we have

$$a_n > \frac{\lambda_n}{n^{\frac{1}{2}}}. \qquad (3.22)$$

On the other hand, we have for $|z| \le 1$,

$$|f(z) - 1| \le \sum_{n=1}^{\infty} |a_n| \le \epsilon \sum_{k=1}^{\infty} 2^{-k} = \epsilon,$$

so that the series for $f(z)$ converges uniformly and so $f(z)$ is continuous in $|z| \le 1$. Also the area, with due count of multiplicity, of the image of $|z| < 1$ by $f(z)$ is

$$\int_0^1 r dr \int_0^{2\pi} |f'(re^{i\theta})|^2 d\theta = \pi \sum_{n=1}^{\infty} n|a_n|^2 = \pi \sum_{k=1}^{\infty} \epsilon^2 2^{-2k} = \frac{\pi \epsilon^2}{3}$$

(see §1.3). If $\pi W(R)$ denotes the amount of this area which lies over $|w| < R$, then $W(R) = 0$ if $R < 1 - \epsilon$, $W(R) < \pi \epsilon^2 / 3$ otherwise. If we choose $\epsilon < \frac{1}{2}$, it follows that

$$\frac{W(R)}{\pi R^2} < \frac{\pi \epsilon^2}{3\pi(\frac{1}{2})^2} = \frac{4\epsilon^2}{3} \quad (0 < R < \infty),$$

and we can make the right-hand side as small as we please by choosing ϵ small enough. Thus $f(z)$, which is continuous in $f(z) \le 1$, can be made mean p-valent there with p as small as we please. Clearly, for any function $f(z)$ bounded and mean p-valent in $|z| < 1$, $\sum_1^{\infty} n|a_n|^2$ converges and so

(3.21) holds. Nevertheless (3.22) shows that nothing stronger than this need be true.

Examples

3.3   If $f(z) = \sum_0^\infty a_n z^n$ is mean p-valent in $|z| < 1$ and

$$M(r,f) = o(1-r)^{-\beta}, \text{ as } r \to 1$$

where $\beta > 0$, prove that

$$I_\lambda(r,f) = o(1-r)^{1-\beta\lambda} \text{ as } r \to 1,$$

provided that $\beta\lambda > 1$. Deduce that, if $\beta > \frac{1}{2}$,

$$I_1(r,f') = o(1-r)^{-\beta} \text{ as } n \to \infty,$$

and

$$|a_n| = o(n^{\beta-1}) \text{ as } n \to \infty.$$

(Use Theorem 3.2 and Lemma 3.1.)

3.4   Suppose that $g(R)$ is positive increasing for $R \geq 0$ and that

$$\int_0^\infty \frac{R\,dR}{g(R)} < \infty.$$

Prove that

$$\sum_{n=0}^\infty \frac{2^{2n}}{g(2^n)} < \infty.$$

Deduce that if $f(z)$ is mean p-valent in $|z| < 1$, then

$$\int_0^1 \rho\,d\rho \int_0^{2\pi} \frac{|f'(\rho e^{i\theta})|^2 d\theta}{g(|f(\rho e^{i\theta})|)} < \infty.$$

3.5   Suppose $G(R)$ to be twice continuously differentiable for $R \geq 0$, $G'(0) = 0$, and that $f(z)$ is regular for $|z| < 1$. If $u = G(|f(z)|)$, prove that

$$\nabla^2 u = \frac{\partial^2 u}{\partial x^2} + \frac{\partial^2 u}{\partial y^2} = |f'(z)|^2 g(|f(z)|), \ |z| < r$$

where $g(R) = G''(R) + R^{-1}G'(R)$, if $R > 0$, and $g(0) = 2G''(0)$. If

$$I_G(r,f) = \frac{1}{2\pi} \int_0^{2\pi} u(re^{i\theta})d\theta,$$

deduce from Green's Theorem that if

$$S_G(r,f) = r\frac{d}{dr}I_G(r,f)$$

then

$$
\begin{aligned}
S_G(r,f) &= \int_0^r \rho d\rho \int_0^{2\pi} |f'(\rho e^{i\theta})|^2 g(|f(\rho e^{i\theta})|)d\theta \\
&= \int_0^\infty Rg(R)p(r,R)dR.
\end{aligned}
$$

By writing $G(R) = (\varepsilon^2 + R^2)^{\frac{1}{2}\lambda}$, and letting ε tend to zero, deduce another proof of Theorem 3.1 (Flett [1954]).

3.5 Coefficients of general mean p-valent functions   Theorems 2.5, 3.3 and 3.4 give immediately

Theorem 3.5   *Suppose that $f(z) = \sum_0^\infty a_n z^n$ is mean p-valent in $|z| < 1$. Then we have for $1 \leq n < \infty$*

$$|a_n| < A(p)\mu_p n^{2p-1}, \quad (p > \tfrac{1}{4}) \tag{3.23}$$

$$|a_n| < A|a_0|n^{-\frac{1}{2}}\log(n+1), \quad (p = \tfrac{1}{4}) \tag{3.24}$$

$$|a_n| < A(p)|a_0|n^{-\frac{1}{2}}, \tag{3.25}$$

and

$$a_n = o(n^{-\frac{1}{2}}), \ as\ n \to \infty \quad (0 < p < \tfrac{1}{4}), \tag{3.26}$$

where $A(p)$ depends only on p and $\mu_p = \sum_{v \leq p} |a_v|$.

For by Theorem 2.3 we may write $C = A(p)\mu_p$, $\beta = 2p$ in Theorems 3.3 and 3.4. The order of magnitude in (3.23), which is due to Biernacki [1936] for p-valent functions, and to Spencer [1941a] for mean p-valent functions, is best possible. The example of the last section shows that the right-hand sides in (3.25), (3.26) cannot at any rate be replaced by $\epsilon_n n^{-\frac{1}{2}}$, where ϵ_n is a fixed sequence which tend to zero as $n \to \infty$.

3.5.1 The case of univalent functions   The example of Section 3.4 shows that (3.12) fails if $\beta < \tfrac{1}{2}$ for bounded mean p-valent functions. However, Clunie and Pommerenke [1967] have shown that for bounded univalent functions the result can be strengthened.

We first give the Clunie–Pommerenke Theorem in a sharpened form recently obtained by Pommerenke [1985a] and then obtain a corresponding extension of Theorem 3.3 due to Baernstein [1986].

Theorem 3.6   *Suppose that $f(z) \in S$. Then if $0 < \lambda < 1$ and $\eta > \eta_0(\lambda) = -\frac{1}{2} + \lambda + \left(\frac{1}{4} - \lambda + 4\lambda^2\right)^{\frac{1}{2}}$ we have*

$$I_\lambda(r, f') < A(\eta, \lambda)(1 - r)^{-\eta} \quad 0 \le r < 1. \tag{3.27}$$

Hence if $\beta > \beta_0 = \{\eta_0(\lambda) + (1 - \lambda)\}/(2 - \lambda)$, we have

$$I_1\left(r, \frac{f'}{f}\right) < A(\beta)(1 - r)^{-\beta}, \quad \frac{1}{2} \le r < 1. \tag{3.28}$$

Choosing $\lambda = .07$ we obtain (3.28) with $\beta_0 < \beta < .4905525 < .5 - \frac{1}{106}$.

Corollary   *If $f(z)$ is univalent and $|f(z)| < 1$ for $|z| < 1$, we have*

$$I_1(r, f') < 2A(\beta)(1 - r)^{-\beta}, \quad \frac{1}{2} \le r < 1, \tag{3.29}$$

$$|a_n| < 2eA(\beta)n^{\beta-1}, \quad n \ge 2. \tag{3.30}$$

We write

$$I(r) = I_\lambda(r, f') = \frac{1}{2\pi} \int_0^{2\pi} |f'(re^{i\theta})|^\lambda d\theta. \tag{3.31}$$

We apply Theorem 3.1 to f' and deduce that

$$rI'(r) = \frac{\lambda^2}{2\pi} \int_0^r \rho d\rho \int_0^{2\pi} |f'(\rho e^{i\theta})|^\lambda \left|\frac{f''(\rho e^{i\theta})}{f'(\rho e^{i\theta})}\right|^2 d\theta,$$

or

$$r\frac{d}{dr}(rI'(r)) = \frac{\lambda^2}{2\pi} \int_0^{2\pi} |f'(\rho e^{i\theta})|^\lambda \left|r\frac{f''(\rho e^{i\theta})}{f'(\rho e^{i\theta})}\right|^2 d\theta.$$

Since $I'(r) \ge 0$, we deduce, writing $z = re^{i\theta}$,

$$r^2 I''(r) \le \frac{\lambda^2}{2\pi} \int_0^{2\pi} |f'(z)|^\lambda \left|\left(z\frac{f''(z)}{f'(z)} - \frac{2r^2}{1 - r^2}\right) + \frac{2r^2}{1 - r^2}\right|^2 d\theta. \tag{3.32}$$

We define

$$z\frac{f''(z)}{f'(z)} - \frac{2r^2}{1 - r^2} = a + ib, \quad \frac{2r^2}{1 - r^2} = c.$$

Then using (1.6) we have

$$
\begin{aligned}
|a + ib + c|^2 &= a^2 + b^2 + c^2 + 2ac \\
&\leq \frac{16r^2 + 4r^4}{(1 - r^2)^2} + \frac{4r^2}{1 - r^2} \Re \left\{ z \frac{f''(z)}{f'(z)} - c \right\} \\
&= \frac{16r^2 - 4r^4}{(1 - r^2)^2} + \frac{4r^2}{(1 - r^2)} \Re \left(z \frac{f''(z)}{f'(z)} \right).
\end{aligned}
$$

To deal with the last term we differentiate (3.31) under the integral sign. We write

$$
\phi(z) = \log f'(z) = u + iv,
$$

so that, with $z = re^{i\theta}$,

$$
r \frac{\partial u}{\partial r} = \Re z \phi'(z) = \Re z \frac{f''(z)}{f'(z)}.
$$

Thus

$$
\begin{aligned}
rI'(r) &= r \frac{d}{dr} \frac{1}{2\pi} \int_0^{2\pi} e^{\lambda u(re^{i\theta})} d\theta = \frac{\lambda}{2\pi} \int_0^{2\pi} r \frac{\partial u}{\partial r} e^{\lambda u(re^{i\theta})} d\theta \\
&= \frac{\lambda}{2\pi} \int_0^{2\pi} |f'(z)|^\lambda \Re \left(z \frac{f''(z)}{f(z)} \right) d\theta. \qquad (3.33)
\end{aligned}
$$

Substituting in (3.32) we obtain finally

$$
r^2 I''(r) \leq \lambda^2 \frac{16r^2 - 4r^4}{(1 - r^2)^2} I(r) + \frac{4r^3 \lambda}{(1 - r^2)} I'(r).
$$

Given a positive ε we can therefore find $r_0 = r_0(\varepsilon)$ such that $r_0 < 1$ and

$$
I''(r) < \frac{2\lambda + \varepsilon}{(1 - r)} I'(r) + \frac{3\lambda^2 + \varepsilon}{(1 - r)^2} I(r), \quad r_0 < r < 1.
$$

Let $\beta = \beta(\varepsilon)$ be the positive solution of

$$
\beta(\beta + 1) = (2\lambda + \varepsilon)\beta + 3\lambda^2 + \varepsilon.
$$

Then the comparison function $v(r) = C(1 - r)^{-\beta}$ satisfies the equation

$$
v''(r) = \frac{2\lambda + \varepsilon}{1 - r} v'(r) + \frac{3\lambda^2 + \varepsilon}{(1 - r)^2} v(r).
$$

It follows for instance from (1.3), (1.6) and (3.33) that

$$
I(r_0) \leq C_0, \quad I'(r_0) \leq \beta C_0, \qquad (3.34)
$$

where C_0 depends on λ, β, r_0 and ε only. Choosing $C = C_0$ and writing

$$
h(r) = I(r) - v(r) = I(r) - C_0 (1 - r)^{-\beta}, \quad r_0 \leq r < 1
$$

we have $h(r_0) \leq 0$, $h'(r_0) \leq 0$, and $h''(r) \leq 0$, $r_0 \leq r < 1$. We deduce that $h'(r) \leq 0$, $r_0 < r < 1$ and hence $h(r) \leq 0$, $r_0 < r < 1$, i.e.

$$I(r) < C_0(1 - r)^{-\beta}, \quad r_0 < r < 1. \tag{3.35}$$

Using (3.32) we see that (3.35) remains valid for $0 \leq r < 1$ possibly with a larger C_0. Here $\beta = \beta(\varepsilon)$ and C_0 depends on β. Also $\beta(\varepsilon) \to \beta(0) = \eta_0(\lambda)$ as $\varepsilon \to 0$. Thus if $\eta > \eta_0(\lambda)$ we can choose ε so small that $\beta(\varepsilon) < \eta$, and now (3.35) implies (3.27).

We note that $\eta_0(\lambda) < \frac{1}{2}\lambda$ for small λ. In fact

$$\eta_0(\lambda) \sim 3\lambda^2 \text{ as } \lambda \to 0.$$

Pommerenke [1985b] has also shown that the best $\eta_0(\lambda)$ satisfies

$$\beta(\lambda) > C\lambda^2$$

for small positive or negative λ, where C is a constant.

We next prove (3.28). We suppose that $0 < \lambda < 1$, $\beta < \beta_0$ and define η by $\beta = (\eta + 1 - \lambda)/(2 - \lambda)$, so that $\eta > \eta_0$. Further we define δ, p, q by

$$\delta = \frac{\lambda}{2 - \lambda}, \quad p = \frac{2}{1 + \delta}, \quad q = \frac{2}{1 - \delta}, \text{ so that } \frac{1}{p} + \frac{1}{q} = 1.$$

We also write

$$a = a(\theta) = |f'(re^{i\theta})|^\delta, \quad b = b(\theta) = \frac{r|f'(re^{i\theta})|^{1-\delta}}{|f(re^{i\theta})|}.$$

We now apply Hölder's inequality (Titchmarsh, 1939, p. 382)

$$\frac{1}{2\pi} \int_0^{2\pi} a(\theta)\, b(\theta)\, d\theta \leq \left(\frac{1}{2\pi} \int_0^{2\pi} a(\theta)^p d\theta\right)^{\frac{1}{p}} \left(\frac{1}{2\pi} \int_0^{2\pi} b(\theta)^q d\theta\right)^{\frac{1}{q}}.$$

We note that $g(z) = f'(z)^2(z/f(z))^q$ is regular in $|z| < 1$ so that, by Theorem 3.1, $I_1(r, g)$ increases. Also since $f \in S$ we deduce from Theorem 1.4 that

$$|f(z)| \geq \tfrac{2}{9} > \tfrac{1}{5}, \quad \tfrac{1}{2} < |z| < 1.$$

Thus, if $r \geq \frac{1}{2}$,

$$
\begin{aligned}
\frac{1}{2\pi} \int_0^{2\pi} b(\theta)^q d\theta &= I_1(r, g) \leq \frac{2}{(1 - r^2)} \int_r^1 I_1(\rho, g)\rho d\rho \\
&= \frac{1}{\pi(1 - r^2)} \int_r^1 \int_0^{2\pi} |g(\rho e^{i\theta})|\rho d\rho d\theta \\
&= \frac{1}{\pi(1 - r^2)} \int\int_{r<|z|<1} \frac{|z|^q |f'(z)|^2 |dz|^2}{|f(z)|^q}
\end{aligned}
$$

$$< \frac{1+5^q}{\pi(1-r^2)} \int \int_{|z|<1} \frac{|f'(z)|^2 |dz|^2}{1+|f(z)|^q} < \frac{A_1}{1-r}$$

by Lemma 3.2, where $A_1, A_2, \ldots$ denote constants depending on λ and β. Again

$$\int_0^{2\pi} a(\theta)^p d\theta = \int_0^{2\pi} |f'(re^{i\theta})|^\lambda d\theta < A_2(1-r)^{-\eta}$$

by (3.27). Thus

$$\int_0^{2\pi} a(\theta)\, b(\theta)\, d\theta = \int_0^{2\pi} r \left| \frac{f'(re^{i\theta})}{f(re^{i\theta})} \right| d\theta < A_3(1-r)^{-\{\eta(1+\delta)/2+(1-\delta)/2\}}$$

$$= A_3(1-r)^{-\beta}.$$

This yields (3.28). Choosing $\lambda = .07$ we have $\beta_0 < \beta < .4905525$.

Finally we apply (3.28) to $\phi(z) = \{f(z) - f(0)\}/f'(0)$. Then $|f'(z)| < 2|\phi'(z)|/|\phi(z)|$. Now (3.29) follows from (3.28) and (3.30) from (3.8). This proves the corollary.

3.5.2 The results of Carleson and Jones

At this stage we ought to mention a recent paper of Carleson and Jones [1992]. Let $\mathfrak{S}_1$ be the class of functions

$$f(z) = \sum_1^\infty a_n z^n$$

considered in the corollary to Theorem 3.6, i.e. such that $f(z)$ is univalent and $|f(z)| < 1$ in $|z| < 1$. Let $\mathfrak{S}_2$ be the class of functions

$$g(z) = \frac{1}{z} + \sum_1^\infty b_n z^n$$

considered in § 1.1, which are univalent in $|z| < 1$. Write

$$A_n = \sup_{f \in \mathfrak{S}_1} |a_n|, \quad B_n = \sup_{g \in \mathfrak{S}_2} |b_n|.$$

The authors prove that there is a positive constant C_0 such that

$$A_n \geq C_0 \sup_{f \in \mathfrak{S}_1} I_1\left(\frac{n-1}{n}, f'\right), \quad B_n \geq C_0 \sup_{g \in \mathfrak{S}_2} I_1\left(\frac{n-1}{n}, g'\right).$$

(The inequality in the opposite direction follows from (3.8)).

They further prove the existence and equality of the limits

$$\gamma = \lim_{n \to \infty} \frac{-\log|A_n|}{\log n} = \lim_{n \to \infty} \frac{-\log|B_n|}{n}.$$

and characterize γ in terms of a conformal dimension of the boundary $\partial\Omega$ of the image Ω of $|z| < 1$ by $f(z)$ or $g(z)$. Further they show the existence of f, g in $\mathfrak{S}_1, \mathfrak{S}_2$ respectively such that

$$\varlimsup_{n\to\infty} \frac{\log |a_n|}{\log n} = \varlimsup_{n\to\infty} \frac{\log |b_n|}{\log n} = -\gamma.$$

Finally they obtain the upper bound $\gamma \leq .76$, sharpening the previous bound $\gamma < .83$ of Pommerenke [1975, p. 133]. Carleson and Jones also conjecture that $\gamma = .75$. Their method however only gives $\gamma \geq .50245$ compared with Pommerenke's $\gamma \geq .59944$ of Theorem 3.6, corollary.

The methods of Carleson and Jones are based on fractals and iterations and are unfortunately outside the scope of this book. Instead we shall in the next section develop a result of Baernstein [1986].

3.5.3 Baernstein's extension of Theorem 3.3 We proved in Section 3.5.1 that (3.27) for the class S with constants η, λ, where $0 < \lambda < 1$ and $\eta < \frac{1}{2}\lambda$, yields (3.29), i.e.

$$I_1(r, f') < A(\eta, \lambda)(1 - r)^{-\frac{1}{2}+\kappa}$$

for functions $f(z)$ univalent and satisfying $|f(z)| < 1$ in $|z| < 1$. Here

$$\tfrac{1}{2} - \kappa = \{\eta + 1 - \lambda\}/(2 - \lambda)$$

i.e.

$$\kappa = \frac{\lambda - 2\eta}{4 - 2\lambda}. \tag{3.36}$$

From this Baernstein [1986] has deduced a corresponding extension of Theorem 3.3. We proceed to prove Baernstein's result.

Theorem 3.7 *Suppose that (3.27) holds for all* $f(z)$ *in* $\mathfrak{S}$ *with constants* η, λ *such that* $0 < \lambda < 1$ *and* $0 < \eta < \frac{1}{2}\lambda$. *Then if*

$$f(z) = \sum_0^\infty a_n z^n$$

is univalent in $|z| < 1$ *and satisfies*

$$M(r, f) \leq C(1 - r)^{-\beta}, \quad 0 < r < 1, \tag{3.37}$$

where $\beta > \frac{1}{2} - \kappa$ *and* κ *is defined by (3.36), and in particular if* $\beta > .4905525$, *we have*

$$I_1(r, f') \leq A_1 C(1 - r)^{-\beta}, \quad 0 \leq r < 1, \tag{3.38}$$

and hence

$$|a_n| \leq A_2 C n^{\beta-1}. \tag{3.39}$$

From now on $A_1, A_2 = eA_1, A_3, \ldots$ will denote constants depending on η, λ, β only. In particular the conclusion holds with $\beta \geq .4905525$ and absolute constants A_1, A_2. We note that if $\beta \geq 1$, the conclusion of Theorem 3.7 follows from Theorem 3.3. Thus we assume from now on that $\beta < 1$.

The proof of Theorem 3.7 is rather complicated and makes essential use of Theorem 2.4 as well as a localized version of Theorem 3.5. We proceed by a number of steps.

3.5.4 Some auxiliary results   We assume now that $f(z)$ is univalent and $f(z) \neq 0$ in $|z| < 1$. Let E be a set of points $z = re^{i\theta}$ lying on $|z| = r$, and suppose that

$$M = \sup_{z \in E} |f(z)|.$$

We proceed to estimate

$$I(E) = I_1(r, f', E) = \int_E |f'(re^{i\theta})| d\theta.$$

Our first result

Lemma 3.3   *We have for* $\frac{1}{2} \leq r < 1$

$$\int_E |f'(re^{i\theta})|^2 d\theta \leq A_3 M^2/(1-r).$$

We note that, since $f(z) \neq 0$ for $|z| < 1$,

$$\phi(z) = \frac{f(z) - f(0)}{f'(0)} \in \mathfrak{S} \tag{3.40}$$

and $\phi(z) \neq -f(0)/f'(0)$. Using Theorem 1.2, we deduce $|f(0)|/|f'(0)| \geq \frac{1}{4}$. We apply this conclusion to $f\{z_0 + (1-r)z\}$, where $z_0 = re^{i\theta}$ and deduce that

$$|f'(z_0)| \leq \frac{4|f(z_0)|}{1-r}, \tag{3.41}$$

and so

$$\left| \frac{\partial}{\partial r} \log |f(re^{i\theta})| \right| \leq \frac{4}{1-r}.$$

Integrating this inequality from $re^{i\theta}$ to $\rho e^{i\theta}$, we deduce that if $(1-r) \le (1-\rho) \le 2(1-r)$, we have

$$\log|f(\rho e^{i\theta})| \le \log|f(re^{i\theta})| + 4\log\frac{1-\rho}{1-r} \le \log|f(re^{i\theta})| + 4\log 2,$$

so that

$$|f(\rho e^{i\theta})| \le 16|f(re^{i\theta})| \le 16M.$$

Let A be the set defined by

$$A : \{z| \ z = \rho e^{i\theta}, \ re^{i\theta} \in E \text{ and } 1-r \le (1-\rho) \le 2(1-r)\}.$$

Since $f(z)$ is univalent and $|f(z)| < 16M$ in A, the area of the image of A by $f(z)$ is at most $\pi(16M)^2 = 2^8\pi M^2$, i.e.

$$\int_{2r-1}^{r} \rho d\rho \int_{E} |f'(\rho e^{i\theta})|^2 d\theta \le 2^8\pi M^2. \tag{3.42}$$

Next we recall the inequality (1.6) which yields

$$\frac{\partial}{\partial\rho}|f'(\rho e^{i\theta})| \le \frac{2\rho+4}{1-\rho^2} \le \frac{4}{1-\rho}.$$

Integrating this we obtain

$$\log|f'(re^{i\theta})| \le \log|f'(\rho e^{i\theta})| + 4\log 2, \ 2r-1 \le \rho \le r.$$

i.e.

$$|f'(re^{i\theta})| \le 16|f'(\rho e^{i\theta})|, \quad 2r-1 \le \rho \le r.$$

Substituting this into (3.42) we obtain

$$\begin{aligned}
\frac{1}{2}\left[r^2 - (2r-1)^2\right]\int_{E}|f'(re^{i\theta})|^2 d\theta &= \int_{2r-1}^{r}\rho d\rho\int_{E}|f'(re^{i\theta})|^2 d\theta \\
&\le 2^8\int_{2r-1}^{r}\rho d\rho\int_{E}|f'(\rho e^{i\theta})|^2 d\theta \\
&\le \pi 2^{16}M^2.
\end{aligned}$$

Thus

$$\int_{E}|f'(re^{i\theta})|^2 d\theta \le \frac{\pi 2^{17}M^2}{(3r-1)(1-r)} \le \frac{\pi 2^{18}M^2}{1-r}, \text{ since } r \ge \frac{1}{2}.$$

This proves Lemma 3.3.

We deduce

Lemma 3.4   *With the hypotheses of Lemma 3.3 we have*

$$\int_E |f'(re^{i\theta})|d\theta \le A_4 |f(0)|^{\lambda/(2-\lambda)} M^{(2-2\lambda)/(2-\lambda)} (1-r)^{\kappa-\frac{1}{2}}.$$

We choose a positive constant B and divide E into the subsets

$$E_1 : |f'(re^{i\theta})| \le B, \text{ and } E_2 : |f'(re^{i\theta})| \ge B.$$

We recall (3.40) and that $|f'(0)| \le 4|f(0)|$ and apply (3.27) to $\phi(z)$. This yields

$$\begin{aligned}
\int_0^{2\pi} |f'(re^{i\theta})|^\lambda d\theta &\le 2\pi |f'(0)|^\lambda A(\eta, \lambda)(1-r)^{-\eta} \\
&\le A_5 |f(0)|^\lambda (1-r)^{-\eta}.
\end{aligned}$$

Thus

$$\int_{E_1} |f'(re^{i\theta})|d\theta \le B^{1-\lambda} \int_{E_1} |f'(re^{i\theta})|^\lambda d\theta \le A_5 B^{1-\lambda} |f(0)|^\lambda (1-r)^{-\eta},$$

while Lemma 3.3 yields

$$\int_{E_2} |f'(re^{i\theta})|d\theta \le \frac{1}{B} \int_{E_2} |f'(re^{i\theta})|^2 d\theta \le \frac{A_3 M^2}{B(1-r)}.$$

Choosing B so that

$$B^{1-\lambda} |f(0)|^\lambda (1-r)^{-\eta} = \frac{M^2}{B(1-r)} = \{f(0)^\lambda (1-r)^{-\eta}\}^{1/(2-\lambda)} \left(\frac{M^2}{1-r} \right)^{(1-\lambda)/(2-\lambda)},$$

i.e.

$$B^{2-\lambda} = \left\{ \frac{M^2}{|f(0)|^\lambda (1-r)^{1-\eta}} \right\},$$

we obtain Lemma 3.4 with $A_4 = A_3 + A_5$.

  We need a localised version of Lemma 3.4. For this purpose we now assume that E lies in an arc $T : \theta_1 \le \theta \le \theta_2$ of $|z| = r$. We write $|I| = \theta_2 - \theta_1$, assume that

$$\tfrac{1}{2}(1-r) < |I| \le e^{-9}, \tag{3.43}$$

and define

$$\phi = \frac{1}{2}(\theta_1 + \theta_2), \quad \rho = 1 - e^8 |I|, \quad z(I) = \rho e^{i\phi}. \tag{3.44}$$

Thus

$$\frac{1}{e} > (1-\rho) = e^8 |I| > e^7 (1-r). \tag{3.45}$$

Lemma 3.5 *With the above hypotheses we have*

$$\int_E |f'(re^{i\theta})|d\theta \le A_6 \left(\frac{1-\rho}{1-r}\right)^{\frac{1}{2}-\kappa} |f(\rho e^{i\phi})|^{\lambda/(2-\lambda)} M^{(2-2\lambda)/(2-\lambda)}.$$

We suppose without loss of generality that $\phi = 0$, so that I is symmetrical about the real axis, since this may be achieved by a rotation. Thus $z(I) = \rho$. We now consider the bilinear map

$$\zeta = \frac{z-\rho}{1-\rho z} \tag{3.46}$$

followed by the radial projection

$$Z = R\frac{\zeta}{|\zeta|}, \text{ where } R = \frac{r-\rho}{1-\rho r}. \tag{3.47}$$

Writing $r = 1 - \varepsilon$, $\rho = 1 - \delta$, we have $\varepsilon < e^{-7}\delta < \frac{1}{2}e^{-7}$ by (3.45). Thus

$$R = \frac{\delta - \varepsilon}{\delta + \varepsilon(1 - \delta)} > \frac{1 - e^{-7}}{1 + e^{-7}} > .998.$$

Let E_1 be the image of E in the ζ plane and let e be the radial projection of E on $|z| = R$. We write $f(z) = g(\zeta)$, where z, ζ are related by (3.46). Then, since $r > \frac{1}{2}$ by (3.45), we have

$$\int_E |f'(z)|d\theta \le 2\int_E |f'(z)||dz| = 2\int_{E_1} |g'(\zeta)||d\zeta|. \tag{3.48}$$

We note that for $|z| = r$, we have

$$1 - |\zeta|^2 = 1 - \left|\frac{z-\rho}{1-\rho z}\right|^2 = \frac{(1-r^2)(1-\rho^2)}{|1-\rho z|^2} \le \frac{(1-r^2)(1-\rho^2)}{(1-\rho r)^2}. \tag{3.49}$$

Thus E_1 lies outside $|\zeta| = R$. If $\zeta = te^{i\psi}$ then, as $z = re^{i\theta}$ varies on I, we have

$$\begin{aligned}
\frac{d\zeta}{\zeta} &= \frac{dt}{t} + id\psi = dz\left\{\frac{1}{z-\rho} + \frac{\rho}{1-\rho z}\right\} \\
&= izd\theta\left\{\frac{1}{z-\rho} + \frac{\rho}{1-\rho z}\right\} \\
&= id\theta\left\{\frac{\rho}{z-\rho} + \frac{1}{1-\rho z}\right\}.
\end{aligned} \tag{3.50}$$

We take imaginary parts in (3.50) and write $z = x+iy, |I| = 2\gamma$. By (3.43), $|\theta| \le \gamma \le \frac{1}{2}e^{-9}$ on I, and $r > 1 - 4\gamma$. Thus

$$x = r\cos\theta > (1-4\gamma)\cos\gamma > (1-4\gamma)(1-\tfrac{1}{2}\gamma) > 1 - 5\gamma > \rho$$

since $\rho = 1 - 2e^8\gamma$ by (3.45). Thus

$$d\psi = d\theta \left\{ \frac{\rho(x - \rho)}{|z - \rho|^2} + \frac{1 - \rho x}{|1 - \rho z|^2} \right\} > \frac{1 - \rho x}{|1 - \rho z|^2} d\theta.$$

We recall that $\rho = 1 - \delta$, $r = 1 - \varepsilon$ so that $\gamma = \frac{1}{2}e^{-8}\delta$ by (3.44). Thus

$$|1 - \rho z| \le 1 - \rho + \rho|z - 1| \le 1 - \rho + (1 - r) + |\theta| \le \delta + \varepsilon + \gamma < 1.01\delta \quad (3.51)$$

and

$$1 - \rho x \ge 1 - \rho = \delta.$$

Hence

$$\frac{d\psi}{d\theta} > \frac{\delta}{(1.01\delta)^2} > \frac{1}{1.03\delta}. \quad (3.52)$$

On the other hand

$$|z - \rho| \ge r - \rho = \delta - \varepsilon > .99\delta, \quad |1 - \rho z| \ge 1 - \rho x \ge \delta.$$

Thus (3.50) yields

$$\left| \frac{d\zeta}{d\theta} \right| \le \frac{1}{\delta} \left\{ 1 + \frac{1}{.99} \right\} < \frac{2.02}{\delta}. \quad (3.53)$$

On combining (3.52) and (3.53) we obtain

$$|d\zeta| < 3 \, d\psi \quad (3.54)$$

in (3.48), where $Z = Re^{i\psi}$ in (3.47).

Next we compare $|g'(\zeta)|$ and $|g'(Z)|$, where ζ, Z are related by (3.47). We have $\zeta = te^{i\psi}$, $Z = Re^{i\psi}$, where by (3.49), (3.51)

$$1 - t^2 = \frac{(1 - r^2)(1 - \rho^2)}{|1 - \rho z|^2} \ge \frac{(1 + r)(1 + \rho)\delta\varepsilon}{(1.01)^2\delta^2} > \frac{3(1 - r)}{(1 - \rho)},$$

since $r > .99$, $\rho > .63$. Also

$$1 - R = \frac{(1 - r)(1 + \rho)}{1 - \rho r} < \frac{2(1 - r)}{1 - \rho} \text{ and } 1 - R > \frac{1 - r}{1 - \rho}. \quad (3.55)$$

Thus

$$(1 - t) > \frac{3}{4}(1 - R). \quad (3.56)$$

The function $h(w) = g\{Re^{i\psi} + (1 - R)w\}$ is univalent in $|w| < 1$. Hence (1.3) in Theorem 1.3 yields for $|w| < 1$

$$|h'(w)| \le \frac{1 + |w|}{(1 - |w|)^3}|h'(0)|.$$

We define w by

$$Re^{i\psi} + (1 - R)w = \zeta = te^{i\psi},$$

so that

$$|w| = \frac{t - R}{1 - R} < \frac{1}{4}$$

by (3.56). Thus

$$|g'(\zeta)| \le \frac{5/4}{(\frac{3}{4})^3} |g'(Z)| < 3|g'(Z)|.$$

On combining this with (3.48) and (3.54) we obtain

$$\int_E |f'(z)|d\theta < A_7 \int_e |g'(Re^{i\psi})|d\psi,$$

with $A_7 = 18$.

We apply Lemma 3.4 with g instead of f, R instead of r, $g(0) = f(\rho)$ and M as before. Thus

$$\int_E |f'(z)|d\theta < A_4 A_7 |f(\rho)|^{\lambda/(2-\lambda)} M^{(2-2\lambda)/(2-\lambda)} (1 - R)^{\kappa - \frac{1}{2}}.$$

Using also (3.55) and $\kappa < \frac{1}{2}$, we deduce Lemma 3.5.

3.5.5 Proof of Theorem 3.7 We can now complete the proof of Theorem 3.7. We assume for the time being that $f(z) \ne 0$, and that $f(z)$ satisfies (3.37) with $C = 1$. We suppose that $\frac{1}{2} \le r < 1$ and write $M = M(r, f)$.

We choose α so that $\frac{1}{2} - \kappa < \alpha < \frac{1}{2}$ and $\alpha < \beta$. We define $\beta - \alpha = \nu > 0$. We recall that λ, η are constants such that (3.27) holds and κ is defined by (3.36). Let k be the smallest nonnegative integer such that

$$2^{-k}M \le e^8(1 - r)^{-\nu}. \tag{3.57}$$

We define

$$B_k = \{\theta \,|\, 0 \le \theta \le 2\pi \text{ and } |f(re^{i\theta})| \le 2^{-k}M\}$$

and, if $k > 0$ and $0 \le j < k$,

$$B_j = \{\theta \,|\, 0 \le \theta \le 2\pi, \text{ and } 2^{-j-1}M < |f(re^{i\theta})| \le 2^{-j}M\}.$$

We apply Lemma 3.4 with $E = B_k$. Since $|f(0)| \le 1$ by (3.37) with $C = 1$, we obtain, since $\alpha > \kappa - \frac{1}{2}$,

$$\int_{B_k} |f(re^{i\theta})|d\theta \le A_4(2^{-k}M)^{(2-2\lambda)/(2-\lambda)} (1 - r)^{-\alpha}.$$

If $2^{-k}M \leq 1$, we deduce that

$$\int_{B_k} |f'(re^{i\theta})|d\theta \leq A_4(1-r)^{-\alpha} \leq A_4(1-r)^{-\beta}.$$

If $2^{-k}M > 1$ we obtain

$$\int_{B_k} |f'(re^{i\theta})|d\theta \leq A_4 2^{-k}M(1-r)^{-\alpha} \leq A_4 e^8(1-r)^{-\nu-\alpha}$$

$$= A_4 e^8(1-r)^{-\beta}. \qquad (3.58)$$

Thus (3.58) holds in all cases. If $k = 0$, $B_k = [0, 2\pi]$ and (3.38) is proved.

Thus we now suppose that $k > 0$, and consider B_j for $0 \leq j < k$. Hence

$$2^{-j}M > e^8(1-r)^{-\nu}.$$

We fix such a value of j and define

$$p = 4(1-r)^{-\beta}2^j M^{-1}, \qquad (3.59)$$

Using also (3.37) with $C = 1$, we deduce that

$$4 \times 2^j \leq p < 4e^{-8}(1-r)^{\nu-\beta} = 4e^{-8}(1-r)^{-\alpha}. \qquad (3.60)$$

Let l be the largest number of the form $2\pi/n$, where n is a natural number, such that

$$l \leq \frac{1}{16}(1-r)\,p^{1/\alpha} < \frac{1}{16}(4e^{-8})^{1/\alpha} < e^{-16} \qquad (3.61)$$

by (3.60) and since $\alpha < \frac{1}{2}$. Thus (3.43) holds if I is an arc of length l.

By (3.61) we have $n > 1$. Thus

$$\frac{2\pi}{(n-1)} > \frac{1}{16}(1-r)p^{1/\alpha},$$

so that using (3.60) we have

$$l = \frac{n-1}{n}\frac{2\pi}{n-1} > \frac{1}{32}(1-r)p^{1/\alpha} > \frac{1}{2}(1-r). \qquad (3.62)$$

Thus (3.43) holds, when $|I| = l$. We divide the interval $[0, 2\pi]$ into n arcs I of length $l = 2\pi/n$ and apply Lemma 3.5 with $E_j = I \cap B_j$. We recall that by (3.37), (3.44) and (3.62)

$$|f(\rho e^{i\phi})| \leq (1-\rho)^{-\beta} = \left(\frac{e^{-8}}{l}\right)^{\beta} < \{32e^{-8}(1-r)^{-1}p^{-1/\alpha}\}^{\beta}.$$

Also by (3.61)

$$1 < \frac{1-\rho}{1-r} = \frac{e^8 l}{1-r} < \frac{e^8}{16}p^{1/\alpha},$$

and $\frac{1}{2} - \kappa < \alpha$. Next we have on E_j

$$|f(re^{i\theta})| \le 2^{-j}M = 4(1-r)^{-\beta}p^{-1}$$

by (3.59). Hence Lemma 3.5 yields

$$\int_{E_j} |f'(re^{i\theta})|d\theta$$

$$\le A_6 \left(\frac{e^8}{16}p^{1/\alpha}\right)^{\alpha} \left\{32e^{-8}\frac{1}{1-r}p^{-\frac{1}{\alpha}}\right\}^{\beta\lambda/(2-\lambda)} \left\{\frac{4}{p(1-r)^{\beta}}\right\}^{(2-2\lambda)/(2-\lambda)}$$

$$< A_8(1-r)^{-\beta}p^{1-(\beta\lambda/\alpha(2-\lambda))-(2-2\lambda)/(2-\lambda)}$$

$$= A_8(1-r)^{-\beta} p^{-\gamma} < A_8 2^{-\gamma j}(1-r)^{-\beta} \tag{3.63}$$

by (3.60), where

$$\gamma = \left(\frac{\beta}{\alpha} - 1\right)\left(\frac{\lambda}{2-\lambda}\right) = \frac{\nu}{\alpha}\frac{\lambda}{2-\lambda}.$$

In order to complete the proof of Theorem 3.7 we need a final Lemma.

Lemma 3.6 *For each fixed j at most A_9 of the sets E_j are nonempty.*

Assuming Lemma 3.6 we deduce from (3.63) that

$$\int_{B_j} |f'(re^{i\theta})|d\theta \le A_8 A_9 (1-r)^{-\beta} 2^{-\gamma j}, \quad 0 \le j < k.$$

On combining this with (3.58) we deduce (3.38) on the assumption that $f(z) \ne 0$.

It remains to prove Lemma 3.6. For this purpose we use Theorem 2.4. We label as I_ν the arcs I which meet B_j and assume the I_ν with increasing ν to be arranged in anticlockwise order along $|z| = r$. Thus on each arc I_ν there is a point z'_ν in B_j, i.e.

$$z'_\nu = re^{i\theta_\nu}, \quad |f(z'_\nu)| > M2^{-j-1} = R_2 \tag{3.64}$$

say, while by (3.37) with $C = 1$

$$z_\nu = \rho e^{i\theta_\nu} \text{ satisfies } |f(z_\nu)| \le (1-\rho)^{-\beta} = R_1 \tag{3.65}$$

say. We define $Q = 19\,000$ so that $Q - 1 > 2\pi e^8$. Also, by (3.61), $(Q-1)l < 1$. Then $(1-\rho) = e^8 l < (Q-1)l/(2\pi) < \frac{1}{2}$, so that $\rho > \frac{1}{2}$. Suppose that, for some integer n_0, there are at least $n_0 Q$ sets E_j. We consider the discs

$$|z - z_\nu| < 1 - \rho, \quad \nu = 1, Q+1, \ldots, (n_0-1)Q+1, \tag{3.66}$$

and note that they are all disjoint. In fact if μ, μ' are 2 distinct values of v in (3.66) we have

$$z_\mu = \rho e^{i\theta}, \quad z_{\mu'} = \rho e^{i\theta'} \text{ where } \theta + (Q-1)l \le \theta' \le \theta + 2\pi - (Q-1)l.$$

This yields

$$|z_\mu - z_{\mu'}| \ge 2\rho \left| \sin \left(\frac{\theta - \theta'}{2} \right) \right| \ge 2\rho \sin \frac{(Q-1)l}{2} > \frac{1}{\pi}(Q-1)l,$$

since $(Q-1)l < \pi$, and $\rho > \frac{1}{2}$. On the other hand

$$2(1 - \rho) = 2e^8 l < \frac{(Q-1)l}{\pi}.$$

Thus the discs (3.66) are all disjoint.

Next we have from (3.45), (3.62), (3.64) and (3.65)

$$
\begin{aligned}
\frac{R_2}{R_1} &= 2^{-(j+1)}M(1-\rho)^\beta = 2^{-(j+1)}M(e^8 l)^\beta \\
&> 2^{-(j+1)}M\left(\frac{e^8}{32}\right)^\beta (1-r)^\beta \, p^{\beta/\alpha} \\
&= 2\left(\frac{e^8}{32}\right)^\beta p^{\frac{\beta}{\alpha}-1}
\end{aligned}
\tag{3.67}
$$

by (3.59). Since $p > 1$, $\alpha > .4$, and $\beta > \alpha$, we obtain

$$\frac{R_2}{R_1} > 2\left(\frac{e^8}{32}\right)^4 = \tfrac{1}{2}e^{3.2} > e^2.$$

Now Theorem 2.4 yields

$$n_0 / \log \left(\frac{A_0(1-\rho)}{1-r} \right) < \frac{2}{\log\left(\frac{R_2}{R_1}\right) - 1} < \frac{4}{\log(R_2/R_1)},$$

i.e.

$$n_0 < \frac{4\log\left\{\frac{A_0(1-\rho)}{1-r}\right\}}{\log\left(\frac{R_2}{R_1}\right)}.$$

By (3.67) we have

$$\log\left(\frac{R_2}{R_1}\right) > 2 + \frac{\beta - \alpha}{\alpha}\log p > \frac{v}{\alpha}\log p,$$

while by (3.61)

$$\log p > \alpha \log \frac{16l}{1-r} = \alpha \log \frac{16e^{-8}(1-\rho)}{1-r} > \alpha\left\{\log \frac{1-\rho}{1-r} - 6\right\}.$$

On the other hand by (3.45)

$$\log\left(\frac{1-\rho}{1-r}\right) > 7, \text{ so that } \log p > \frac{\alpha}{7}\log\frac{1-\rho}{1-r},$$

Thus

$$\log\frac{R_2}{R_1} > \frac{v}{\alpha}\log p > \frac{v}{7}\log\frac{1-\rho}{1-r},$$

and

$$n_0 < \frac{4\log\frac{1-\rho}{1-r}}{\log\frac{R_2}{R_1}} + \frac{4\log A_0}{\log\frac{R_2}{R_1}} < \frac{28}{v} + 2\log A_0.$$

This proves Lemma 3.6 with $A_9 = 19\,000\{2\log A_0 + 28v^{-1} + 1\}$, since otherwise we obtain a contradiction from our assumption that at least $n_0 Q$ of the E_j are non-empty.

To complete the proof of (3.38) with $C = 1$, we need to eliminate the assumption $f(z) \neq 0$. It follows from (3.37) with $C = 1$, $\beta < 1$, that $|f(z)| < 2$ for $|z| < \frac{1}{2}$, so that by Cauchy's inequality $|a_1| = |f'(0)| \leq 4$. Suppose that the image of $|z| < 1$ by $w = f(z)$ contains the disc $|w| < R$. Then $z = f^{-1}(w)$ gives a univalent map of $|w| < R$ onto a subdomain of $|z| < 1$ and now Cauchy's inequality yields

$$\frac{1}{4} \leq \left|\frac{1}{a_1}\right| = \left|\frac{dz}{dw}\right| \leq \frac{1}{R}, \text{ i.e. } R \leq 4.$$

Thus there exists w_0, such that $|w_0| \leq 4$, and $f(z) \neq w_0$ in $|z| < 1$. Hence if

$$F(z) = \frac{f(z) - w_0}{5}$$

then $F(z)$ is univalent, $F(z) \neq 0$ in $|z| < 1$ and

$$|F(z)| \leq \frac{|f(z)|}{5} + \frac{4}{5} \leq \frac{C}{5}(1 - |z|)^{-\beta} + \frac{4}{5} \leq C(1 - |z|)^{-\beta}.$$

Thus we can apply what we have just proved to $F(z)$ and obtain (3.38) with

$$I_1(r, F') = \frac{1}{5}I_1(r, f')$$

instead of $I_1(r, f')$. This proves (3.38) with $5A_1$ instead of A_1, when $C = 1$. We obtain the result for general C by considering $f(z)/C$ instead of $f(z)$. Finally (3.39) follows from (3.8) and (3.38). This proves Theorem 3.7.

3.6 Growth and omitted values   We give next an application of Theorem
3.4 to a problem raised by Dvoretzky [1950]. We show that if a univalent
function $f(z)$ omits a set of values which is fairly dense in the plane, then
the effect on the coefficients is nearly as strong as if $f(z)$ is bounded.[†]

Theorem 3.8   *Suppose that* $f(z) = \sum_{0}^{\infty} a_n z^n$ *is univalent in* $|z| < 1$ *and
maps* $|z| < 1$ *(1,1) conformally onto a domain* D *in the* w *plane. Let* $d(R)$
be the radius of the largest disc, $|w - w_0| < \rho$, *whose centre lies on* $|w_0| = R$
and which lies entirely in D. *Then if for some constant* $\alpha < .245$ *we have*

$$d(R) \leq \alpha R \quad (0 < R < \infty), \qquad (3.68)$$

we deduce

$$|a_n| < A(\alpha)|a_0|n^{-.509} \quad (n = 1, 2, \ldots). \qquad (3.69)$$

We define $z_0 = re^{i\theta}$ and

$$\phi(z) = f\left(\frac{z_0 + z}{1 + \bar{z}_0 z}\right) = b_0 + b_1 z + \ldots$$

Then $\phi(z)$ is univalent in $|z| < 1$ and maps $|z| < 1$ onto D.
Hence there exists w_1 outside D, such that

$$|w_1 - b_0| \leq \alpha|b_0|.$$

Thus

$$\frac{\phi(z) - b_0}{b_1} = z + \ldots$$

is univalent in $|z| < 1$ and omits the value $(w_1 - b_0)/b_1$. Now Theorem
1.2 gives

$$\left|\frac{w_1 - b_0}{b_1}\right| \geq \frac{1}{4},$$

and so $|b_1| \leq 4\alpha|b_0|$, i.e.

$$(1 - r^2)|f'(re^{i\theta})| \leq 4\alpha|f(re^{i\theta})|.$$

We deduce

$$\frac{\partial}{\partial r} \log|f(re^{i\theta})| \leq \frac{4\alpha}{1 - r^2} \quad (0 < r < 1),$$

and integrating with respect to r, we obtain

$$\left|\frac{f(re^{i\theta})}{f(0)}\right| \leq \left(\frac{1+r}{1-r}\right)^{2\alpha}.$$

[†] For generalizations to mean p-valent and other functions see Hayman [1952].

Since $\alpha < \frac{1}{4}$ this gives

$$M(r,f) < \sqrt{2}|a_0|(1-r)^{-2\alpha} \quad (0 < r < 1),$$

and now Theorem 3.8 follows from Theorem 3.7, with .491 instead of β. If $0 < \lambda \leq 1$, then the function

$$w = f(z) = \left(\frac{1+z}{1-z}\right)^{\lambda}$$

maps $|z| < 1$ (1,1) conformally onto the angle $|\arg w| < \frac{1}{2}\pi\lambda$ and satisfies $d(R) = R\sin(\frac{1}{2}\pi\lambda)$. It does not satisfy (3.69) if $\lambda > .491$. Thus in (3.69) .245 cannot at any rate be replaced by .7 > sin(.246π).

3.7 k-symmetric functions and Szegö's conjecture If $f_k(z)$ is regular in $|z| < 1$ and of the form

$$f_k(z) = z + a_{k+1}z^{k+1} + a_{2k+1}z^{2k+1}\ldots, \tag{3.70}$$

we say that $f_k(z)$ is k-symmetric. If further $f_k(z) \in \mathfrak{S}$ where k is a positive integer then also

$$f(z) = \left\{f_k(z^{\frac{1}{k}})\right\}^k \in \mathfrak{S}.$$

In fact since $f_k(z)$ is univalent

$$\frac{f_k(z)}{z} = 1 + a_{k+1}z^k + a_{2k+1}z^{2k}$$

is regular and non zero in $|z| < 1$, and so is

$$\phi(z) = 1 + a_{k+1}z + a_{2k+1}z^2 + \ldots = \frac{f_k(z^{1/k})}{z^{1/k}}.$$

Hence so is

$$f(z) = z\phi(z)^k = z + ka_{k+1}z^2 + \ldots$$

It remains to show that $f(z)$ is univalent. Suppose then that z_1, z_2 are distinct numbers in $|z| < 1$, such that $f(z_1) = f(z_2) = w$. Since $f(z) = 0$ only for $z = 0$, we may assume that $w \neq 0$. Let ζ_1, ζ_2 be k th roots of z_1, z_2, so that $\zeta_1^k = z_1$, $\zeta_2^k = z_2$. Then

$$w = \{f_k(\zeta_1)\}^k = f(\zeta_1^k) = f(\zeta_2^k) = \{f_k(\zeta_2)\}^k.$$

Thus

$$f_k(\zeta_1) = w_1, \quad f_k(\zeta_2) = w_2$$

where $w_2^k = w_1^k = w$, so that $w_2 = w_1\omega$, where $\omega = \exp\left(\frac{2\pi i p}{k}\right)$.
Now by (3.70)

$$f_k(\omega\zeta_1) = \omega f_k(\zeta_1).$$

Since f_k is univalent we deduce that $\zeta_2 = \omega\zeta_1$. Thus $z_2 = \zeta_2^k = \zeta_1^k = z_1$,
so that $f(z)$ is also univalent, i.e. $f(z) \in \mathfrak{S}$. We can now prove

Theorem 3.9 *(Szegö's conjecture). If $f_k(z) \in \mathfrak{S}$, where $k = 1, 2, 3, 4$ and*
$f(z)$ *has the expression (3.70) then*

$$|a_n| \le A n^{2/k-1}, \quad n \ge 1. \tag{3.71}$$

The result was obtained by Littlewood [1925] for $k = 1$, by Littlewood
and Paley [1932] for $k = 2$, by V. I. Levin [1934] for $k = 3$ and by
Baernstein [1986] for $k = 4$. Littlewood [1938] showed that (3.69) is false
for large k and an example of Pommerenke [1967, and 1975, p. 133]
proves that it is false for $k \ge 12$. The conjecture (3.71) is attributed by
Levin [1934] to an oral communication of Gabor Szegö.

We note that since

$$f_k(z) = f(z^k)^{1/k}$$

we have

$$M(r, f_k) = M(r^k, f)^{1/k} \le r/(1 - r^k)^{2/k}, \quad 0 < r < 1$$

by Theorem 1.3. Now (3.71) follows from Theorem 3.3 for $k = 1, 2, 3$ and
from Theorem 3.7 for $k = 4$. We see that the cases $5 \le k \le 11$ remain
open. To prove (3.71) for $k = 5$ the range of Theorem 3.7 would need
to be extended to $\beta \ge .4$ instead of $\beta \ge .491$. This seems well beyond
techniques available at present.

If $f_k(z)$ is k-symmetric and p-valent a similar argument, based upon
Theorem 3.3, shows that

$$|a_{nk+1}| < A(p,k)\mu_p n^{(2p/k)-1}, \tag{3.72}$$

provided that $1 \le k < 4p$ (Robertson [1938]) and the proof was modified
so as to apply to mean p-valent functions by Spencer [1940b, 1941a].
For this latter case (3.72) breaks down when $k > 4p$, as is shown by a
suitable form of the examples in §3.4.

For this type of proof is essential that *all* the coefficients vanish, except
those whose suffixes form an arithmetic progression of common difference

k. We proceed to prove (3.72) under somewhat weaker assumptions, basing ourselves on the fact that the functions $f_k(z)$ satisfy

$$|f_k[re^{i(\theta + 2\pi i v/k)}]| = |f_k(re^{i\theta})| \quad (0 \le v \le k - 1).$$

3.7.1   We need the following preliminary result:

Theorem 3.10   *Suppose that* $f(z) = \sum_0^\infty a_n z^n$ *is mean p-valent in* $|z| < 1$ *and that there exist k points* $z_1', z_2', \ldots, z_k'$ *on* $|z| = r$, *where* $0 < r < 1$ *and* $k \ge 2$, *such that*

(i) $|z_i' - z_j'| \ge \delta \quad (1 \le i < j \le k)$

and

(ii) $|f(z_i')| \ge R \quad (1 \le i \le k).$

Then we have

$$R < A(p)\mu \delta^{2p(1/k - 1)}(1 - r)^{-2p/k},$$

where μ *is given by (2.8).*

We suppose that

$$\delta > 4^{p+2}(1 - r). \tag{3.73}$$

For if this is false, we have by Theorem 2.3

$$
\begin{aligned}
R &\le M(r, f) < A(p)\mu(1 - r)^{-2p} \\
&\le A(p)\mu(1 - r)^{-2p/k} \left(\frac{4^{p+2}}{\delta} \right)^{2p - 2p/k} \\
&< A(p)\mu(1 - r)^{-2p/k} \delta^{2(p/k) - 2p},
\end{aligned}
$$

so that Theorem 3.10 holds.

At least one of the annuli

$$1 - 4^{-v}\delta < |z| < 1 - 4^{-(v+1)}\delta \quad (1 \le v \le [p] + 1)$$

is free from zeros of $f(z)$, since $f(z)$ has $q \le p$ zeros in $|z| < 1$. For such a value of v we choose

$$r_1 = 1 - \tfrac{1}{2}4^{-v}\delta = 1 - \delta_0,$$

say, and note that $f(z)$ has no zeros in $1 - 2\delta_0 < |z| < 1 - \tfrac{1}{2}\delta_0$. Using (3.73) we deduce further that

$$\frac{\delta}{8} \ge \delta_0 = \tfrac{1}{2}4^{-v}\delta > \tfrac{1}{2}4^{p+2-v}(1 - r) \ge 2(1 - r), \tag{3.74}$$

since $1 \leq v \leq p+1$, and hence that $r_1 < r$.

Suppose now that

$$z'_j = re^{i\theta_j} \quad (1 \leq j \leq k).$$

We write

$$z_j = r_1 e^{i\theta_j} \quad (1 \leq j \leq k),$$

and apply Theorem 2.4 with $R_1 = 2^p M(r_1, f)$, using the number R of Theorem 3.10 (ii) instead of R_2, and δ_0 instead of r_n. This gives

$$\delta_j = \frac{\delta_0 - |z'_j - z_j|}{\delta_0} = \frac{1-r}{\delta_0} \quad (1 \leq j \leq k).$$

By construction the circles $|z - z_j| < \frac{1}{2}\delta_0$ contain no zeros of $f(z)$. Also the circle $|z - z_j| < \delta_0$ contains the point z'_j and has diameter $2\delta_0 \leq \frac{1}{4}\delta$ by (3.74). In view of hypothesis (i) of Theorem 3.10 it now follows that the circles $|z - z_j| < \delta_0$ $(1 \leq j \leq k)$ are non-overlapping and Theorem 2.4 is applicable. We obtain

$$k \left[\log \frac{A_0 \delta_0}{1-r} \right]^{-1} \leq 2p \left[\log \left(\frac{R_2}{eR_1} \right) \right]^{-1},$$

$$\frac{R_2}{eR_1} < \left\{ \frac{A_0 \delta_0}{1-r} \right\}^{2p/k} < A(p) \left(\frac{\delta}{1-r} \right)^{2p/k}$$

by (3.74).[†] Since $r_1 = 1 - \delta_0$ Theorem 2.3 gives further

$$R_1 = 2^p M(r_1, f) < A(p)\mu \delta_0^{-2p} = A(p)\mu \left(\frac{2 \times 4^v}{\delta} \right)^{2p} < A(p)\mu \delta^{-2p},$$

and we deduce Theorem 3.10 on writing R instead of R_2.

3.8 Power series with gaps   It is easy to see that the theorems quoted in §3.7 follow at once from Theorems 2.3, 3.3 and 3.10. We can, however, prove more.

Theorem 3.11   *Suppose that*

$$f(z) = \sum_0^{N-1} a_n z^n + a_N z^N + a_{N+k} z^{N+k} + a_{N+2k} z^{N+2k} + \cdots$$

[†] We may assume that $R_2 > eR_1$. Since $\delta/(1-r) \geq 16$ by (3.74), our result is trivial otherwise.

is mean p-valent in $|z| < 1$. *Then*

$$M(r, f) < A(p, k, N)\mu(1 - r)^{-2p/k} \quad (0 < r < 1), \tag{3.75}$$

and if $1 \le k < 4p$ *we have further*

$$|a_n| < A(p, k, N)\mu n^{(2p/k)-1} \quad (n = 1, 2, \ldots). \tag{3.76}$$

Write

$$g(z) = \sum_0^{N-1} a_n z^n, \quad h(z) = \sum_{n=0}^{\infty} a_{N+kn} z^{N+kn},$$

and for $r \ge \frac{1}{2}$ choose θ_0 so that $|f(re^{i\theta_0})| = M(r, f)$. Then if

$$\theta_v = \theta_0 + \frac{2\pi v}{k} \quad (0 \le v \le k - 1),$$

we have

$$h(re^{i\theta_v}) = h(re^{i\theta_0}) \exp\left(\frac{2\pi i v N}{k}\right).$$

Again for $|z| = r < 1$, we have from Theorem 3.5

$$|g(z)| \le \sum_{n=0}^{N-1} |a_n| \le \sum_{n=0}^{N-1} A(p)\mu n^{2p-1} \le A(p, N)\mu$$

if $p \ge \frac{1}{2}$ and

$$|g(z)| \le \sum_{n=0}^{N-1} A|a_0| = AN|a_0|$$

if $p < \frac{1}{2}$. Thus

$$
\begin{aligned}
|f(re^{i\theta_v})| &\ge |h(re^{i\theta_v})| - |g(re^{i\theta_v})| \\
&\ge |h(re^{i\theta_0})| - A(p, N)\mu \\
&\ge M(r, f) - 2A(p, N)\mu \quad (0 \le v \le k - 1).
\end{aligned}
$$

We may therefore apply Theorem 3.10 with

$$\delta = |re^{2\pi i/k} - r| \ge \tfrac{1}{2}|e^{2\pi i/k} - 1| = A(k),$$

and

$$R = M(r, f) - A(p, N)\mu,$$

and obtain

$$M(r, f) < A(p, N)\mu + A(p)\mu A(p, k)(1 - r)^{-2p/k},$$

and this gives (3.75). The inequality, proved on the assumption $r \geq \frac{1}{2}$, clearly remains valid if $r < \frac{1}{2}$, since $M(r, f)$ increases with r. Also (3.76) follows from (3.75) and Theorem 3.3. This proves Theorem 3.11.

3.8.1   If $k = 2$, we can sharpen Theorem 3.11.

Theorem 3.12   *Suppose that* $f(z) = \sum_0^\infty a_n z^n$ *is mean p-valent in* $|z| < 1$ *and that* $a_n = 0$ *whenever* $n = bm + c$, *where* b, c *are fixed positive integers and m goes from* 1 *to* ∞. *Then*

$$M(r, f) < A(p, b, c)\mu(1 - r)^{-p} \quad (0 < r < 1), \tag{3.77}$$

and so, if $p > \frac{1}{2}$,

$$|a_n| < A(p, b, c)\mu n^{p-1} \quad (n \geq 1). \tag{3.78}$$

To prove Theorem 3.12, consider

$$f_v(z) = \sum_{m=0}^\infty a_{bm+v} z^{bm+v} \quad (0 \leq v \leq b - 1),$$

and write $\omega = \exp[2\pi i/b]$. Then we have

$$f_v(z) = \frac{1}{b} \sum_{\mu=0}^{b-1} \omega^{-\mu v} f(\omega^\mu z).$$

In fact the coefficient of z^n on the right-hand side is

$$\frac{a_n}{b} \sum_{\mu=0}^{b-1} \omega^{-\mu v} \omega^{\mu n} = \frac{a_n}{b} \sum_{\mu=0}^{b-1} [\omega^{n-v}]^\mu,$$

and this is a_n or 0, according as $n - v$ is or is not a multiple of b.

Hence if v is so chosen that $c - v$ is a multiple of b, we have from the hypothesis of Theorem 3.12 for $|z| < 1$

$$\left| \sum_{\mu=0}^{b-1} \omega^{-\mu v} f(\omega^\mu z) \right| = b|f_v(z)| \leq b \sum_{n=0}^{b+c} |a_n| \leq A(p, b, c)\mu = K,$$

say.

Choose now, for $\frac{1}{2} \leq r \leq 1$, z_0 so that

$$|z_0| = r, \quad |f(z_0)| = M(r, f).$$

If $M(r, f) \leq 2K$, then (3.77) follows. Otherwise we obtain

$$\left| \sum_{\mu=1}^{b-1} \omega^{-\mu\nu} f(\omega^{\mu} z_0) \right| \geq |f(z_0)| - \left| \sum_{\mu=0}^{b-1} \omega^{-\mu\nu} f(\omega^{\mu} z_0) \right|$$

$$\geq |f(z_0)| - K \geq \tfrac{1}{2}|f(z_0)| = \tfrac{1}{2}M(r, f).$$

Hence we can find μ $(1 \leq \mu \leq b - 1)$ such that

$$|f(\omega^{\mu} z_0| \geq \frac{M(r, f)}{2(b-1)},$$

and

$$|\omega^{\mu} z_0 - z_0| = r|1 - \omega^{\mu}| \geq \tfrac{1}{2}|1 - \omega| = \sin\left(\tfrac{\pi}{2b}\right) \geq \tfrac{1}{b}.$$

We can thus apply Theorem 3.10, with $k = 2$, $z_1' = z_0$, $z_2' = \omega^{\mu} z_0$, $\delta = b^{-1}$ and $R = \tfrac{1}{2}M(r, f)/(b - 1)$. We obtain

$$M(r, f) < 2(b - 1)A(p)\mu b^p (1 - r)^{-p},$$

and this proves (3.77) if $r \geq \tfrac{1}{2}$. Since $M(r, f)$ increases with r, the result for $0 \leq r < \tfrac{1}{2}$ also follows, and (3.78) follows from (3.77) and Theorem 3.3. Thus Theorem 3.12 is proved.

It is worth noting that the above argument does not require the full strength of our hypotheses. It would be sufficient to assume that on $|z| = r$

$$|f_{\nu}(z)| = \left| \sum_{m=0}^{\infty} a_{bm+\nu} z^{bm+\nu} \right| = O(1 - r)^{-p} \text{ as } r \to 1,$$

in order to obtain $M(r, f) = O(1 - r)^{-p}$. A similar remark applies to Theorem 3.11.

The conclusion of Theorem 3.11 continues to hold if $a_n = 0$, except for a sequence n_{ν} such that $n_{\nu+1} - n_{\nu} \geq k$ finally (instead of $n_{\nu+1} - n_{\nu} = k$ finally) [Hayman 1967]. Similarly that of Theorem 3.12 holds if $a_n = 0$ for a sequence $n = n_{\nu}$, such that $n_{\nu+1} - n_{\nu} \leq b$ [Hayman 1969]. These results involve some Fourier series arguments to show that the hypotheses of Theorem 3.10 are satisfied and are beyond the scope of this book.

Examples

3.6   Show that if $f(z)$ is univalent and satisfies the hypotheses of Theorem 3.11 for $k = 4$, then $|a_n| < A(N)\mu n^{-\frac{1}{2}}$. (Use Theorem 3.7.) Note that this conclusion extends Szegö's conjecture for $k = 4$.

3.7   If $f(z)$ is univalent as well as mean p-valent in Theorem 3.12, show that (3.78) continues to hold for $p \geq .491$.

3.8   If $k > 4p$ in Theorem 3.11, or $p < \frac{1}{2}$ in Theorem 3.12, show that $a_n = o(n^{-\frac{1}{2}})$ and that this conclusion is sharp. (Use Theorem 3.4 and the example of Section 3.4.)

4

Symmetrization

4.0 Introduction   In this chapter we develop the theory of symmetrization in the form due to Pólya and Szegö [1951] as far as it is necessary
for our function-theoretic applications.

Given a domain D, we can, by certain types of lateral displacement
called symmetrization, transform D into a new domain D^* having some
aspects of symmetry. The precise definition will be given in §4.5. Pólya
and Szegö showed that while area for instance remains invariant under
symmetrization, various domain constants such as capacity, inner radius, principal frequency, torsional rigidity, etc., behave in a monotonic
manner.

We shall here prove this result for the first two of these concepts
in order to deduce Theorem 4.9, the principle of symmetrization. If
$f(z) = a_0 + a_1 z + \dots$ is regular in $|z| < 1$, and something is known about
the domain D_f of values assumed by $f(z)$, this principle allows us to
assert that in certain circumstances $|a_1|$ will be maximal when $f(z)$ is
univalent and D_f symmetrical. Applications of this result will be given
in Sections 4.10–4.12. Some of these will in turn form the basis of further
studies of p-valent functions in Chapter 5. Some of these results can
also be proved in another manner by a consideration of the transfinite
diameter (Hayman [1951]). The chapter ends with a recent proof of
Bloch's Theorem by Bonk [1990].

Since the early part of the chapter is used only to prove Theorem
4.9, the reader who is prepared to take this result for granted on a first
approach may start with §4.5 and then go straight to §4.9.

We shall need to refer to Ahlfors [1979] (which we continue to denote
by C. A.) for a number of results which space does not permit us to
consider in more detail here. A good set-theoretic background is provided
by C. A., Chapter 3, §1, and we use generally the notation there given.

103

We shall, however, in accordance with common English usage, call an open connected set a *domain* and not a region. The closure of a domain D will be denoted by $\bar{D}$. A domain D has connectivity n if its complement in the extended plane has exactly n components; if $n = 1$, D is simply connected, if $n = 2$ doubly connected, etc. (C. A. pp. 139 and 146). If the boundary of D consists of a finite number n of analytic simple closed curves (C. A. pp. 68 and 234), no two of which have common points, we shall call D an analytic domain.

We shall assume a right-handed system OX, OY of rectangular Cartesian axes in the plane. Points in the plane will be denoted in terms of their coordinates x, y either by (x, y) or by $z = x + iy$, whichever is more convenient. Accordingly, functions u will be written as either $u(z)$ or $u(x, y)$.

4.1 Lipschitzian functions   Let E be a plane set and let $P(z)$ be a function defined on E. We shall say that $P(z)$ is *Lipschitzian* or Lip on E if there is a constant C such that

$$|P(z_1) - P(z_2)| \le C|z_1 - z_2| \qquad (4.1)$$

whenever z_1, z_2 lie in E. It is clear that a Lip function is continuous, and further that if P, Q are Lip and bounded on E, then PQ is Lip on E.

Suppose that E is compact and that $P(z)$ is Lip in some neighbourhood of every point z_0 of E. Then P is Lip on E. For if not, we could find sequences of distinct pairs of points z_n, z_n' $(n \ge 1)$ on E such that

$$\frac{|P(z_n) - P(z_n')|}{|z_n - z_n'|} \to \infty \quad (n \to \infty). \qquad (4.2)$$

By taking subsequences if necessary we may assume that $z_n \to z_0$, $z_n' \to z_0'$, where z_0, z_0' lie in E. If z_0, z_0' are distinct, we at once obtain a contradiction from our local hypothesis and (4.2), since P is bounded near any point of E. If $z_0 = z_0'$, then z_n, z_n' finally lie in that neighbourhood of z_0 where P is Lip and this again contradicts (4.2).

If $P(z) = P(x, y)$ is defined in a disc γ (C. A. p. 52) and has bounded partial derivatives there, then $P(x, y)$ is Lip in γ. Suppose first that $P(x, y)$ is real. If (x_0, y_0), $(x_0 + h, y_0 + k)$ both lie in γ, then either $(x_0, y_0 + k)$ or $(x_0 + h, y_0)$ also lies in γ. Suppose, for example, the former. Then if M is a bound for the absolute values of the partial derivatives in γ we have from the mean-value theorem

$$|P(x_0 + h, y_0 + k) - P(x_0, y_0)|$$

$$\leq |P(x_0 + h, y_0 + k) - P(x_0, y_0 + k)| + |P(x_0, y_0 + k) - P(x_0, y_0)|$$

$$= \left| h \left(\frac{\partial P}{\partial x} \right)_{(x_0 + \theta h, y_0 + k)} \right| + \left| k \left(\frac{\partial P}{\partial y} \right)_{(x_0, y_0 + \theta' k)} \right|$$

$$\leq M(|h| + |k|) \leq 2M \sqrt{(h^2 + k^2)}.$$

Thus P is Lip in y. For complex P we can prove the corresponding result by considering real and imaginary parts.

If follows that if P has continuous partial derivatives in a domain containing a compact set E, then P is Lip on E. For in this case P has continuous partial derivatives in some neighbourhood and so bounded partial derivatives in some smaller neighbourhood of every point of E.

Conversely, we note that if $P(x, y)$ is Lip on a segment $a \leq y \leq b$ of the line $x = $ constant, then P is an absolutely continuous function of y on this segment and so $\partial P/\partial y$ exists almost everywhere on the segment and is uniformly bounded.[†] Thus

$$P(x, b) - P(x, a) = \int_a^b \frac{\partial P(x, y)}{\partial y} dy. \tag{4.3}$$

4.2 The formulae of Gauss and Green   We proceed to prove these formulae in the form in which we shall require them in the sequel.

Lemma 4.1 (Gauss formula)   *Suppose that D is a bounded analytic domain in the plane and that its boundary γ is described so as to leave D on the left. Then if $P(x, y)$, $Q(x, y)$ are Lip in $\bar{D}$, we have*

$$\int_\gamma (P dx + Q dy) = \int \int_D \left(\frac{\partial Q}{\partial x} - \frac{\partial P}{\partial y} \right) dx dy.$$

Let $z = \alpha(t)$ ($u \leq t \leq b$) give an arc of γ. Then $\alpha(t)$ is a regular function of t and $\alpha'(t) \neq 0$. The tangent is parallel to OY at those points where $\alpha'(t)$ is pure imaginary, and this can be true only at a finite number of points, since otherwise $\alpha'(t)$ would be identically pure imaginary and so this arc of γ would reduce to a straight line.[‡] This is, however, impossible, since γ consists of a finite number of analytic closed curves. Thus there are only a finite number of tangents to γ which are parallel to OY, and

---

[†] Burkill [1951], Chapter IV.

[‡] The real part of $\alpha'(t)$ is a regular function of t for $a \leq t \leq b$ and so has only isolated zeros or vanishes identically.

we assume that these are

$$x = x_m \quad (1 \le m \le M),$$

where $x_1 < x_2 < \ldots < x_M$.

For $x_{m-1} < \xi < x_m$, the line $x = \xi$ meets γ in $2n$ points

$$y = y_1(\xi), \quad \ldots, \quad y = y_{2n}(\xi),$$

where n depends only on m and the $y_\nu(\xi)$ ($1 \le \nu \le 2n$) are differentiable functions of ξ for $x_{m-1} < \xi < x_m$. Also the part D_m of D lying in $x_{m-1} < x < x_m$ consists of n domains

$$D_{m,\nu} : y_{2\nu-1}(x) < y < y_{2\nu}(x), \quad x_{m-1} < x < x_m \quad (1 \le \nu \le n),$$

since at the intersections of $x = \xi$ with γ, the line $x = \xi$ alternately enters and leaves D.

Consider now

$$I = \int_\gamma P(x, y) dx$$

taken along γ so as to keep D on the left. Then $dx > 0$ on the curves $y = y_{2\nu-1}(x)$ and $dx < 0$ on the curves $y = y_{2\nu}(x)$. Thus if I_m is the integral taken over those points of γ which lie in $x_{m-1} < x < x_m$, then

$$
\begin{aligned}
I_m &= \int_{x_{m-1}}^{x_m} \sum_{\nu=1}^{n} \{P[x, y_{2\nu-1}(x)] - P[x, y_{2\nu}(x)]\} dx \\
&= -\int_{x_{m-1}}^{x_m} dx \sum_{\nu=1}^{n} \int_{y_{2\nu-1}}^{y_{2\nu}} \frac{\partial P(x,y) dy}{\partial y} = \sum_{\nu=1}^{n} \int \int_{D_{m,\nu}} -\frac{\partial P}{\partial y} dx dy
\end{aligned}
$$

by (4.3), since $P(x, y)$ is Lip. Adding over the separate ranges $x_{m-1} < x < x_m$ and all the domains $D_{m,\nu}$ we obtain

$$\int_\gamma P dx = \int \int_D -\frac{\partial P}{\partial y} dx dy.$$

Similarly we prove

$$\int_\gamma Q dy = \int \int_D \frac{\partial Q}{\partial x} dx dy,$$

and Lemma 4.1 follows.

We deduce

Lemma 4.2 (Green's formula) *Suppose that D is a bounded analytic domain with boundary γ, that u is Lip in $\bar{D}$, and that v possesses continuous* *second partial derivatives near every point of* $\bar{D}$. *Then*

$$\int_\gamma u\frac{\partial v}{\partial n}ds = -\int\int_D \left[u\left(\frac{\partial^2 v}{\partial x^2} + \frac{\partial^2 v}{\partial y^2}\right) + \left(\frac{\partial u}{\partial x}\frac{\partial v}{\partial x} + \frac{\partial u}{\partial y}\frac{\partial v}{\partial y}\right)\right]dxdy,$$

where $\partial/\partial n$ *denotes differentiation along the normal into D, and ds denotes arc length along* γ.

In fact u, $\partial v/\partial x$, $\partial v/\partial y$ are Lip in $\bar{D}$ in this case and hence so are $u(\partial v/\partial x)$, $u(\partial v/\partial y)$. Let (x_0, y_0) be a point of γ and let θ be the angle the tangent to γ makes with OX, where γ is described so as to keep D on the left. Then the normal into D makes an angle $\theta + \frac{1}{2}\pi$ with OX and

$$\frac{\partial v}{\partial n} = \lim_{h \to 0} \frac{v[x_0 + h\cos(\theta + \frac{1}{2}\pi), y_0 + h\sin(\theta + \frac{1}{2}\pi)] - v[x_0, y_0]}{h}$$

$$= \cos\theta\frac{\partial v}{\partial y} - \sin\theta\frac{\partial v}{\partial x}.$$

If ds is an element of length of γ at (x_0, y_0), its projections on OX, OY are $dx = ds\cos\theta$ and $dy = ds\sin\theta$. Thus

$$\frac{\partial v}{\partial n}ds = \frac{\partial v}{\partial y}dx - \frac{\partial v}{\partial x}dy.$$

We now apply Lemma 4.1 with $P = u(\partial v/\partial y)$, $Q = -u(\partial v/\partial x)$ and Lemma 4.2 follows.

4.3 Harmonic functions and the problem of Dirichlet   A function $u(x, y)$ is harmonic in a domain D if u has continuous second partial derivatives in D which satisfy Laplace's equation

$$\nabla^2(u) \equiv \frac{\partial^2 u}{\partial x^2} + \frac{\partial^2 u}{\partial y^2} = 0.$$

It follows from this that $\phi(z) = \partial u/\partial x - i(\partial u/\partial y)$ has continuous partial derivatives which satisfy the Cauchy–Riemann equations, and so $\phi(z)$ is regular in D. Also in any disc with centre z_0 that lies in D, u is the real part of the regular function

$$f(z) = \int_{z_0}^z \phi(\zeta)d\zeta + u(z_0).$$

Thus u possesses continuous partial derivatives of all orders in D. Also, by the maximum modulus theorem applied to $\exp\{\pm f(z)\}$, u can have no local maximum or minimum in D unless u is constant in D.

Suppose now that D is a domain in the open plane and that $u(\zeta)$ is a continuous function of ζ, defined on the boundary γ of D considered in the extended plane. Thus, if D is unbounded, γ includes the point at infinity. The problem of Dirichlet consists in finding a function $u(z)$ harmonic in D, continuous in $\bar{D}$ and coinciding with $u(\zeta)$ on γ. If u_1, u_2 are two functions satisfying these conditions, then $u_1 - u_2$ is harmonic in D, continuous in $\bar{D}$ and zero on γ, and so by the maximum principle $u_1 - u_2$ vanishes identically in D. Thus a solution to the problem of Dirichlet is unique, if it exists.

A solution does not always exist. If D consists of the annulus $0 < |z| < 1$, then any functions bounded and harmonic in D can be extended to be harmonic also at $z = 0$ and so in $|z| < 1$. Thus boundary values assigned on $|z| = 1$ determine $u(0)$.

The question of existence can often be settled by means of the following result, for whose proof we must refer the reader to C. A. p. 250.

Theorem 4.1   *Suppose that for every boundary point z_0 of D there exists $\omega(z)$, harmonic in D, continuous in D, and positive in $\bar{D}$, except at z_0, where $\omega(z_0) = 0$. Then the problem of Dirichlet possesses a solution for any assigned continuous boundary values on the boundary of D.*

From this we deduce the following criterion:

Theorem 4.2   *The problem of Dirichlet always possesses a solution for the domain D if, given any boundary point of $\bar{D}$, there exists an arc of a straight line or circle containing z_0 and lying outside D.*

We shall deduce Theorem 4.2 from Theorem 4.1. Suppose first that z_0 is finite. Let the arc c have end-points z_0 and z_1. The function

$$\zeta = e^{i\alpha} \frac{z - z_0}{z_1 - z}$$

maps the exterior of this arc c onto the ζ plane cut along a ray through the origin, so that $z = z_0$, z_1 correspond to $\zeta = 0$, ∞. By suitably choosing α, we may arrange that the ray is the negative real axis. Then

$$w = 1 + \frac{\zeta^{\frac{1}{2}} - 1}{\zeta^{\frac{1}{2}} + 1}$$

maps this cut plane onto the circle $|w - 1| < 1$. Also $z = z_0$ corresponds to $w = 0$, and D maps onto a domain whose closure lies in $|w - 1| \leq 1$, and such that only $z = z_0$ correspond to $w = 0$. Thus $\omega = \Re w$ is the harmonic function whose existence is required in Theorem 4.1.[†] If z_0 is infinite and some ray from a finite point to infinity lies outside D, the argument is similar. This completes the deduction. A plane domain satisfying the criterion of Theorem 4.2 will be called *admissible* (for the problem of Dirichlet).

4.4 The Dirichlet integral and capacity Historically Dirichlet attempted to base a proof of the existence of a solution to his problem on the problem of finding the minimum for the Dirichlet integral

$$I_D(u) = \int\int_D \left[\left(\frac{\partial u}{\partial x} \right)^2 + \left(\frac{\partial u}{\partial y} \right)^2 \right] dxdy,$$

for all functions having the assigned boundary values and satisfying certain smoothness conditions. In favourable circumstances this minimum is attained by the required harmonic function. However, even for continuous boundary values on the unit circle, the harmonic function inside the circle with these boundary values may have an infinite Dirichlet integral, so that the minimum problem has no solution.

We shall need the Dirichlet minimum principle only in a special case, where its validity can be proved without difficulty.

Theorem 4.3 *Let D be an admissible domain in the open plane whose complement consists of a compact set E_1 and a closed unbounded set E_0, not meeting E_1. Let $v(z)$, $\omega(z)$ both be continuous in the extended plane, $0, 1$ on E_0, E_1 respectively and Lip on every compact subset of D. Suppose, further, that $\omega(z)$ is harmonic in D. Then*

$$I_D[v(z)] \geq I_D[\omega(z)] = \int_{\gamma_a} \left| \frac{\partial \omega}{\partial n} \right| ds,$$

where γ_a is the set $\{z : \omega(z) = a\}$, and a is any number such that $0 < a < 1$ and $\partial \omega / \partial x - i(\partial \omega / \partial y) \neq 0$ on γ_a. In this case γ_a consists of a finite number of analytic Jordan curves.

The system consisting of the domain D and the sets E_0, E_1 will be called a *condenser*, and $I_D[\omega(z)]$ will be called the *capacity* of the condenser for evident physical reasons. In the special case when E_0, E_1 are continua, so that D is doubly connected, a simple function-theoretic interpretation

---

[†] A point of $\bar{D}$ on the arc through z_0, z_1 corresponds to a pair of complex conjugate values of w. Thus ω is uniquely defined and continuous even on this arc.

is possible. We may then map D (1,1) conformally onto an annulus $\Delta\{z : 1 < |z| < R\}$. (For a construction of the mapping function in terms of $\omega(z)$ see C. A. p. 255.) The Dirichlet integral is clearly invariant under this transformation so that Δ has the same capacity as D. The harmonic function satisfying our boundary-value problem in Δ is

$$\Omega(z) = \frac{\log|R/z|}{\log R} \text{ in } \Delta,$$

and hence the capacity of Δ is

$$\int_{|z|=\rho} \frac{1}{\log R} \frac{\rho d\theta}{\rho} = \frac{2\pi}{\log R}.$$

The quantity $\log R$ is frequently called the *modulus* of the doubly connected domain D. It is thus a multiple of the reciprocal of the capacity of D considered as a condenser. For our purposes it is, however, essential that E_0, E_1 need not be connected.

4.4.1 *Proof of Theorem 4.3*

Let $\omega(z)$ be the function of Theorem 4.3. Then $\phi(z) = \partial\omega/\partial x - i(\partial\omega/\partial y)$ is regular and not identically zero in D, and so $\phi(z)$ vanishes at most on a countable set of points in D, which we shall call the branch points of $\omega(z)$. Suppose that $0 < a < 1$ and that the set γ_a, where $\omega(z) = a$, does not pass through any branch point of $\omega(z)$. Clearly γ_a is compact and lies in D by our hypothesis. Also, near any point z_0 of γ_a, ω is the real part of a regular function

$$w = f(z) = \omega + i\omega_1 = a + ib + a_1(z - z_0) + \dots.$$

Here $a_1 \neq 0$, since $\phi(z) \neq 0$ on γ_a. Thus the inverse function $z = f^{-1}(w)$ is also regular and the part of γ_a near z_0 is the image by a regular function of a straight line segment, i.e. an analytic Jordan arc. If we continue along this arc we must finally return to our starting point, since otherwise γ_a would possesses a limit point not of the kind described above. Thus γ_a consists of analytic Jordan curves, no two of which have common points. There can only be a finite number of these curves, since otherwise there would be a point of γ_a any neighbourhood of which meets an infinite number of curves. This again contradicts the local behaviour of γ_a established above.

Suppose again that γ_a, γ_b contain no branch points, that $0 < a < b < 1$, and that $D_{a,b} = \{z : a < \omega(z) < b\}$. Then $D_{a,b}$ consists of a finite number of bounded analytic domains, of which γ_a, γ_b form the complete boundary. Also if $\partial/\partial n$ denotes differentiation along the normal into $D_{a,b}$

then $\partial\omega/\partial n > 0$ on γ_a and $\partial\omega/\partial n < 0$ on γ_b. Now, on applying Green's formula to the regions of $D_{a,b}$ separately and adding, we obtain

$$\int_{\gamma_a} \frac{\partial\omega}{\partial n} ds + \int_{\gamma_b} \frac{\partial\omega}{\partial n} ds = 0,$$

and since $\partial\omega/\partial n$ has constant sign on γ_a, γ_b, we deduce

$$\int_{\gamma_a} \left| \frac{\partial\omega}{\partial n} \right| ds = \int_{\gamma_b} \left| \frac{\partial\omega}{\partial n} \right| ds = I,$$

say. Also a second use of Green's formula gives

$$(b - a)I = -\int_{\gamma_a} \omega \frac{\partial\omega}{\partial n} ds - \int_{\gamma_b} \omega \frac{\partial\omega}{\partial n} ds = I_{D_{a,b}}[\omega(z)].$$

Since every point of D belongs to some $D_{a,b}$, we deduce, on making a tend to 0 and b tend to 1 that

$$I_D[\omega(z)] = \lim_{a \to 0, b \to 1} I_{D_{a,b}}[\omega(z)] = I.$$

Suppose now that $v(z)$ is the function of Theorem 4.3 and set $h = v - \omega$. Then $h(z)$ is continuous in the plane and vanishes on E_0, E_1 and so at infinity. Hence, given $\epsilon > 0$, the set

$$E = \{z : |h(z)| \geq \epsilon\}$$

is a compact subset of D. Let a_0, b_0 be the lower and upper bounds of $\omega(z)$ on E, so that $0 < a_0 < b_0 < 1$. If $0 < a < a_0$, $b_0 < b < 1$, then γ_a, γ_b do not meet E and so $|h(z)| < \epsilon$ on γ_a, γ_b. Now h is Lip on $\bar{D}_{a,b}$ so that Green's formula gives

$$\left| \int\int_{D_{a,b}} \left[\frac{\partial h}{\partial x} \frac{\partial\omega}{\partial x} + \frac{\partial h}{\partial y} \frac{\partial\omega}{\partial y} \right] dxdy \right| = \left| \int_{\gamma_a + \gamma_b} h \frac{\partial\omega}{\partial n} ds \right|$$

$$\leq \epsilon \int_{\gamma_a + \gamma_b} \left| \frac{\partial\omega}{\partial n} \right| ds = 2\epsilon I.$$

Thus

$$I_{D_{a,b}}(h, \omega) = \int\int_{D_{a,b}} \left[\frac{\partial h}{\partial x} \frac{\partial\omega}{\partial x} + \frac{\partial h}{\partial y} \frac{\partial\omega}{\partial y} \right] dxdy \to 0 \quad (a \to 0, b \to 1).$$

Also

$$I_{D_{a,b}}(v) = I_{D_{a,b}}(\omega) + I_{D_{a,b}}(h) + 2I_{D_{a,b}}(h, \omega).$$

Making a tend to 0, b tend to 1 we deduce $I_D(v) = I_D(\omega) + I_D(h)$. This proves the inequality of Theorem 4.3 and completes the proof of that theorem. We note also that equality is possible only if $I_D(h) = 0$. In this case it is not difficult to see that $h(z)$ vanishes identically so that $v(z)$ must coincide with $\omega(z)$. We shall not, however, make any use of this.

4.4.2 Capacity decreases with expanding domain   We make an immediate deduction from Theorem 4.3, which has independent interest:

Theorem 4.4   *Let E_0, E_1; E_0', E_1' be pairs of closed sets characterizing two condensers as in Theorem 4.3. Let I, I' be their capacities and suppose that $E_0' \subset E_0$, $E_1' \subset E_1$. Then $I' \le I$.*

Let $\omega(z)$ be the potential function corresponding to the first condenser as in Theorem 4.3. If $0 < a < b < 1$, we define a function $v(z)$ as follows: $v(z) = 0$ where $\omega(z) \le a$, $v(z) = 1$ where $\omega(z) \ge b$, and

$$v(z) = \frac{\omega(z) - a}{b - a} \text{ in } D_{a,b}.$$

Then $v(z)$ is Lip in the plane and $v = 0$, 1 on E_0', E_1' respectively. Thus if D' is the complement of E_0' and E_1', Theorem 4.3 gives

$$I' \le I_{D'}(v) = -\int_{\gamma_a, \gamma_b} v \frac{\partial v}{\partial n} ds = \frac{1}{b-a} I_D[\omega(z)] = \frac{I}{b-a}.$$

Making a tend to 0, b tend to 1, we obtain $I' \le I$ as required.

4.5 Symmetrization   We shall now consider the process of symmetrization introduced in the middle of last century by Steiner and developed by Pólya and Szegö [1951].

Let O be an arbitrary open set in the open plane. We shall define a symmetrized set O^* in two ways as follows.

4.5.1 Steiner symmetrization   In this case we symmetrize with respect to a straight line, which we can take to be the x axis of Cartesian coordinates. For any real ξ, the line $x = \xi$ meets O in a set of mutually disjoint open intervals of total length $l(\xi)$ say, where $0 \le l(\xi) \le \infty$. The symmetrized set is to be the set

$$O^* = \{(x, y) : |y| < \tfrac{1}{2} l(x)\}.$$

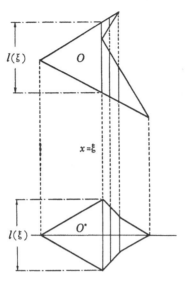

Fig. 3. Steiner symmetrization

4.5.2 Circular (Pólya) symmetrization[†]   In this case we symmetrize with respect to a half-line or ray, which we take to be the line $\theta = 0$ of polar coordinates r, θ. We define the symmetrized set O^* as follows. Consider the intersection of O with the circle $r = \rho$, $0 \leq \rho < \infty$. If this intersection includes the whole circle or is null, then the intersection of O^* with $r = \rho$ is also to include the whole circle, or to be null respectively. Otherwise let O meet the circle $r = \rho$ in a set of open arcs of total length $\rho l(\rho)$, where $0 < l(\rho) \leq 2\pi$. Then O^* is to meet the circle $r = \rho$ in the single arc

$$|\theta| < \tfrac{1}{2}l(\rho).$$

We proceed to discuss some properties of symmetrization. Unless the contrary is stated results are true for either kind of symmetrization.

4.5.3 The symmetrized set O^* is open   We prove this only for circular symmetrization. The proof for Steiner symmetrization is similar.

Suppose that (ρ_0, π) lies in O^*. Then O contains the whole circle $r = \rho_0$. Since O is open, O therefore contains some annulus $\rho_0 - \delta < r < \rho_0 + \delta$, and this annulus also lies in O^*. Thus (ρ_0, π) is an interior point of O^*. A slightly modified argument applies if the origin lies in O^*.

[†] Pólya [1950]

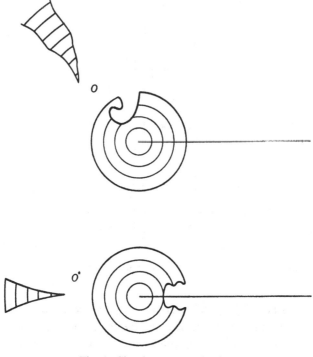

Fig. 4. Circular symmetrization

Suppose next that (ρ_0, θ_0) lies in O^*, with $0 < \rho_0 < \infty$, $|\theta_0| < \pi$. Then $l(\rho_0) > 2|\theta_0|$. Choose x so that $2|\theta_0| < x < l(\rho_0)$. Then we can find a finite number of open arcs of total length greater than $\rho_0 x$ on the circle $r = \rho_0$ and in O. By diminishing these arcs slightly, we may take them to be closed and still lying in O and of total length greater than $\rho_0 x$. If $\alpha_v \leq \theta \leq \beta_v$ is such an arc and δ_v is its distance from the complement of O, let δ be the smallest of the δ_v. Then the arc $\alpha_v \leq \theta \leq \beta_v$ of every circle $r = \rho$ for

$$\rho_0 - \delta < \rho < \rho_0 + \delta$$

lies in O. Thus

$$l(\rho) \geq x > 2|\theta_0| \quad \text{for} \quad \rho_0 - \delta < \rho < \rho_0 + \delta,$$

and so (r, θ) lies in O^* for

$$\rho_0 - \delta < r < \rho_0 + \delta \quad \text{and} \quad |\theta| < \tfrac{1}{2}x,$$

where $\frac{1}{2}x > |\theta_0|$. Thus (ρ_0, θ_0) is an interior point of O^* and so O^* is open.

4.5.4 The Steiner symmetrized set D^* of a domain D is either the whole plane, or a simply connected admissible domain   Suppose that the lines $x = \xi_1, \xi_2$, where $\xi_1 < \xi_2$, meet D^* and so D. Then $x = \xi$ meets D for $\xi_1 < \xi < \xi_2$, since otherwise D could be expressed as the union of the disjoint non-null open subsets lying in the half-planes $x < \xi$ and $x > \xi$ respectively. Thus the segment $\xi_1 \leq x \leq \xi_2$ of the real axis lies in D^* in this case. Two points (ξ_1, η_1), (ξ_2, η_2) in D^* can be joined in D^* along the polygonal arc (ξ_1, η_1), $(\xi_2, 0)$, $(\xi_2, 0)$, (ξ_2, η_2). Thus D^* is a domain.

Suppose next that D^* is not the whole plane, so that $l(\xi) < +\infty$ for at least one ξ. Then the complement of D^* meets $x = \xi$ in the pair of rays $|y| > \frac{1}{2}l(\xi)$. Thus the condition for admissibility in Theorem 4.2 is satisfied for every point (ξ, η) in the complement of D^* and also at ∞. Further, every point in the complement of D^* can be joined to infinity by a ray, and so this complement is connected in the extended plane. Thus D^* is simply connected and is admissible.

The results for circular symmetrization are not so simple. If, for instance, D consists of an annulus $\rho_1 < r < \rho_2$ except for the single point (ρ, π), then D^* coincides with D and so is neither simply connected nor admissible.

4.5.5 The circularly symmetrized set of a domain D is a domain D^*. If D is simply connected, so is D^*. If D is admissible or D^* is simply connected, D^* is admissible or D^* is the whole plane   The proof that D^* is connected and is a domain is similar to that for Steiner symmetrization. If D is simply connected and contains the circle $r = \rho$, then D must contain the interior of this circle, since otherwise the complement of D would contain points inside and outside the circle (e.g. infinity) but no points on the circle. Hence if D contains $r = \rho$, D and D^* also contain the disc $r < \rho$ in this case. Let ρ_0 be the upper bound of all ρ for which this holds. If $\rho_0 = +\infty$, D and D^* contain the whole open plane. Otherwise the complement of D contains some point of every circle $r = \rho \geq \rho_0$, and so the complement of D^* contains the ray $r \geq \rho_0$, $\theta = \pi$.

Any boundary point of D^* on this ray and the point at infinity clearly satisfy the criterion of admissibility. If (ρ, θ_0) is any other point in the complement of D^*, then $\rho \geq \rho_0$, $|\theta_0| < \pi$, and the arc $|\theta_0| \leq \theta \leq \pi$ of the circle $r = \rho$ lies in the complement of D^* and joins (ρ, θ_0) to the ray $\theta = \pi$, $r \geq \rho_0$ in this complement. Thus the complement of D^* is connected and D^* is simply connected. The criterion of admissibility is satisfied in all cases for every boundary point of D^* not on the line $\theta = \pi$. The above argument shows that if D^* is simply connected the criterion is also satisfied for boundary points (ρ, π).

It remains to show that if D is an admissible domain, so is D^*. By the above argument it suffices to establish the criterion of Theorem 4.2 for the boundary points (ρ_0, π) of D^*, and for ∞. Since D is admissible the complement of D contains a ray, which must meet every sufficiently large circle $r = \rho$. Thus (ρ, π) lies in the complement of D^* for large ρ, and so this complement contains a ray $r > \rho$, $\theta = \pi$. Thus our criterion is satisfied at ∞. Suppose next that (ρ_0, π) is a boundary point of D^*. If an arc $\alpha \leq \theta \leq \beta$ of $r = \rho_0$ lies in the complement of D, then an arc of $r = \rho_0$ bisected by (ρ_0, π) lies in the complement of D^* and the criterion is satisfied. If not, let (ρ_0, θ) be a frontier point of D. Since D is admissible there is an arc of a straight line or circle containing (ρ_0, θ) but lying outside D and so by our hypothesis not on $r = \rho_0$. This arc must meet $r = \rho$ for all ρ either in some interval $\rho_0 - \delta \leq \rho \leq \rho_0$ or $\rho_0 \leq \rho \leq \rho_0 + \delta$. Thus the interval $[\rho_0 - \delta, \rho_0]$ or $[\rho_0, \rho_0 + \delta]$ of $\theta = \pi$ lies in the complement of D^* and contains (ρ_0, π), and so D^* is admissible.

4.6 Symmetrization of functions   Let $u(z)$ be a function, real, continuous and bounded in the plane. We symmetrize u to obtain a new function $u^*(z)$ by simultaneously symmetrizing all the sets

$$D_a = \{z : u(z) > a\}(-\infty < a < +\infty).$$

These sets are open since u is continuous.

More precisely, let D_a^* be the symmetrized set of D_a with respect to some straight line or ray. For any point z in the plane we define $u^*(z)$ as the least upper bound of all a for which z lies in D_a^*.

In practice we shall be concerned only with the case where $u(z)$ is non-negative and vanishes continuously at infinity. In this case the sets D_a are bounded for $a > 0$ and their closures $\bar{D}_a$ are compact. Now a function $u(z)$ continuous on a compact set E is uniformly continuous on E (C. A. p. 65). In other words if $\Omega(\delta)$ is the upper bound of $|u(z_1) - u(z_2)|$ for z_1, z_2 on E and $|z_1 - z_2| \leq \delta$, then $\Omega(\delta)$ is finite and $\Omega(\delta) \to 0$ as $\delta \to 0$. The quantity $\Omega(\delta)$ is called *the modulus of continuity of $u(z)$*. Clearly $u(z)$ is Lip on E if and only if $\Omega(\delta) \leq C\delta$ for some positive C and $0 < \delta < \infty$.

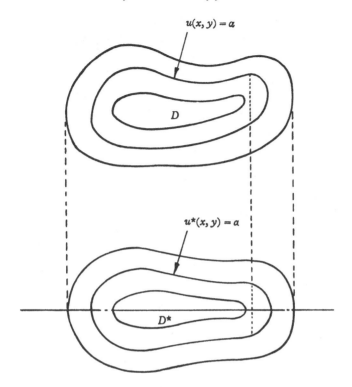

Fig. 5. Symmetrization of functions

4.6.1 Symmetrization decreases the modulus of continuity We shall be
able to show that $u^*(z)$ is 'at least as continuous as $u(z)$' by proving

Theorem 4.5 *Suppose that, with the above definitions of $u(z)$, $u^*(z)$, the
set $E = \{z : a \leq u(z) \leq b\}$ is bounded and that $u(z)$ has modulus of
continuity $\Omega(\delta)$ on E. If $E^* = \{z : a \leq u^*(z) \leq b\}$, then $u^*(z)$ is continuous
on E^* with modulus of continuity $\Omega^*(\delta) \leq \Omega(\delta)$. In particular if u is Lip
on E, u^* is Lip on E^*.*

We shall prove this result for Steiner symmetrization. The proof
of circular symmetrization is similar, but the details are a little more
complicated. We need

Lemma 4.3 *With the above hypotheses let $l(x,t)$ denote the total length
of the y intervals on the line $x = $ constant where $u(x,y) > t$. Suppose that*

$a \leq t_2 < t_1 - \Omega(\delta) \leq b - \Omega(\delta)$ *and* $|x_2 - x_1| \leq \delta$. *Then*

$$l(x_2, t_2) \geq l(x_1, t_1) + 2\sqrt{[\delta^2 - (x_2 - x_1)^2]}.$$

Let F_1 be the set of all y such that $u(x_1, y) > t_1$ and let F_2 be the set of y for which $u(x_2, y) > t_2$. Then

$$y_2 \in F_2 \text{ if } y_1 \in F_1 \text{ and } (x_2 - x_1)^2 + (y_2 - y_1)^2 \leq \delta^2.$$

For otherwise we could find, on the line joining (x_1, y_1) and (x_2, y_2), two points at distance at most δ from each other at which the continuous function u takes the values t_1, t_2 respectively, and this would contradict the definition of $\Omega(\delta)$.

Taking $y_2 = y_1$ we deduce at once that F_2 contains F_1. Further, let y' be the upper bound of F_1. Then $u(x_1, y') = t_1$. Hence it follows that

$$u(x_2, y) > t_2 \text{ if } (y - y')^2 + (x_2 - x_1)^2 \leq \delta^2,$$

and in particular

$$y \in F_2 \text{ if } y' \leq y \leq y' + \sqrt{[\delta^2 - (x_2 - x_1)^2]}.$$

Similarly, if y'' is the lower bound of F_1,

$$y \in F_2 \text{ if } y'' - \sqrt{[\delta^2 - (x_2 - x_1)^2]} \leq y \leq y''.$$

Thus F_2 contains the whole of F_1 together with two intervals, each of length $\sqrt{[\delta^2 - (x_2 - x_1)^2]}$, which do not belong to F_1. Since $l(x_1, t_1), l(x_2, t_2)$ are the lengths of F_1, F_2 respectively, the lemma follows.

Suppose now that Theorem 4.5 is false. Then we can find (x_1, y_1), (x_2, y_2) on E^*, such that for some positive δ

$$(x_2 - x_1)^2 + (y_2 - y_1)^2 \leq \delta^2,$$

and

$$a \leq u^*(x_2, y_2) < u^*(x_1, y_1) - \Omega(\delta) \leq b - \Omega(\delta).$$

Choose t_1, t_2 so that

$$u^*(x_2, y_2) < t_2 < t_1 - \Omega(\delta) < u^*(x_1, y_1) - \Omega(\delta).$$

Then it follows from Lemma 4.3 that

$$l(x_2, t_2) \geq l(x_1, t_1) + 2\sqrt{[\delta^2 - (x_2 - x_1)^2]}.$$

Also by the definition of $u^*(x, y)$ we have

$$l(x_1, t_1) > 2|y_1|, \quad l(x_2, t_2) \leq 2|y_2|,$$

since (x_1, y_1) lies in $D_{t_1}^*$, but (x_2, y_2) does not lie in $D_{t_2}^*$. Thus

$$|y_2 - y_1| \geq |y_2| - |y_1| > \sqrt{[\delta^2 - (x_2 - x_1)^2]}$$

and we have the contradiction which proves our theorem.

Examples

4.1 Prove Theorem 4.5 for circular symmetrization.

4.7 Symmetrization of condensers Consider a condenser satisfying the conditions of Theorem 4.3. Let $\omega(z)$ be the potential function of Theorem 4.3. We symmetrize the function $\omega(z)$, and write

$$E_i^* = \{z : \omega^*(z) = i\} \quad (i = 0, 1).$$

By Theorem 4.5 $\omega^*(z)$ is continuous and so E_0^*, E_1^* are closed. It also follows that the complement of E_0^* is the symmetrized set of the complement of E_0, and that E_1^* is obtained by symmetrizing just as in the definitions 4.5.1 and 4.5.2 except that open set and open interval must be replaced by closed set and closed interval.

The complementary set to E_0^*, E_1^* consists of the open set

$$D^* = \{z : 0 < \omega^*(z) < 1\};$$

and this is again a domain. Suppose, for example that we are symmetrizing with respect to a half-line $\theta = 0$. Let r_0, r_1 be the lower and upper bounds of r in D. Then every circle $r = \rho$ $(r_0 < \rho < r_1)$ meets D^* in a single symmetrical arc $\gamma(\rho)$ of one of the forms $|\theta| < l(\rho)$, $|\theta| \leq \pi$, $l(\rho) < |\theta| \leq \pi$ or in a pair of arcs $\gamma_+(\rho)$, $\gamma_-(\rho)$ given by $l_1(\rho) < |\theta| < l_2(\rho)$. One of the former cases certainly holds unless $r = \rho$ meets both E_0 and E_1, and so holds if ρ is sufficiently near r_0 or r_1. It follows from the openness of D^*, that if ρ is sufficiently near to ρ_0, $\gamma_+(\rho)$ and $\gamma_+(\rho_0)$ (or $\gamma(\rho)$ and $\gamma(\rho_0)$) can be joined by a straight line segment in D^*, and the connectedness of D^* follows.

We can further show just as in 4.5.4 and 4.5.5 that D^* is admissible, since D is admissible. We have thus obtained a new condenser, which is called the symmetrized condenser of the original condenser. We now prove the following result of Pólya and Szegö [1951]:

Theorem 4.6 *Suppose that a condenser C and its symmetrized condenser C^* have capacities I, I^* respectively. Then $I^* \leq I$.*

By Theorem 4.5 $\omega^*(z)$ is continuous in the plane, and since $\omega(z)$ is Lip on the set $\bar{D}_{a,b} = \{z : a \leq \omega(z) \leq b\}$ if $0 < a < b < 1$, $\omega^*(z)$ is Lip on $\bar{D}_{a,b} = \{z : a \leq \omega^*(z) \leq b\}$. On any compact set E^* in D^*, $\omega^*(z)$ has lower and upper bounds a, b where $0 < a < b < 1$, and so $\omega^*(z)$ is Lip on E^*. Also $\omega^*(z) = 0$, 1 on E_0^*, E_1^* respectively. Thus we have by Theorem 4.3

$$I^* \leq I_{D^*}[\omega^*(z)], \quad I = I_D[\omega(z)].$$

To complete the proof of Theorem 4.6 it thus remains to prove that

$$I_{D^*}[\omega^*(z)] \leq I_D[\omega(z)]. \tag{4.4}$$

This result represents the basic tool of the Pólya–Szegö theory.

4.7.1 Symmetrization decreases the Dirichlet integral   We shall prove the inequality (4.4) for circular symmetrization. The proof for Steiner symmetrization is similar and even a little simpler. We express the Dirichlet integral in polar coordinates:

$$I_D(\omega) = \int_0^\infty \rho\, d\rho \int_{\mathfrak{E}_\rho} \left[\left(\frac{\partial\omega}{\partial\rho}\right)^2 + \frac{1}{\rho^2}\left(\frac{\partial\omega}{\partial\theta}\right)^2 \right] d\theta.$$

Here $\mathfrak{E}_\rho = \{\theta : 0 < \omega(\rho e^{i\theta}) < 1\}$

Consider the expression

$$J(\rho) = \int_{\mathfrak{E}_\rho} \left[\left(\frac{\partial\omega}{\partial\rho}\right)^2 + \frac{1}{\rho^2}\left(\frac{\partial\omega}{\partial\theta}\right)^2 \right] d\theta.$$

If ω is constant on every circle with centre the origin, then D must be an annulus $r_1 < \rho < r_2$ and D^* coincides with D and $\omega(z)$ with $\omega^*(z)$. We ignore this case. Otherwise, since $\omega(z)$ is harmonic in D, $\omega(z)$ can be constant and different from $0, 1$ on the circle $r = \rho$ only for isolated values of ρ,[†] which we may omit from the range of integration. If $\omega(z)$ is not constant on $r = \rho$, $\omega(z)$ cannot be constant in any interval of $\mathfrak{E}_\rho$. For the end-points of such an interval would have to lie outside $\mathfrak{E}_\rho$, and so $\omega(z)$ would be 0 or 1 in the interval, contrary to the definition of $\mathfrak{E}_\rho$. We thus assume that $\omega(z)$ is not constant in any interval of $\mathfrak{E}_\rho$, and so $\partial\omega/\partial\theta = 0$ only at isolated points in $\mathfrak{E}_\rho$. The values of $\omega(z)$ at these points will be called stationary values of $\omega(z)$. They can have at most 0, 1 as limit points. We can thus arrange the stationary values in order of magnitude as a sequence t_m, where the lower bound of m is finite or $-\infty$

[†] Otherwise $\partial\omega/\partial\theta$ would be identically zero.

and the upper bound of m is finite or $+\infty$. If 0 is the lower bound of $\omega(z)$ on $\mathfrak{E}_\rho$ and 0 is not a limit point of the t_m, we include 0 as the smallest t_m; similarly for the value 1. Let t_m, t_{m+1} be successive stationary values. Then there will be n open intervals in $\mathfrak{E}_\rho$ in which $\omega(\rho, \theta)$ increases from t_m to t_{m+1} and n intervals where ω decreases from t_{m+1} to t_m, where n is a positive integer, and these intervals occur alternately on the circle $r = \rho$, since ω is continuous on the circle.

Let us denote these intervals in the order in which they occur on the circle $r = \rho$ by $T_{m,v}$ $(1 \le v \le 2n)$. We suppose that $\omega(\rho, \theta)$ increases in $T_{m,v}$ for odd v and decreases in $T_{m,v}$ for even v. Also the totality of intervals $T_{m,v}$ for varying m, v make up $\mathfrak{E}_\rho$, except for isolated points. Thus

$$J(\rho) = \sum_m \sum_{v=1}^{2n(m)} \int_{T_{m,v}} \left[\left(\frac{\partial \omega}{\partial \rho} \right)^2 + \frac{1}{\rho^2} \left(\frac{\partial \omega}{\partial \theta} \right)^2 \right] d\theta.$$

We now change the variable of integration from θ to $t = \omega(\rho, \theta)$ in each interval $T_{m,v}$.

Since $\partial \omega / \partial \theta \ne 0$ in $T_{m,v}$ the relation between θ in $T_{m,v}$ and t is (1,1). We regard θ as a function $\theta = \theta_v(t, \rho)$ of t, ρ. The partials of this function are given in terms of the original $\partial \omega / \partial \theta$, $\partial \omega / \partial \rho$ by the formulae

$$\frac{\partial \theta}{\partial t} = 1 / \frac{\partial \omega}{\partial \theta}, \quad \frac{\partial \theta}{\partial \rho} = -\frac{\partial \omega}{\partial \rho} / \frac{\partial \omega}{\partial \theta} \quad \text{and} \quad d\theta = \frac{\partial \theta}{\partial t} dt.$$

Thus if $\theta_v(t, \rho)$ denotes the value of θ in $T_{m,v}$ such that $\omega(\rho, \theta) = t$, we have

$$J(\rho) = \sum_m \int_{t_m}^{t_{m+1}} \sum_{v=1}^{2n(m)} \left\{ \frac{(\partial \theta_v / \partial \rho)^2}{|\partial \theta_v / \partial t|} + \frac{1}{\rho^2} \frac{1}{|\partial \theta_v / \partial t|} \right\} dt. \qquad (4.5)$$

If $t_m < t < t_{m+1}$, the set of intervals on $r = \rho$ where $\omega(\rho, \theta) < t$ is given by $\theta_{2v-1}(t) < \theta < \theta_{2v}(t)$ $(1 \le v \le n)$. Thus the length of this set is given by $\rho l(\rho, t)$, where

$$l(\rho, t) = \sum_{v=1}^n [\theta_{2v}(t) - \theta_{2v-1}(t)].$$

Clearly $l(\rho, t)$ decreases strictly with increasing t in $t_m < t < t_{m+1}$. Thus $\omega^*(\rho, \phi)$ satisfies

$$\omega^*(\rho, \phi) = t, \quad \text{where } \phi = \pm \tfrac{1}{2} l(\rho, t) \quad \text{if } t_m < t < t_{m+1}.$$

Thus if $J^*(\rho)$ corresponds to $\omega^*(z)$ as $J(\rho)$ corresponds to $\omega(z)$ we have[†]

$$J^*(\rho) = \sum_m 2 \int_{t_m}^{t_{m+1}} \left\{ \frac{(\partial\phi/\partial\rho)^2}{|\partial\phi/\partial t|} + \frac{1}{\rho^2} \frac{1}{|\partial\phi/\partial t|} \right\} dt, \qquad (4.6)$$

where

$$\phi = \tfrac{1}{2}[(\theta_2 - \theta_1) + \ldots + (\theta_{2n} - \theta_{2n-1})].$$

Thus

$$\frac{\partial\phi}{\partial t} = \frac{1}{2}\sum_{v=1}^{2n}(-1)^v\frac{\partial\theta_v}{\partial t} = -\frac{1}{2}\sum_1^{2n}\left|\frac{\partial\theta_v}{\partial t}\right|, \qquad \frac{\partial\phi}{\partial\rho} \le \frac{1}{2}\sum_1^{2n}\left|\frac{\partial\theta_v}{\partial\rho}\right|.$$

Thus the arithmetic–harmonic mean theorem gives

$$\frac{2}{\left|\dfrac{\partial\phi}{\partial t}\right|} = \frac{4}{\displaystyle\sum_1^{2n}\left|\dfrac{\partial\theta_v}{\partial t}\right|} \le \frac{4}{(2n)^2}\sum_1^{2n}\frac{1}{\left|\dfrac{\partial\theta_v}{\partial t}\right|}.$$

Also Schwarz's inequality gives $(\sum a)^2 \le (\sum a^2/|b|)(\sum |b|)$ and so

$$\frac{2(\partial\phi/\partial\rho)^2}{|\partial\phi/\partial t|} \le \frac{(\sum |\partial\theta_v/\partial\rho|)^2}{\sum |\partial\theta_v/\partial t|} \le \sum \left\{ \frac{|\partial\theta_v/\partial\rho|^2}{|\partial\theta_v/\partial t|} \right\}.$$

Substituting these inequalities in the expressions (4.5) and (4.6) for $J(\rho)$ and $J^*(\rho)$ we deduce

$$J^*(\rho) \le J(\rho),$$

and hence on integrating with respect to ρ we deduce (4.4). This completes the proof of Theorem 4.6.

4.8 Green's function and the inner radius Suppose that D is a domain in the complex z plane, z_0 a point of D, and that there exists a function $g[z, z_0, D]$, continous in the closed plane and harmonic in D except at z_0, vanishing outside D and such that

$$g[z, z_0, D] + \log|z - z_0|$$

remains harmonic at $z = z_0$. Then $g[z, z_0, D]$ is called the (classical) *Green's function* of D.

Clearly $g[z, z_0, D]$ is unique if it exists, for if $g_1(z)$ is another function with the same properties, then $g - g_1$ is harmonic in the whole of D,

---

[†] In estimating the integral $I_{D^*}[\omega^*(z)]$ we may ignore the sets where $\omega^* = t_m$. For such a set either has zero area or else $\partial\omega^*/\partial\rho, \partial\omega^*/\partial\theta$ vanish almost everywhere on it.

continuous in $\bar{D}$, and zero outside D, and so $g - g_1$ vanishes identically by the maximum principle.

If D is admissible and bounded $g[z, z_0, D]$ exists. For let $h(z, z_0)$ be harmonic in D and have boundary values $\log|z - z_0|$ on the boundary of D. Then

$$g(z, z_0) = h(z, z_0) - \log|z - z_0|$$

is the required Green's function in D.

Suppose now that $g(z, z_0)$ exists. Then $g > 0$ in D, since otherwise g would have a minimum in D at a point other than z_0, and this is impossible. Since

$$g(z, z_0) + \log|z - z_0|$$

remains harmonic at z_0, the limit

$$\gamma = \lim_{z \to z_0} g(z, z_0) + \log|z - z_0|$$

exists. We write

$$\gamma = \log r_0$$

and call r_0 the *inner radius* of D at z_0.

To explain this terminology, suppose that D is simply connected and that

$$w = \psi(z) = z - a_0 + b_2(z - a_0)^2 + \ldots$$

maps D (1,1) conformally onto a disc $|w| < r_0$, so that $\psi(a_0) = 0$, $\psi'(a_0) = 1$. Then

$$g(z, a_0) = \log \frac{r_0}{|\psi(z)|}$$

is the Green's function of D at a_0, and the inner radius of D at a_0 is r_0. For $g(z, a_0)$ is harmonic in D except at a_0 and as z approaches the boundary of D in any manner $|\psi(z)| \to r_0$ and so $g(z) \to 0$. Also near $z = a_0$ we have

$$g(z, a_0) = \log \frac{r_0}{|z - a_0||1 + o(1)|} = \log \frac{1}{|z - a_0|} + \log r_0 + o(1),$$

as required.

We also note that if

$$z = f(w) = a_0 + a_1 w + \ldots$$

maps $|w| < 1$ onto D (1,1) conformally, then

$$w = f^{-1}(z) = \frac{z - a_0}{a_1} + \dots$$

near $z = a_0$, so that $a_1 f^{-1}(z)$ has properties of $\psi(z)$ in the above analysis. In particular, $|a_1|$ is the inner radius of D at a_0 in this case. We remark specifically, however, that our definition of the inner radius applies also to multiply connected domains D, which cannot be mapped onto a disc, and even in the simply connected case we do not need to assume the existence of the mapping.

We note that if the domains D, D_1 possess Green's functions $g(z, z_0)$, $g_1(z, z_0)$, and if $D \subset D_1$, then the difference

$$g_1(z, z_0) - g(z, z_0)$$

is harmonic in D and non-negative on the boundary of D. Thus the difference is non-negative in D, and so if r, r_1 are the inner radii of D, D_1 at z_0, then $\log r_1 \geq \log r$ *and so* $r \leq r_1$. *Thus the inner radius increases with expanding domain.*

We accordingly define the inner radius r_0 at a point a_0 of an arbitrary domain D, containing a_0, as the least upper bound of the inner radius at a_0 of all domains containing a_0, contained in D, and possessing (classical) Green's functions. We shall have $0 < r_0 \leq \infty$.

4.8.1 *The inner radius and conformal mapping* Our application of the inner radius to function theory derives from

Theorem 4.7 *Suppose that* $w = f(z) = a_0 + a_1 z + \dots$ *is regular in* $|z| < 1$ *and takes there values* w, *which lie in a domain* D *having inner radius* r_0 *at* a_0. *Then* $|a_1| \leq r_0$. *Equality holds if* $f(z)$ *maps* $|z| < 1$ *(1,1) conformally onto* D.[†]

We discussed in the previous section the case when $f(z)$ maps $|z| < 1$ (1,1) conformally onto D. In the general case take $0 < \rho < 1$ and choose ρ so that $f'(z) \neq 0$ on $|z| = \rho$. This can be achieved by increasing ρ slightly if necessary. Let $C(\rho)$ be the curve which is the image of $|z| = \rho$ by $w = f(z)$. Thus $C(\rho)$ is a closed analytic curve which may cross itself. The set of values w assumed by $w = f(z)$ inside $|z| < \rho$ forms a domain $D(\rho)$ whose boundary consists of certain arcs of $C(\rho)$. If two such arcs

---

[†] This is, in fact, the only case of equality. See, for example, Hayman [1951] for the limiting case $a_0 = \infty$.

touch at a point P, we add the interior of a small circle of centre P to $D(\rho)$. In this way we construct a domain D_0 which contains $D(\rho)$, is contained in D, and whose boundary consists of a finite number of analytic arcs no two of which touch each other.

Thus D_0 satisfies the condition of admissibility of Theorem 4.2. Further, D_0 is bounded and so possesses a Green's function

$$g_\rho(w, a_0) = \log \left| \frac{1}{w - a_0} \right| + \log r(\rho) + o(1)$$

near $w = a_0$, where $r(\rho)$ is the inner radius of D_0 at a_0. Since D_0 lies in D, we have $r(\rho) \leq r_0$,

Consider now the function

$$h(z) = g_\rho(f(z), a_0) + \log \frac{|z|}{\rho}.$$

Then $h(z)$ is non-negative on $|z| = \rho$ and harmonic in $|z| \leq \rho$ except possibly at points where $f(z) = a_0$. At all such points, except possibly the origin, $h(z)$ becomes positively infinite. At the origin we have

$$
\begin{aligned}
h(z) &= \log \left| \frac{1}{f(z) - a_0} \right| + \log r(\rho) + \log \frac{|z|}{\rho} + o(1) \\
&= \log \frac{r(\rho)}{|a_1|\rho} + o(1).
\end{aligned}
$$

Thus $h(z)$ remains harmonic at the origin, unless $a_1 = 0$, in which case Theorem 4.7 is trivial, and

$$h(0) = \log \frac{r(\rho)}{|a_1|\rho}.$$

Thus $h(z) \geq 0$ in $|z| < \rho$, since otherwise $h(z)$ would have a negative minimum somewhere in $|z| < \rho$, and at such a point $h(z)$ would be harmonic, which is impossible. Thus

$$h(0) = \log \frac{r(\rho)}{|a_1|\rho} \geq 0,$$

$$|a_1| \leq \frac{r(\rho)}{\rho} \leq \frac{r_0}{\rho}.$$

When $\rho \to 1$, we have the inequality of Theorem 4.7.

4.8.2 The inner radius and symmetrization Theorem 4.7 becomes a powerful tool in the theory of functions, when combined with the following result of Pólya and Szegö [1951]:

Theorem 4.8   *Suppose that a_0 is a point of a domain D in the w plane
and that D^* is obtained from D by symmetrizing with respect to a line or
half-line containing a_0. Let r_0, r_0^* be the inner radii of D, D^* at a_0. Then
$r_0 \leq r_0^*$.*[†]

Following Polya and Szegö we shall deduce Theorem 4.8 from Theorem
4.6. Suppose first that D is bounded and admissible so that D has a
Green's function $g(w, a_0)$, satisfying

$$g(w) = \log \left| \frac{1}{w - a_0} \right| + \log r_0 + o(1)$$

near $w = a_0$. We have

$$\frac{\partial}{\partial r} g(a_0 + re^{i\theta}) = \frac{-1}{r} + O(1) \quad (r \to 0).$$

Thus if K is sufficiently large, there is exactly one value of r for each
θ such that $g(a_0 + re^{i\theta}) = K$, and the set of these points $a_0 + re^{i\theta}$ forms
an analytic Jordan curve γ_K surrounding a_0. We define $r = |w - a_0|$. On
γ_K we have

$$r = e^{-K}[1 + o(1)]r_0$$

for large K. In other words, given $\epsilon > 0$, we have for sufficiently large K

$$r_0 e^{-(K+\epsilon)} < r < r_0 e^{-(K-\epsilon)} \text{ on } \gamma_K. \tag{4.7}$$

Consider now the condenser formed by keeping γ_K and its interior at
unit potential and the boundary of D at zero potential. The corresponding
potential function is $g(w)/K$, and hence by Theorem 4.3 the capacity of
the condenser is

$$-\frac{1}{K} \int_{\gamma_{K'}} \frac{\partial g}{\partial n} ds,$$

where $0 < K' < K$. Since g is harmonic it follows from Lemma 4.2 that
if r is small

$$-\int_{\gamma_{K'}} \frac{\partial g}{\partial n} ds = -\int_{|w-a_0|=r} \frac{\partial g}{\partial r} ds = 2\pi r \left[\frac{1}{r} + O(1) \right] \to 2\pi \quad (r \to \infty).$$

Thus the capacity of our condenser is $2\pi/K$.

Consider next the capacity $c(r)$ of the condenser formed on replacing γ_K
by the circle $|w - a_0| = r$. By Theorem 4.4 we have $c(r) \geq 2\pi K^{-1}$ or $c(r) \leq
2\pi K^{-1}$ according as $|w - a_0| \leq r$ includes or is included in the interior of

[†] Jenkins [1955] has shown that when D is simply connected $r_0 < r_0^*$ unless D^* coincides
with D or is obtained from D by rigid rotation.

γ_K. Thus we have from (4.7) for large K

$$c(r_0 e^{-K-\epsilon}) < \frac{2\pi}{K} < c(r_0 e^{-K+\epsilon}),$$

or

$$\log \frac{r_0}{r} - \epsilon < \frac{2\pi}{c(r)} < \log \frac{r_0}{r} + \epsilon$$

for all sufficiently small r. Thus

$$\frac{1}{c(r)} = \frac{1}{2\pi} \log \frac{r_0}{r} + o(1) \quad (r \to 0).$$

If we now symmetrize our condenser with respect to a line or half-line through a_0, then the disc $|w - a_0| \leq r$, being already symmetrical, is unaltered. The domain D is replaced by its symmetrized domain D^*. If $c^*(r)$ denotes the capacity of the new condenser, we have by Theorem 4.6

$$\frac{1}{c(r)} \leq \frac{1}{c^*(r)}.$$

On applying the above asymptotic expression for $1/c(r)$ we deduce

$$\log \frac{r_0}{r} + o(1) \leq \log \frac{r_0^*}{r} + o(1),$$

and this gives $r_0 \leq r_0^*$.

We have thus proved Theorem 4.8 when D is bounded and admissible. For a general domain D in the open plane we argue as follows. Let r_0 be the inner radius of D at a_0. Let D_1 be a domain in D, having a Green's function $g(w, a_0)$ and inner radius r_1 at a_0, where $r_1 < r_0 - \epsilon$. Suppose that ϵ is so chosen that

$$\frac{\partial g}{\partial u} - i \frac{\partial g}{\partial v} \neq 0$$

on the curves $\gamma_\epsilon = \{w : g(w) = \epsilon\}$, where $w = u + iv$. Let D_2 be that domain in D_1, bounded by some of the analytic Jordan curves γ_ϵ, which contains a_0. Then D_2 is analytic and so admissible, the Green's function of D_2 is $g - \epsilon$, and the inner radius, r_2, of D_2 at a_0 is $r_1 e^{-\epsilon}$.

Thus if D^*, D_2^* are the symmetrized domains of D, D_2 and r_0^*, r_2^* their inner radii at a_0, we have $r_0^* \geq r_2^* \geq r_2 > e^{-\epsilon}(r_0 - \epsilon)$. Since ϵ is arbitrary, the inequality $r_0 \leq r_0^*$ of Theorem 4.8 is proved.

4.9 The principle of symmetrization[†] We may combine Theorems 4.7 and 4.8 in the following result, which is in a form suitable for applications:

[†] For the remaining results in Sections 4.9 to 4.12 see Hayman [1950].

Theorem 4.9   *Suppose that $w = f(z) = a_0 + a_1 z + \ldots$ is regular in $|z| < 1$, and that $D = D_f$ is the domain of all values w assumed by $w = f(z)$ at least once in $|z| < 1$. Suppose further that the symmetrized domain D^* of D with respect to a line or ray through a_0 lies in a simply connected domain D_0 and that*

$$w = \phi(z) = a_0 + a_1' z + a_2' z^2 + \ldots$$

maps $|z| < 1$ (1,1) conformally onto D_0. Then $|a_1| \leq |a_1'|$.

Let r, r^* and r_0 be the inner radii of D, D^* and D_0 at a_0. Then we have

$$|a_1| \leq r \leq r^* \leq r_0 = |a_1'|.$$

The first inequality follows from Theorem 4.7, as does the equation $|a_1'| = r_0$. Again we have $r \leq r^*$ by Theorem 4.8 and $r^* \leq r_0$, since, as was shown in Section 4.8, the inner radius increases with expanding domain. This proves Theorem 4.9.

Using the work of Jenkins [1955] and Hayman [1951] quoted above, one can show that strict inequality holds in Theorem 4.9, unless D coincides with D_0 and D^*, and $f(z)$ with $\phi(ze^{i\lambda})$. However, we shall not need to use this result.

In Theorem 4.9 we may allow $f(z) = a_0 + a_1 z + \ldots$ to vary over a class $\mathfrak{F}$ of functions for which a_0 is fixed and D_f^* remains inside D_0. Then if $\phi(z) \in \mathfrak{F}$, Theorem 4.9 shows that $|a_1'|$ gives the exact upper bound for $|a_1|$, when $f(z)$ varies over the class $\mathfrak{F}$. We shall give some examples of this type of result, which may have independent interest, in the next two sections. Frequently the extremal functions $\phi(z)$ give also the exact upper bounds for $M(r, f)$ and $M(r, f')$ when $f(z) \in \mathfrak{F}$, and in the last section of the chapter we shall give a proof of this in a typical case, which is important for Chapter 5.

4.10 Applications of Steiner symmetrization   We suppose that the hypotheses of Theorem 4.9 are satisfied and that a_0 is real, and we symmetrize with respect to the real axis $v = 0$ in the $w = u + iv$ plane. Let $\theta(u)$ be the total measure of the intervals in which D_f meets the line $u = \text{constant}$. Then D^* is the domain

$$\{w : v < \tfrac{1}{2}\theta(u), -\infty < u < +\infty\}.$$

As we saw in § 4.5.4, D^* is necessarily simply connected in this case, and so we might choose $D_0 = D^*$ in Theorem 4.9, unless D^* is the whole plane. For by the Riemann mapping theorem (C. A. p. 230) any other simply connected domain may be mapped (1,1) conformally onto $|z| < 1$ by a function $\phi(z) = a_0 + a_1' z + \ldots$, where a_0 is an assigned point of the domain.

We shall not appeal to this general result, but simply give two examples of the method where the extremal function $\phi(z)$ can easily be calculated explicity so that numerical inequalities result. Evidently many other examples could be given.

Theorem 4.10[†]   *Suppose that* $f(z) = a_0 + a_1 z + \ldots$ *is regular in* $|z| < 1$ *and* D_f *intersects each line* $u = constant$ *in the* $w = u + iv$ *plane in a set of intervals of total length at most* l. *Then*

$$|a_1| \leq \frac{2l}{\pi}.$$

Equality holds for $f(z) = a_0 + \frac{l}{\pi} \log\{(1+z)/(1-z)\}$, *which maps* $|z| < 1$ *(1,1) conformally onto the strip* $|v - \Im a_0| < \frac{1}{2} l$.

We may, without loss of generality, suppose a_0 real by subtracting an imaginary constant if necessary. We then symmetrize with respect to the real axis and note that D_f^* lies in the strip $|v| < \frac{1}{2} l$. Since

$$\phi(z) = a_0 + \frac{l}{\pi} \log \frac{1+z}{1-z} = a_0 + \frac{2l}{2\pi} z + \ldots$$

maps $|z| < 1$ (1,1) conformally onto this strip, the inequality $|a_1| \leq 2l/\pi$ follows from Theorem 4.9. This inequality is sharp, since $\phi(z)$ satisfies the hypotheses for $f(z)$.

As another example we prove

Theorem 4.11   *Suppose that* $f(z) = a_0 + a_1 z + \ldots$ *is regular in* $|z| < 1$ *and that* D_f *meets the imaginary axis in the* w *plane in a set of intervals of total length at most* l. *Then*

$$|a_1| \leq [4|a_0|^2 + l^2]^{\frac{1}{2}}.$$

Equality holds if a_0 *is real and* $f(z)$ *maps* $|z| < 1$ *(1,1) conformally onto the domain* D_0 *consisting of the* w *plane cut from* $-\frac{1}{2} il$ *to* $-i\infty$ *and from* $\frac{1}{2} il$ *to* $+i\infty$ *along the imaginary axis.*

[†] For univalent $f(z)$ this an unpublished result of Rogosinski. For the general case see Hayman [1950, 1951].

Suppose first that a_0 is real. We may then symmetrize with respect to the real axis and then D_f^* lies in D_0. Thus if

$$\phi(z) = a_0 + a_1'z + \ldots$$

maps $|z| < 1$ onto D_0 we shall have $|a_1| \leq |a_1'|$. It remains to find a_1'. We note that

$$\zeta = z + \frac{1}{z}$$

maps $|z| < 1$ onto the closed ζ plane cut along the segment $(-2, 2)$ of the real axis. Thus

$$w = \frac{il}{\zeta}$$

maps $|z| < 1$ onto D_0. We now choose r so that $-1 < r < 1$ and put

$$iz = \frac{z_1 + r}{1 + rz_1}.$$

Thus we obtain a more general map of $|z_1| < 1$ onto D_0, given by

$$w = \frac{l(z_1 + r)(1 + rz_1)}{(1 - z_1^2)(1 - r^2)} = \frac{lr}{(1 - r^2)} + \frac{l(1 + r^2)}{(1 - r^2)}z_1 + \ldots.$$

We may choose r so that $lr/(1 - r^2) = a_0$. Then

$$a_1' = \frac{l(1 + r^2)}{(1 - r^2)} = \sqrt{(4a_0^2 + l^2)},$$

and Theorem 4.11 follows for real a_0.

If, finally, $a_0 = \alpha + i\beta$, we consider $f(z) - i\beta$ instead of $f(z)$ and deduce

$$|a_1| \leq \sqrt{(4\alpha^2 + l^2)} \leq \sqrt{(4|a_0|^2 + l^2)}.$$

This completes the proof.

4.11 Applications of circular symmetrization Circular symmetrization is more powerful than Steiner symmetrization, and any result obtainable by the latter method can also be obtained by the former on taking exponentials, though this may be less direct. When we use circular symmetrization the domain D_f^* is not necessarily simply connected. On the other hand, D_f^* cannot reduce to the whole plane unless D_f does, whereas in Steiner symmetrization D_f^* consists of the whole plane as soon as D_f meets every line perpendicular to the line of symmetrization in a set of intervals of infinite total length.

We proceed to give some examples

Theorem 4.12 *Suppose that $0 < \alpha < 2$, that*

$$f(z) = a_0 + a_1 z + \dots$$

is regular in $|z| < 1$ and that D_f meets each circle $|w| = \rho$ $(0 < \rho < \infty)$ in a set of arcs of total length $\pi\alpha\rho$ at most. Then $|a_1| \leq 2\alpha|a_0|$, with equality when $f(z) = a_0\{(1 + z)/(1 - z)\}^\alpha$.

We may suppose that a_0 is real and positive, since this may be achieved by considering $e^{i\lambda}f(z)$ instead of $f(z)$. With this assumption we symmetrize D_f with respect to the positive real axis. Then D_f^* lies in the domain

$$D_0 = \{w : |\arg w| < \tfrac{1}{2}\alpha\pi, \quad 0 < |w| < \infty\}.$$

The function

$$w = \phi(z) = a_0 \left(\frac{1 + z}{1 - z}\right)^\alpha = a_0(1 + 2\alpha z + \dots)$$

maps $|z| < 1$ onto D_0 and now Theorem 4.12 follows from Theorem 4.9.

Our key result for further applications is

Theorem 4.13 *Let $f(z) = a_0 + a_1 z + \dots$ be regular in $|z| < 1$, and let $R = R_f$ be the least upper bound, supposed finite, of all ρ for which D_f contains the whole circle $|w| = \rho$. If there are no such ρ, we put $R = 0$. Then $|a_1| \leq 4(|a_0| + R)$.*

Equality holds when $a_0 \geq 0$ and

$$f(z) = a_0 + \frac{4(a_0 + R)z}{(1 - z)^2},$$

which maps $|z| < 1$ onto D_0 consisting of the w plane cut from $-\infty$ to $-R$ along the negative real axis.

We again suppose that $a_0 \geq 0$ and symmetrize with respect to the positive real axis. Then, since D_f does not contain the whole circle $|w| = \rho$ for $\rho > R$, D_f^* does not contain the point $w = -\rho$ for $\rho > R$ and so D_f^* lies in D_0. Since

$$\phi(z) = a_0 + \frac{4(a_0 + R)z}{(1 - z)^2} = a_0 + 4(a_0 + R)z + \dots$$

evidently maps $|z| < 1$ onto D_0 (see §1.1), Theorem 4.13 follows.

Various special cases of Theorem 4.13 are of interest. Thus if

$$f(z) = z + a_2 z^2 + \dots \quad (|z| < 1),$$

we may take $a_0 = 0$, $a_1 = 1$ and obtain $R \geq \tfrac{1}{4}$. This is a sharp form of a theorem of Landau [1922]. The result also includes Theorem 1.2 of Koebe–Bieberbach. In fact if $f(z)$ is, in addition, univalent, D_f is simply connected and so its complement is connected. Thus if this complement contains a point w_0, it must meet every circle $|w| = \rho$ for $\rho > |w_0|$, and so $|w_0| \geq R \geq \frac{1}{4}$. Thus D_f contains the disc $|w| < \frac{1}{4}$.

Again if $R = 0$, we obtain $|a_1| \leq 4|a_0|$, a limiting case of Theorem 4.12.

It is easy to obtain inequalities when in addition to the other hypotheses $f(z)$ is bounded above. As an example we state

Theorem 4.14   *Suppose that $f(z)$ satisfies the hypotheses of Theorem 4.13 and that in addition $|f(z)| < M$ in $|z| < 1$. Then*

$$|a_1| \leq \frac{4(M - |a_0|)(M^2 + |a_0|R)(|a_0| + R)}{(M + R)^2(M + |a_0|)}.$$

Equality holds when $a_0 > 0$, and $f(z)$ maps $|z| < 1$ (1,1) conformally onto D_0, consisting of $|w| < M$, cut from $-M$ to $-R$ along the negative real axis.

The proof is left to the reader (see Example 4.2 at the end of the chapter). Clearly D_f^* lies in D_0, and the determination of the map of $|z| < 1$ onto D_0, which gives the extremal value for $|a_1|$, is an elementary exercise in conformal mapping.

Finally, we obtain the following corollary of Theorem 4.13:

Theorem 4.15   *Suppose that $w = f(z) = a_0 + a_1z + \ldots$ is regular in $|z| < 1$, that $0 < l < \infty$, and that the complement of D_f in the w plane contains a sequence of points*

$$u + i(v + nl) \quad (-\infty < n < +\infty),$$

where v may depend on u, for every real u. Then $|a_1| \leq 2l/\pi$. Equality holds for $f(z) = a_0 + \frac{l}{\pi}\log\{(1 + z)/(1 - z)\}$ as in Theorem 4.10.

Consider

$$g(z) = e^{2\pi f(z)/l} = e^{2\pi a_0/l}\left(1 + \frac{2\pi a_1}{l}z + \ldots\right).$$

Then $g(z)$ is regular in $|z| < 1$, and since

$$\log g(z) \quad \neq \quad \frac{2\pi}{l}(u + iv + inl) \quad (-\infty < n < +\infty),$$

$$g(z) \quad \neq \quad \exp\left[\frac{2\pi}{l}(u + iv)\right].$$

Since u is arbitrary subject to $-\infty < u < +\infty$, we can for every positive ρ find a number w, such that $|w| = \rho$ and $g(z) \neq w$ in $|z| < 1$. Thus we can apply Theorem 4.13 to $g(z)$ with $R = R_g = 0$, and obtain

$$\left|\frac{2\pi a_1}{l}\right| \leq 4.$$

Equality holds for real a_0, when

$$f(z) = a_0 + \frac{l}{\pi} \log\left(\frac{1+z}{1-z}\right),$$

which maps $|z| < 1$ onto the strip $|v| < \frac{1}{2}l$. This proves Theorem 4.15.

A comparison of Theorems 4.10 and 4.15 illustrates the relative power of Steiner and circular symmetrization. It may be seen that Theorem 4.15, while containing Theorem 4.10 as a special case, is significantly stronger. For, under the hypotheses of Theorem 4.15, D_f may well meet each line $u = $ constant in a set of intervals of infinite length, provided that a single sequence of points in arithmetic progression of common difference l is on the line and outside D_f. The results also suggest that the hypotheses of Theorem 4.15 might be further weakened to the assumption that D_f meets no line $u = $ constant in any *interval* of length greater than l. Dubinin [1993] has proved that the inequality $|a_1| \leq 2l/\pi$ still holds under this assumption.

4.12 Bounds for $|f(z)|$ and $|f'(z)|$ The functions which give the maximum for $|a_1|$ under the hypotheses of Theorems 4.10–4.15 also give the maximum values for $M(r,f)$ and $M(r,f')$ for $0 < r < 1$, and we can even prove in some cases that they are essentially the only functions that do so. The proofs are similar and we confine ourselves to the case of Theorem 4.13, which has most applications.

Theorem 4.16   *Suppose that $f(z)$ satisfies the hypotheses of Theorem 4.13. Then we have, for $|z| = r \ (0 < r < 1)$,*

$$|f(z)| < |a_0| + \frac{4(|a_0| + R)r}{(1-r)^2}, \tag{4.8}$$

$$|f'(z)| \leq \frac{4(R + |f(z)|)}{1-r^2} < \frac{4(|a_0| + R)(1+r)}{(1-r)^3}, \tag{4.9}$$

unless

$$f(z) = a_0 + \frac{4(Re^{i\lambda} + a_0)ze^{-i\theta}}{(1 - ze^{i\theta})^2},$$

where θ, λ are real and $\arg a_0 = \lambda$ or $a_0 = 0$.

We suppose that $|z| < 1$ and consider

$$\phi(z) = f\left(\frac{z_0 + z}{1 + \bar{z}_0 z}\right) = f(z_0) + (1 - |z_0|^2)f'(z_0)z + \dots,$$

instead of $f(z)$. Then $D_\phi = D_f$, since ϕ is obtained from f by a map of the unit disc onto itself, and so we can apply Theorem 4.13 to $\phi(z)$ with the same value of R. This gives

$$(1 - |z_0|^2)|f'(z_0)| \le 4(R + |f(z_0)|),$$

and, dropping the suffix, we obtain the left inequality of (4.9). We next write $z_0 = re^{i\theta}$, $f(z_0) = T(r)e^{i\lambda(r)}$ where θ is fixed. Then

$$e^{i\theta}f'(re^{i\theta}) = e^{i\lambda(r)}[T'(r) + i\lambda'(r)T(r)].$$

Thus we have

$$T'(r) \le \frac{4(R + T(r))}{1 - r^2}, \tag{4.10}$$

and equality is possible only if $\lambda'(r) = 0$ or $T(r) = 0$. We may write (4.10) as

$$d\log[R + T(r)] \le d\left[2\log\left(\frac{1 + r}{1 - r}\right)\right].$$

Thus

$$\psi(r) = \left(\frac{1 - r}{1 + r}\right)^2 [R + T(r)] \tag{4.11}$$

cannot increase with increasing r. If $\psi(r_1) = \psi(r_2)$, where $0 \le r_1 < r_2 < 1$, then $\psi(r)$ is constant for $r_1 \le r \le r_2$, and so equality holds in (4.10). Then $\lambda'(r) = 0$ in this range, so that

$$\lambda(r) = \lambda = \text{constant}.$$

This gives

$$e^{i\theta}f'(z) = \frac{4[Re^{i\lambda} + f(z)]}{1 - (ze^{-i\theta})^2}$$

for $z = re^{i\theta}$ $(r_1 < r < r_2)$, and hence by analytic continuation throughout $|z| < 1$. Thus

$$\frac{d}{dz}\log[Re^{i\lambda} + f(z)] = \frac{4e^{-i\theta}}{1 - (ze^{-i\theta})^2},$$

$$\frac{Re^{i\lambda} + f(z)}{Re^{i\lambda} + a_0} = \left(\frac{1 + ze^{-i\theta}}{1 - ze^{-i\theta}}\right)^2,$$

and since $\arg f(re^{i\theta}) \equiv \lambda$, we must have $\arg a_0 = \lambda$ unless $a_0 = 0$. Thus in this case $f(z)$ reduces to one of the extremals given in Theorem 4.16. They clearly satisfy the hypotheses of that theorem and give equality in the inequalities (4.8) and (4.9), when $z = re^{i\theta}$.

In all other cases the function $\psi(r)$ in (4.11) decreases strictly with increasing r for $0 < r < 1$, and in particular

$$\psi(r) < \psi(0) = R + |a_0|,$$

i.e.

$$|f(re^{i\theta})| < (R + |a_0|) \left(\frac{1+r}{1-r}\right)^2 - R.$$

This proves (4.8), and by substitution we obtain the second inequality of (4.9). This completes the proof of Theorem 4.16.

Two special cases of Theorem 4.16 are worth noting. If $a_0 = 0$, then (4.8) gives

$$M(r, f) < \frac{4Rr}{(1-r)^2}.$$

We thus obtain the following sharp form of a theorem of Bohr [1923]:

If $f(z)$ is regular in $|z| < 1, f(0) = 0$, and $M(r, f) = 1$ for some r such that $0 < r < 1$, then D_f includes a circle $|w| = \rho$, where

$$\rho > \frac{(1-r)^2}{4r},$$

except when

$$f(z) = \frac{(1-r)^2}{r} \frac{ze^{i\lambda}}{(1 - ze^{-i\theta})^2}.$$

For, except in this case, (4.8) yields

$$R > \frac{(1-r)^2}{4r} M(r, f) = \frac{(1-r)^2}{4r},$$

and we may choose ρ so that $\frac{1}{4}r^{-1}(1-r)^2 < \rho < R$, and $|w| = \rho$ lies in D_f.

We also collect together the results when $R = 0$:

Theorem 4.17 *Suppose that $f(z) = a_0 + a_1 z + \dots$ is regular in $|z| < 1$ and that for each $\rho \; (0 \le \rho < \infty)$ we can find $w = \rho e^{i\phi(\rho)}$ such that $f(z) \ne w$ in $|z| < 1$. Then unless $f(z) = a_0[(1 + ze^{i\theta})/(1 - ze^{i\theta})]^2$ for some real θ we* *have for* $|z| = r$ $(0 < r < 1)$

$$|a_0| \left(\frac{1-r}{1+r}\right)^2 < |f(z)| < |a_0| \left(\frac{1+r}{1-r}\right)^2, \qquad (4.12)$$

$$|f'(z)| \le \frac{4}{1-r^2} |f(z)| < 4|a_0| \frac{1+r}{(1-r)^3}, \qquad (4.13)$$

and $\{(1-r)/(1+r)\}^2 M(r,f)$ *decreases strictly with increasing* r $(0 < r < 1)$.

In this case we take $R = 0$ in Theorem 4.16, and (4.8) and (4.9) give the right-hand inequality of (4.12) and (4.13). Also $[f(z)]^{-1}$ satisfies the same hypotheses as $f(z)$, and we obtain the left-hand inequality in (4.12) by applying (4.8) to $[f(z)]^{-1}$ instead of $f(z)$.

To prove the last statement of Theorem 4.17, suppose that $0 \le r_1 < r_2 < 1$ and choose θ so that

$$|f(r_2 e^{i\theta})| = M(r_2, f).$$

Then unless $f(z) \equiv a_0 [(1 + z e^{i\theta})/(1 - z e^{i\theta})]^2$, the function $\psi(r)$ of (4.11) decreases strictly with increasing r and so

$$\begin{aligned}
\left(\frac{1-r_2}{1+r_2}\right)^2 M(r_2, f) &= \left(\frac{1-r_2}{1+r_2}\right)^2 |f(r_2 e^{i\theta})| \\
&< \left(\frac{1-r_1}{1+r_1}\right)^2 |f(r_1 e^{i\theta})| \le \left(\frac{1-r_1}{1+r_1}\right)^2 M(r, f).
\end{aligned}$$

This completes the proof of Theorem 4.17.

4.13 Bloch's Theorem   Although rather on the boundary of the area of multivalent functions the following proof of Bloch's Theorem recently obtained by Bonk [1990] is too elegant to be missed out.

Theorem 4.18   *Let*

$$f(z) = z + a_2 z^2 + \dots \qquad (4.14)$$

be regular in $|z| < 1$. *Then* $f(z)$ *maps some subdomain* Δ_0 *of* $|z| < 1$ *(1,1) conformally onto a disc of radius* B, *where* $B \ge \frac{\sqrt{3}}{4} = .433\dots$

The result with some absolute constant B is due to Bloch [1926]. The bound $B \ge \frac{1}{4}\sqrt{3}$ is due to Ahlfors [1938]. This was sharpened by Heins [1962] to $B > \frac{1}{4}\sqrt{3}$. Bonk [1990] proved that $B > \frac{1}{4}\sqrt{3} + 10^{-14}$. In the opposite direction Ahlfors and Grunsky [1937] showed that $B \le .472$ and their upper bound is widely conjectured to give the correct value for

B. We follow Bonk's argument to show that $B \geq \frac{1}{4}\sqrt{3}$. The proof that $B > \frac{1}{4}\sqrt{3} + 10^{-14}$ lies rather deeper and we omit it.

Lemma 4.4 *Let $\mathfrak{B}_0$ be the class of functions $f(z)$ regular in $|z| < 1$, having an expansion (4.14), and satisfying*

$$(1 - |z|^2)|f'(z)| \leq 1, \quad |z| < 1. \tag{4.15}$$

Then if $f(z) \in \mathfrak{B}_0$

$$\Re\{f'(z)\} \geq \frac{1 - |z|\sqrt{3}}{(1 - |z|/\sqrt{3})^3}, \text{for } |z| \leq \frac{1}{\sqrt{3}}, \tag{4.16}$$

and $w = f(z)$ maps $|z| < \frac{1}{\sqrt{3}}$ onto a domain D, containing the disc $|w| < \frac{1}{4}\sqrt{3}$.

We define

$$F(w) = \frac{1}{\sqrt{3}} \frac{1 - w}{1 - w/3}$$

and

$$G(w) = \frac{9}{4}w\left(1 - \frac{1}{3}w\right)^2.$$

Then

$$|G(w)|(1 - |F(w)|^2) = 1, \text{ when } |w| = 1. \tag{4.17}$$

In fact if $w = e^{i\theta}$, we have

$$|F(w)|^2 = \frac{3(1 - w)(1 - \bar{w})}{(3 - w)(3 - \bar{w})} = \frac{6(1 - \cos\theta)}{10 - 6\cos\theta} = \frac{3 - 3\cos\theta}{5 - 3\cos\theta},$$

so that

$$1 - |F(w)|^2 = \frac{2}{5 - 3\cos\theta},$$

and

$$|G(w)| = \frac{9}{4}\left(1 - \frac{w}{3}\right)\left(1 - \frac{\bar{w}}{3}\right) = \frac{1}{2}(5 - 3\cos\theta).$$

This proves (4.17).

We now consider

$$h(z) = \left\{\frac{f'(F(z))}{G(z)} - 1\right\}\frac{z}{(z - 1)^2}. \tag{4.18}$$

Since $f(z) \in \mathfrak{B}_0$, we have

$$|f'(z)| = |1 + zf''(0) + \ldots| \leq \frac{1}{1 - |z|^2}.$$

If $f''(0) = \rho e^{i\alpha}$, we set $z = t e^{-i\alpha}$, and let t tend to zero. This yields

$$1 + \rho t + O(t^2) \leq 1 + t^2 + O(t^4),$$

and we deduce that $\rho = 0$, i.e $f''(0) = 0$.

The functions $F(w), G(w)$ are regular for $|w| \leq 1$ and, by (4.17), $|F(w)| < 1$ there. Hence $h(z)$, given by (4.18), is regular for $|z| \leq 1$, except for possible poles at $z = 0, 1$. The zero of $G(z)$ at $z = 0$ is cancelled by the factor z. To study the behaviour of $h(z)$ at $z = 1$, we put $z = 1 + \eta$ and see that, for small η,

$$F(z) = \frac{-\eta\sqrt{3}}{2 - \eta}, \quad f'\{F(z)\} = 1 + O\{F(z)\}^2 = 1 + O(\eta^2),$$

since $f''(0) = 0$, and

$$G(z) = \tfrac{1}{4}(1 + \eta)(2 - \eta)^2 = 1 + O(\eta^2).$$

Thus $h(z)$ remains bounded and so regular in $|z| \leq 1$. Also by (4.15) and (4.17) we have for $z = e^{i\theta}$, $\quad 0 < \theta < 2\pi$

$$\left| \frac{f'(F(z))}{G(z)} \right| \leq \frac{1}{(1 - |F(z)|^2)|G(z)|} = 1$$

while

$$\frac{z}{(z - 1)^2} = -\frac{1}{4}\mathrm{cosec}^2\frac{\theta}{2} < 0.$$

Thus

$$\Re h(z) \geq 0.$$

By continuity this remains true for $z = 1$, and so for $|z| \leq 1$, by the maximum principle. In particular we obtain for $0 < w < 1$

$$\Re \frac{f'(F(w))}{G(w)} \geq 1; \quad \Re f'\{F(w)\} \geq G(w), \text{ since } G(w) > 0.$$

Setting $z = F(w)$, we see that z decreases from $\frac{1}{\sqrt{3}}$ to 0, as w increases from 0 to 1, while

$$w = \frac{1 - z\sqrt{3}}{1 - z/\sqrt{3}}, \quad G(w) = \frac{1 - z\sqrt{3}}{(1 - z/\sqrt{3})^3}.$$

This yields (4.16) when $z > 0$. If $z = t e^{i\alpha}$, we apply the conclusion to $e^{-i\alpha} f(z e^{i\alpha})$ instead of $f(z)$. This completes the proof of (4.16).

We deduce that, with the hypotheses of Lemma 4.4, $f(z)$ is univalent in $|z| \leq 1/\sqrt{3}$. For if z_1, z_2 are two distinct points in $|z| \leq 1/\sqrt{3}$, write $z_2 = z_1 + \rho e^{i\alpha}$. The segment

$$T : z = z_1 + te^{i\alpha}, \quad 0 < t < \rho$$

lies in $|z| < 1/\sqrt{3}$, and so by (4.16) $\Re f'(z) > 0$ in T. Thus

$$\Re e^{-i\alpha}\{f(z_2) - f(z_1)\} = \Re e^{-i\alpha} \int_{z_1}^{z_2} f'(z)dz$$

$$= \Re \int_0^{\rho} f'(z_1 + te^{i\alpha})dt > 0. \qquad (4.19)$$

Thus $f(z)$ maps $|z| < 1/\sqrt{3}$ $(1,1)$ conformally onto a domain D.

The boundary ∂D of D is the image by $f(z)$ of $|z| = 1/\sqrt{3}$. If $w = f\left((1/\sqrt{3})e^{i\alpha}\right)$ is a point of this image, (4.16) and (4.19) yield

$$|w| \geq \Re e^{-i\alpha} w = \Re \int_0^{\frac{1}{\sqrt{3}}} f'(te^{i\alpha})dt \geq \Re \int_0^{\frac{1}{\sqrt{3}}} \frac{(1 - t\sqrt{3})dt}{(1 - t/\sqrt{3})^3} = \frac{\sqrt{3}}{4}.$$

Thus D contains the disc $|w| < \frac{\sqrt{3}}{4}$ and Lemma 4.4 is proved and so is Theorem 4.18 if $f(z) \in \mathfrak{B}_0$.

Suppose next that $f(z)$ is not in $\mathfrak{B}_0$. Then there exists z_0, such that

$$(1 - |z_0|^2)|f'(z_0)| > 1$$

somewhere in $|z| < 1$. We choose ρ such that $0 < \rho < 1$, and

$$(1 - |z_0|^2)\rho|f'(\rho z_0)| > 1.$$

This is possible by continuity. We now define $\psi(z) = f(\rho z)$ and

$$\mu(z) = (1 - |z|^2)\,|\psi'(z)| = (1 - |z|^2)\rho\,|f'(\rho z)|, \quad |z| \leq 1. \qquad (4.20)$$

Then $\mu(z) = 0$ for $|z| = 1$, since $\rho < 1$, and so $\mu(z)$ attains its maximum value μ_0 in $|z| \leq 1$ at z_1 where $|z_1| < 1$. Also $\mu_0 \geq \mu(z_0) > 1$.

We suppose that $\arg \psi'(z_1) = \alpha$ and consider

$$\phi(z) = \frac{e^{-i\alpha}}{\mu_0} \left\{ \psi\left(\frac{z_1 + z}{1 + \bar{z}_1 z}\right) - \psi(z_1) \right\}.$$

We note that $\phi(z)$ is regular in $|z| < 1$ and

$$\phi(0) = 0, \quad \phi'(0) = \frac{e^{-i\alpha}(1 - |z_1|^2)\psi'(z_1)}{\mu_0} = 1.$$

Also we have for $|z| < 1$

$$
\begin{aligned}
(1 - |z|^2)|\phi'(z)| &= \frac{(1 - |z|^2)}{\mu_0} \psi'\left(\frac{z_1 + z}{1 + \bar{z}_1 z}\right) \frac{1 - |z_1|^2}{|1 + \bar{z}_1 z|^2} \\
&= \left(1 - \left|\frac{z_1 + z}{1 + \bar{z}_1 z}\right|^2\right)\left|\psi'\left(\frac{z_1 + z}{1 + \bar{z}_1 z}\right)\right| \leq 1.
\end{aligned}
$$

by (4.20). Thus $\phi(z) \in \mathfrak{B}_0$, and so, by Lemma 4.4, $\phi(z)$ maps $|z| < 1/\sqrt{3}$ onto a domain containing the disc $|w| < \sqrt{3}/4$. Hence $\psi(z)$ maps some subdomain of $|z| < 1$ onto a disc of radius $\mu_0\sqrt{3}/4$, and so $f(z)$ maps some subdomain of $|z| < \rho$ onto a disc of radius $\mu_0\sqrt{3}/4 > \sqrt{3}/4$. This completes the proof of Theorem 4.18.

4.14 Some other results We close the chapter by mentioning some further developments. Baernstein [1975] has proved that, with the hypotheses of Theorem 4.9, we have further

$$
I_\lambda(r, f) \leq I_\lambda(r, \phi), \quad 0 \leq r < 1, \quad 0 < \lambda \leq \infty \tag{4.21}
$$

where

$$
I_\lambda(r, f) = \left\{\frac{1}{2\pi}\int_0^{2\pi} |f(re^{i\theta})|^\lambda\right\}^{1/\lambda}, \text{ if } \lambda < \infty
$$

and

$$
I_\infty(r, f) = M(r, f).
$$

If $f(z) \in \mathfrak{S}$ he proved that

$$
I_\lambda(r, f) \leq I_\lambda(r, k),
$$

where $k(z)$ is the Koebe function (1.1).

Next Weitsman [1986] proved that Theorem 4.9 extends to arbitrary domains D, whose complement in the extended plane contains at least 3 points. If $D^* = D_0$ and D^* is multiply connected, $\phi(z)$ maps $|z| < 1$ onto the universal covering surface over D_0. It is not known whether the analogue of (4.21) holds in this more general case.

The case when the complement of D^* consists of the points $(0, -1, \infty)$ is of particular interest, since it leads to sharp forms of the theorems of Landau [1904] and Schottky [1904]. This special case had previously been dealt with by Hempel [1979] and Lai [1979].

For proofs of the above results the interested reader is referred to the original papers. A connected account is also given in Hayman [1989, Chapter 9].

Examples

4.2   If
$$\zeta = \left(\frac{1+w}{1-w}\right)^2 - \left(\frac{1-r}{1+r}\right)^2 = K\left(\frac{1+z}{1-z}\right)^2,$$

show that the ζ plane cut along the negative real axis corresponds (1,1) conformally to $|z| < 1$, and to $|w| < 1$ cut from -1 to $-r$ along the real axis.

Hence complete the proof of Theorem 4.14, first when $M = 1$, and then generally by considering $f(z)/M$, R/M, 1 instead of $f(z)$, R, M respectively.

The following examples are generalisations of Koebe's Theorem 1.2.

Examples

4.3   We denote by $\mathfrak{M}$ the class of functions meromorphic in $|z| < 1$ and having an expression (4.14) near $z = 0$. Show that if $f \in \mathfrak{M}$ and a, b are positive numbers such that $1/a - 1/b > 4$, then D_f contains a circle $|w| = r$, such that $a < r < b$, (Hayman [1951]). Show by an example that this result fails whenever $1/a - 1/b = 4$. (Assume the result is false and apply Theorem 4.9.)

4.4   If $w = f(z) \in \mathfrak{M}$ and $0 < r < 1$, show that D_f meets $|w| = r$ in a set of linear measure at least $2r\alpha$, where $r = \cos(\alpha/2)$, and for each r give an example for which this lower bound is attained. (Show that the map

$$z = \frac{\{1 - w\,e^{i\alpha}\}^{\frac{1}{2}} - (1 - w\,e^{-i\alpha})^{\frac{1}{2}}}{(e^{-i\alpha} - w)^{\frac{1}{2}} - (e^{i\alpha} - w)^{\frac{1}{2}}}$$

maps the complement in the closed w plane of the arc $w = e^{i\theta}$, $\alpha \le |\theta| \le \pi$, onto $|z| < 1$.)

4.5   If $W = f(z) \in \mathfrak{S}$ where $\mathfrak{S}$ is the class of Chapter 1, prove that, for $\frac{1}{4} < r < 1$, $D = D_f$ meets $|w| = r$ in a set of linear measure at least $2r\alpha$ where $r = \cos^4(\alpha/4)$. (If the conclusion is false, the complement of D^* contains the ray $-\infty$, $-r$ of the real axis as well as the arc $\alpha \le |\arg z| \le \pi$ of $|w| = r$. To find the extremal map, set

$$\frac{z}{(1+z)^2} = 4k\,\frac{Z}{(1+Z)^2},$$

where

$$k = \frac{\cos\frac{\alpha}{2}}{(1 + \cos\frac{\alpha}{2})^2},$$

and $W = rw, w = f(z)$, is the map of Example 4.4. (Netanyahu [1969] has shown that a function of this type yields the extremal value $|a_2| d = 2/3$ where d is the radius of the largest disc with centre 0 contained in the image of $|z| < 1$).

4.6   If $w = f(z) \in \mathfrak{M}$ and $f(z)$ is regular in $|z| < 1$, show that for $0 < r < 1$, D_f meets $|w| = r$ in a set of linear measure at least $2r\alpha$, where α is the (unique) solution, in $0 < \alpha < \pi$, of

$$r = \frac{\sin^2 \frac{\alpha}{2}}{2 \log \left(\sec \frac{\alpha}{2} \right)},$$

This example needs the full strength of Weitsman's Theorem. The symmetrized domain D^* lies in the domain D_0, whose complement in the closed plane consists of the arc $w = re^{i\theta}$, $\alpha \leq |\theta| \leq \pi$ and the point at ∞. Thus, by Weitsman's Theorem the extremal map is $w = \psi(z) = r A_1 z + \dots$, where $\psi(z)$ maps $|z| < 1$ onto the universal covering surface over this doubly connected domain D_0. In the map $w = f(z)$ of Example 4.4, $|z| < 1$, punctured at $z = c$, corresponds to D_0 $(1,1)$ conformally, and $z = c = \cos(\alpha/2)$ is mapped onto $w = \infty$. We now set

$$Z = \frac{\log z_1 + K}{\log z_1 - K}, \text{ where } z_1 = \frac{c - z}{1 - cz}, \quad K = \log \frac{1}{c}.$$

Then $|Z| < 1$ is mapped onto the universal covering surface of the unit disc $|z| < 1$ punctured at $z = c$. Thus the combined map $w = \psi(Z) = f(z(Z))$ gives the required extremal map with $\psi(0) = 0$,

$$|\psi'(0)| = \frac{r}{c} \frac{2 \log 1/c}{(1/c - c)} = \frac{2r \log 1/c}{1 - c^2}.$$

If $\psi(z) \in \mathfrak{M}$, then, as required,

$$r = \frac{1 - c^2}{2 \log 1/c}.$$

The function

$$\frac{1 - c^2}{2 \log 1/c}$$

increases monotonically from 0 to 1 as c increases from 0 to 1 and so as α decreases from π to 0. Thus the equation

$$r = \frac{\sin^2(\alpha/2)}{2 \log \sec(\alpha/2)}$$

has exactly one solution in the range $0 < \alpha < \pi$.

We denote by $\mathfrak{B}$ the class of functions regular in $|z| < 1$ and such that

$$\mu_0 = \sup_{|z|<1}(1 - |z|^2)|f'(z)| < \infty. \tag{4.22}$$

The functions in $\mathfrak{B}$ are called *Bloch functions* [Hayman 1952] and have been widely studied. (See e.g. Pommerenke 1975, p. 269 *et seq.*)

Examples

4.7   If $f(z) \in \mathfrak{B}$ show that $f(z)$ maps some subdomain of $|z| < 1$ *(1, 1)* conformally onto a disc of radius R, provided that $R < \mu_0 B$, where B is Bloch's constant, i.e. the quantity of Theorem 4.18.

4.8   Suppose that $f(z)$ is regular in $|z| < 1$ and that R_0 is the upper bound of the radii R of discs $|w - w_0| < R$ onto which some subdomain of $|z| < 1$ is mapped by $w = f(z)$. Then R_0 is finite if and only if μ_0 is finite and

$$\mu_0 B \le R_0 \le \mu_0. \tag{4.23}$$

(To obtain the right-hand inequality consider the map $z = f^{-1}(w)$ from a disc $|w - w_0| < R$ into $|z| < 1$.)

4.9   If $f(z) = \sum_0^\infty a_n z^n \in \mathfrak{B}$, show that $|a_n| \le \frac{1}{2}e\mu_0$, $n \ge 1$. (Prove that the exact bound for $|a_n|$ is $\frac{1}{2}\mu_0 A_n$, where $A_1 = 2$, $\log A_n = \frac{n}{2}\log\left(\frac{1+1/n}{1-1/n}\right) + \frac{1}{2}\log\left(1 - \frac{1}{n^2}\right)$, $n \ge 2$).)

4.10   (Landau's Theorem). If $f(z) = a_0 + a_1 z + \dots$ is regular in $|z| < 1$, and $f(z) \ne 0, 1$ there, we define

$$g(z) = \frac{1}{2\pi i}\log f(z), \quad h(z) = g(z) + \sqrt{(g(z)^2 - 1)}, \quad b(z) = \log h(z).$$

Show that, with a suitable choice of the square roots and logarithms at $z = 0$, the functions $g(z)$, $h(z)$ and $b(z)$ are regular in $|z| < 1$, and

$$b(z) \ne \pm\log\{n + \sqrt{(n^2 - 1)}\} + 2m\pi i$$

for any positive integer n, or integer m. Deduce that $b(z) \in \mathfrak{B}$ and that $R_0 < 2\sqrt{3}$. Hence show that $|a_1| < 8|a_0|\{|\log|a_0|| + 2\pi\}$. (For the best result $|a_1| < 2|a_0|\{|\log|a_0|| + 4.37\dots\}$ see Hempel [1979] and Lai [1979] or Hayman [1989, p. 702].)

5

Circumferentially mean p–valent functions

5.0 Introduction We consider again functions $f(z)$ regular and not constant in a domain Δ, define $n(w) = n(w, \Delta, f)$ as the number of roots of the equation $f(z) = w$ in Δ, and write as in (2.4)

$$p(R) = p(R, \Delta, f) = \frac{1}{2\pi} \int_0^{2\pi} n(Re^{i\phi})d\phi.$$

In Chapters 2 and 3 we considered functions satisfying the condition

$$\int_0^R p(\rho)d(\rho^2) \leq pR^2 \quad (R > 0),$$

where p is a positive number. In what follows we shall call such functions *areally mean p–valent* (a.m.p–valent). We shall consider in the first part of this chapter some consequences of the more restrictive hypothesis

$$p(R) \leq p \quad (R > 0). \tag{5.1}$$

Functions satisfying this condition were introduced by Biernacki [1946] and are generally called *circumferentially mean p–valent* (c.m.p–valent).

We first prove some sharp inequalities restricting the growth of functions c.m.p–valent in $|z| < 1$ and such that either $f(z) \neq 0$, or $f(z)$ has a zero of order p at $z = 0$, where p is a positive integer, basing ourselves on Theorems 4.13 and 4.17 of the last chapter.

In the second part of the chapter we prove some regularity theorems for a.m.p–valent functions in $|z| < 1$. We proved in Theorem 2.10 that if $f(z) = \sum_0^\infty a_n z^n$ is such a function, then

$$\alpha = \lim_{r \to 1-} (1 - r)^{2p} M(r, f)$$

144

exists. We shall prove that, if further $p > \frac{1}{4}$, then also

$$\lim_{n \to \infty} \frac{|a_n|}{n^{2p-1}} = \frac{\alpha}{\Gamma(2p)}.$$

A generalization to positive powers of mean p–valent functions and functions of the type $\sum_{n=0}^{\infty} a_{kn+\nu} z^{kn+\nu}$ will also be proved. As an application of these results, we shall prove that if

$$f(z) = z + a_2 z^2 + \cdots$$

is a.m. 1–valent in $|z| < 1$, then $|a_n| \le n$ for $n > n_0(f)$ (cf. Example 2.12).

5.1 Functions without zeros   In this section we prove[†]

Theorem 5.1   *Suppose that* $f(z) = a_0 + a_1 z + \cdots$ *is c.m.p–valent and* $f(z) \neq 0$ *in* $|z| < 1$. *Then*

$$|a_1| \le 4p|a_0|.$$

Further, unless $f(z) = a_0[(1 + z e^{i\theta})/(1 - z e^{i\theta})]^{2p}$ *for a real* θ, *we have for* $|z| = \rho$ $(0 < \rho < 1)$

$$|a_0| \left(\frac{1 - \rho}{1 + \rho} \right)^{2p} < |f(z)| < |a_0| \left(\frac{1 + \rho}{1 - \rho} \right)^{2p}$$

and

$$|f'(z)| \le \frac{4p}{1 - \rho^2} |f(z)| < \frac{4|a_0| p (1 + \rho)^{2p-1}}{(1 - \rho)^{2p+1}}.$$

Also $[(1 - \rho)/(1 + \rho)]^{2p} M(\rho, f)$ *decreases strictly with increasing* ρ $(0 < \rho < 1)$, *and so approaches a limit* α_0 *as* $\rho \to 1$ *where* $\alpha_0 < |a_0|$.

We need two lemmas:

Lemma 5.1   *Suppose that* $\eta > 0$, *that* $f(z)$ *is regular in a domain* Δ *and that* $\psi(z) = f(z)^\eta$ *is single valued there.*

(a) If $f(z)$ *is c.m.p–valent in* Δ, *then* $\psi(z)$ *is c.m.(ηp)–valent in* Δ.

(b) If $f(z)$ *is a.m.p–valent in* Δ, *then* $\psi(z)$ *is a.m. $(\eta_0 p)$–valent in* Δ, *where* $\eta_0 = \max(\eta, \eta^2)$.

[†] Hayman [1950, 1951, 1955]

The result is sharp (see Example 5.1 at the end of the chapter).

We write $\psi(z) = W = Re^{i\Phi}$, $f(z) = w = re^{i\phi}$, so that $W = w^\eta$. Thus we may write $R = r^\eta$, $\Phi = \eta\phi$. If on passing along an arc of a level curve $|f(z)| = r = $ constant, $\phi = \arg f(z)$ increases strictly by an amount $\delta\phi$, then this level curve contributes $\frac{1}{2\pi}\delta\phi$ to $p(r, \Delta, f)$. Also the corresponding contribution to $p(R, \Delta, \psi)$ where $R = r^\eta$, is $\frac{\eta}{2\pi}\delta\phi$. We call $\delta\phi$ the variation of $\arg f(z)$ on the arc.

We may express the set of level curves $|f(z)| = r$ in Δ as the union of a finite or ennumerable set of Jordan arcs, no two of which have more than end-points in common and in each of which $\arg f(z)$ varies monotonically. Then $2\pi p(r, \Delta, f)$ is the sum of the variations of $\arg f(z)$ on these Jordan arcs, and so we obtain by addition

$$p(R, \Delta, \psi) = \eta p(r, \Delta, f). \tag{5.2}$$

Now (a) follows from (5.1). To prove (b) we write

$$p(r) = p(r, \Delta, f), \quad P(R) = p(R, \Delta, \psi),$$

$$w(r) = \int_0^r p(\rho)d(\rho^2), \quad W(R) = \int_0^R P(s)d(s^2).$$

Thus (5.2) yields $P(s) = \eta p(s^{1/\eta})$ and

$$\begin{aligned}
W(R) &= \int_0^R \eta p(s^{1/\eta})2s\,ds = 2\eta^2 \int_0^{R^{1/\eta}} p(\rho)\rho^{2\eta-1}d\rho \\
&= \eta^2 \int_0^{R^{\frac{1}{\eta}}} \rho^{2\eta-2}dw(\rho) \\
&= \eta^2 \left[\rho^{2\eta-2}w(\rho)\right]_0^{R^{1/\eta}} - \eta^2(2\eta-2)\int_0^{R^{1/\eta}} \rho^{2\eta-3}w(\rho)d\rho. \tag{5.3}
\end{aligned}$$

Since f is a.m.p–valent we have $0 \leq w(\rho) \leq p\rho^2$, $0 < \rho < \infty$. Thus if $\eta \leq 1$ we obtain

$$W(R) \leq p\eta^2 R^2 + \eta^2(2 - 2\eta) \int_0^{R^{1/\eta}} p\rho^{2\eta-1}d\rho = p\eta R^2,$$

so that (b) holds in this case. If $\eta > 1$ we obtain at any rate

$$W(R) \leq p\eta^2 R^2,$$

since the integral on the right-hand side of (5.3) is nonnegative. This completes the proof of Lemma 5.1.

Lemma 5.2 *If $f(z)$ is c.m. 1–valent in $|z| < 1$, then there exists a number* $l = l_f \geq 0$, *with the following property. If* $|w| < l$ *then the equation* $f(z) = w$ *has exactly one root in* $|z| < 1$, *while if* $R \geq l$, *we can find* $w_R = Re^{i\phi}$, *such that* $f(z) \neq w_R$ *in* $|z| < 1$.

Let $n(w)$ be the number of roots of the equation $f(z) = w$ in $|z| < 1$. We note that the sets $\{w : n(w) \geq K\}$ in the w plane are open for every finite positive K. In fact if $n(w_0) \geq K$, then there exist a finite number of distinct points $z_1, z_2, \ldots, z_q$ in $|z| < 1$ at which $f(z_v) = w_0$ with multiplicity K_v and $\sum_{v=1}^{q} K_v \geq K$. We can then, given a small positive δ, choose ε so small that for $0 < |w - w_0| < \varepsilon$ the equation $f(z) = w$ has exactly K_v roots in $|z - z_v| < \delta$.[†] If δ is sufficiently small all these roots are distinct and so

$$n(w) \geq \sum_{v=1}^{q} K_v \geq K \quad \text{if } |w - w_0| < \varepsilon.$$

Thus the set where $n(w) \geq K$ is open.

It follows that if $n(Re^{i\phi_0}) > 1$, then $n(Re^{i\phi}) \geq 2$ in some range $|\phi - \phi_0| < \varepsilon$. Since also

$$\frac{1}{2\pi} \int_0^{2\pi} n(Re^{i\phi}) d\phi \leq 1,$$

by hypothesis, we deduce that $n(Re^{i\phi}) < 1$, i.e. $n(Re^{i\phi}) = 0$, for some other value of ϕ. Thus if $n(Re^{i\phi}) \geq 1$ $(0 \leq \phi \leq 2\pi)$, then $n(Re^{i\phi}) = 1$ in this range. In this case the circle $|w| = R$ corresponds $(1,1)$ continuously to a set γ in $|z| < 1$, which is therefore a simple closed curve. Take $|w_0| < R$. Then since $|f(z)| = R$ on γ, it follows from Rouché's Theorem[‡] that the functions $f(z)$ and $f(z) - w_0$ have equally many zeros inside γ, N_0 say. Since $f(z)$ is not constant by the introductory hypotheses of this chapter, $f(z)$ assumes some values w such that $|w| < R$ inside γ, and so $N_0 \geq 1$. Thus $n(w) \geq N_0 \geq 1$ $(|w| < R)$, and so $n(w) = 1$ for $|w| < R$.

Let now $l = l_f$ be the least upper bound of all R, if any, which satisfy the above hypotheses. If there are no such R, we put $l = 0$. Then $n(w) = 1$ for $|w| < l$. This shows that $l < +\infty$, since otherwise the inverse function $z = f^{-1}(w)$ would provide a map of the whole w plane onto $|z| < 1$, which contradicts Liouville's Theorem. Also for $R > l$, we can find ϕ such that $n(Re^{i\phi}) = 0$. This remains true for $R = l$. In fact the set where $n(w) \geq 1$ is open, as we saw above, and so if it includes the circle

[†] C. A. Theorem 11, p. 131.
[‡] C. A., p.153

$|w| = l$ it will also include $|w| = l + \varepsilon$ for sufficiently small ε, contradicting the definition of l. This proves Lemma 5.2.

We can now prove Theorem 5.1. We write

$$\psi(z) = [f(z)]^{1/p} = a_0^{1/p}\left[1 + \frac{a_1}{pa_0}z + \cdots\right].$$

Since $f(z)$ is regular and $f(z) \neq 0$ in $|z| < 1$, we can choose a single-valued branch of $\psi(z)$, which is therefore c.m. 1–valent by Lemma 5.1. Also $\psi(z) \neq 0$ in $|z| < 1$, and so $l = l_\psi = 0$ in Lemma 5.2. Thus for $R > 0$ we can find $w = w_R$ such that $|w_R| = R$ and $\psi(z) \neq w_R$ in $|z| < 1$.

We may therefore apply Theorem 4.17 to $\psi(z)$. (4.13) gives

$$\left|\frac{\psi'(z)}{\psi(z)}\right| \leq \frac{4}{1 - |z|^2},$$

and, putting $z = 0$, we deduce $|a_1| \leq 4p|a_0|$. Also unless

$$\psi(z) = a_0^{1/p}\left(\frac{1 + ze^{i\theta}}{1 - ze^{i\theta}}\right)^2,$$

we deduce from (4.12)

$$|a_0|^{1/p}\left(\frac{1 - \rho}{1 + \rho}\right)^2 < |\psi(z)| < |a_0|^{1/p}\left(\frac{1 + \rho}{1 - \rho}\right)^2 \quad (|z| = \rho),$$

and for $0 < \rho < 1$, $[(1 - \rho)/(1 + \rho)]^2 M(\rho, \psi)$ decreases strictly with increasing ρ. Writing $f(z) = [\psi(z)]^p$ we deduce Theorem 5.1. The extremal functions

$$a_0[(1 + ze^{i\theta})/(1 - ze^{i\theta})]^{2p}$$

are c.m.p–valent and not zero by Lemma 5.1, since their pth roots are univalent and map $|z| < 1$ onto the plane cut along a ray from the origin to infinity.

We remarked in §2.3 that if $f(z)$ is a.m.p–valent and hence *a fortiori* if $f(z)$ is c.m.p–valent then $f(z)$ has $q \leq p$ zeros. Thus the condition $f(z) \neq 0$ of Theorem 5.1 is a consequence of mean p–valency if $p < 1$.

5.2 Functions with a zero of order p at the origin In this section we consider a function

$$f(z) = z^p + a_{p+1}z^{p+1} + \cdots,$$

c.m.*p*–valent in $|z| < 1$, where *p* is a positive integer. In this case $f(z)$ has no zeros in $0 < |z| < 1$ and so

$$[f(z)]^{1/p} = z\left(1 + \frac{a_{p+1}}{p} + \cdots\right)$$

is single–valued and hence c.m. 1–valent in $|z| < 1$ and vanishes only at $z = 0$. Thus we can reduce our problem to the case $p = 1$. We show that the methods of §1.2.1 are applicable.

Theorem 5.2   *Let $f(z) = z + a_2 z^2 + \cdots$ be c.m. 1–valent in $|z| < 1$. Then $f(z)$ belongs to the class $\mathfrak{S}_0$ defined in §1.2.1.*

We suppose that $|z_0| < 1$ and that $z = \omega(\zeta)$ gives a univalent map of $|\zeta| < 1$ into $|z| < 1$ in such a way that $\omega(\zeta) \neq z_0$. Consider

$$\psi(\zeta) = f[\omega(\zeta)].$$

Then the equation $\psi(\zeta) = w$ never has more roots in $|\zeta| < 1$ than the equation $f(z) = w$ has in $|z| < 1$, since $\omega(\zeta)$ is univalent. Thus $\psi(\zeta)$ is c.m. 1–valent in $|\zeta| < 1$. We write $f(z_0) = Re^{i\phi_0}$. Then it follows from Lemma 5.2 that if the equation $f(z) = Re^{i\phi_0}$ has a root in $|z| < 1$ other than z_0, then we can find ϕ' such that $f(z) \neq Re^{i\phi'}$ in $|z| < 1$. In this case $\psi(\zeta) \neq Re^{i\phi'}$ in $|\zeta| < 1$. If, on the other hand, $f(z) \neq Re^{i\phi_0}$ for $z \neq z_0$, then $\psi(\zeta) \neq Re^{i\phi_0}$ in $|\zeta| < 1$.

Thus in either case we can find $\phi = \phi(R)$ such that $\psi(\zeta) \neq Re^{i\phi}$ in $|\zeta| < 1$. Now Lemma 5.2 shows that if $\rho \geq R$ there exists $\phi = \phi(\rho)$ such that $\psi(\zeta) \neq \rho e^{i\phi(\rho)}$ in $|\zeta| < 1$. Hence Theorem 4.13 gives

$$|\psi'(0)| \leq 4[|\psi(0)| + R] = 4[|\psi(0)| + |f(z_0)|].$$

Hence $f(z) \in \mathfrak{S}_0$ and Theorem 5.2 is proved.

We deduce immediately[†]

Theorem 5.3   *Suppose that $f(z) = z^p + a_{p+1} z^{p+1} + \cdots$ is c.m.p–valent in $|z| < 1$, where p is a positive integer. Then $|a_{p+1}| \leq 2p$. Further we have for $|z| = r$ $(0 < r < 1)$*

$$\frac{r^p}{(1+r)^{2p}} \leq |f(z)| \leq \frac{r^p}{(1-r)^{2p}},$$

[†] In this form the result appears in Hayman [1950]. The inequality $|a_{p+1}| \leq 2p$ has been proved for a.m.*p*–valent functions by Spencer [1941b], and the last statement of the theorem was extended to a.m.*p*–valent functions by Garabedian and Royden [1954].

$$|f'(z)| \le \frac{p(1+r)}{r(1-r)}|f(z)| \le \frac{pr^{p-1}(1+r)}{(1-r)^{2p+1}}.$$

Finally the equation $f(z) = w$ *has exactly p roots in* $|z| < 1$ *if* $|w| < 4^{-p}$.

Theorem 5.4[†]   *With the hypotheses of Theorem 5.3*

$$r^{-p}(1-r)^{2p}M(r,f)$$

decreases strictly with increasing r $(0 < r < 1)$, *and so tends to* $\alpha < 1$ $r \to$ 1, *unless* $f(z) = z^p(1 - ze^{i\theta})^{-2p}$. *Hence the upper bounds for* $|f(z)|, |f'(z)|$, *given in Theorem 5.3 are only attained by these functions.*

We write

$$\psi(z) = [f(z)]^{1/p} = z + \frac{a_{p+1}}{p}z^2 + \cdots.$$

By Theorem 5.2 $\psi(z) \in \mathfrak{S}_0$, and so Theorems 1.4 and 1.5 are applicable with $\psi(z)$ instead of $f(z)$. We deduce the results of Theorems 5.3 and 5.4 by taking pth powers.

5.3 Regularity Theorems: the case $\alpha = 0$ For the rest of this chapter we concentrate on the more general hypothesis that $f(z)$ is a.m.p–valent and shall say that $f(z)$ is mean p–valent in this case. We start with the regularity Theorems 2.10 and 2.11 for the maximum modulus and radius of greatest growth of $f(z)$ to obtain corresponding results for the coefficients of $f(z)^\lambda$ if $p\lambda > \frac{1}{4}$. The results fail for $p\lambda < \frac{1}{4}$, as the example of Section 3.4 shows. The extension to a.m.p–valent functions is due to Eke [1967b] and was made possible by his regularity Theorems 2.10 and 2.11. For c.m.p–valent functions the results are contained in Hayman [1955].

Suppose now that $f(z)$ is mean p–valent and $f(z) \ne 0$ in an annulus $1 - 2\delta < |z| < 1$. Put $\phi(z) = [f(z)]^\lambda$. Then $\phi(z)$ may be analytically continued throughout the annulus $1 - 2\delta < |z| < 1$. Also if $\phi_2(z)$ is obtained from a branch $\phi_1(z)$ of $\phi(z)$ by analytic continuation once around the annulus in the positive direction, then

$$|\phi_2(z)/\phi_1(z)| = 1,$$

and so by the maximum modulus principle $\phi_2(z)/\phi_1(z) = e^{i\mu}$, where μ is a real constant.

[†] For this and subsequent results of this chapter see Hayman [1955].

Thus $\phi(z)/z^{\mu}$ remains one–valued in the annulus and possesses a Laurent expansion, so that

$$\phi(z) = z^{\mu} \sum_{n=-\infty}^{+\infty} b_n z^n. \tag{5.4}$$

We shall prove

Theorem 5.5   *Suppose that $f(z)$ is mean p–valent in $0 < |z| < 1$, that $p\lambda > \frac{1}{4}$, and that $\phi(z) = [f(z)]^{\lambda}$ possesses a power series expansion (5.4) in an annulus $1 - 2\delta < |z| < 1$. Then*[†]

$$\lim_{n \to +\infty} \frac{|b_n|}{n^{2p\lambda - 1}} = \frac{\alpha^{\lambda}}{\Gamma(2p\lambda)},$$

where α is the constant of Theorem 2.10.

We shall use the formula

$$
\begin{aligned}
(n + \mu)b_n &= \frac{1}{2\pi i} \int_{|z|=\rho} \frac{\phi'(z)dz}{z^{n+\mu}} \\
&= \frac{\rho^{1-n-\mu}}{2\pi} \int_{-\pi}^{\pi} \phi'(\rho e^{i\theta})e^{-i(n+\mu-1)\theta}d\theta,
\end{aligned} \tag{5.5}
$$

where $1 - 2\delta < \rho < 1$. In this section we prove Theorem 5.5 when $\alpha = 0$. We choose a positive constant t such that

$$0 < t < 2, \quad t > 2 - \frac{2}{\lambda} \quad \text{and} \quad t > \frac{1}{2p\lambda}. \tag{5.6}$$

This is possible since $p\lambda > \frac{1}{4}$. Then Schwarz's inequality gives

$$\frac{1}{2\pi} \int_{-\pi}^{\pi} |\phi'(\rho e^{i\theta})|d\theta$$

$$\leq \left(\frac{1}{2\pi} \int_{-\pi}^{\pi} |\phi'(\rho e^{i\theta})|^2 |\phi(\rho e^{i\theta})|^{-t} d\theta \right)^{\frac{1}{2}} \left(\frac{1}{2\pi} \int_{-\pi}^{\pi} |\phi(\rho e^{i\theta})|^t d\theta \right)^{\frac{1}{2}}$$

$$= \lambda \left(\frac{1}{2\pi} \int_0^{2\pi} |f'(\rho e^{i\theta})|^2 |f(\rho e^{i\theta})|^{(2-t)\lambda - 2} d\theta \right)^{\frac{1}{2}} \times \left(\frac{1}{2\pi} \int_0^{2\pi} |f(\rho e^{i\theta})|^{\lambda t} d\theta \right)^{\frac{1}{2}}.$$

Since $\alpha = 0$ we can, given $\varepsilon > 0$, find $r_0(\varepsilon) < 1$ such that

$$M(r,f) \leq \varepsilon(1-r)^{-2p} \quad (r_0(\varepsilon) < r < 1). \tag{5.7}$$

Since $f(z)$ is mean p–valent and (5.6) holds, we may apply Lemma 3.1

[†] Here and subsequently $\Gamma(x)$ denotes the Gamma function. Relevant properties will be found for instance in Titchmarsh [1939] pp. 55–8. We need assume only that $f(z)/z^{\nu}$ remains one–valued for some real ν (cf. Example 2.8).

with $(2-t)\lambda$ instead of λ. Hence if $r_0(\varepsilon) < r < 1$ there exists ρ such that $2r - 1 \le \rho \le r$ and

$$\frac{1}{2\pi} \int_0^{2\pi} |f'(\rho e^{i\theta})|^2 |f(\rho e^{i\theta})|^{(2-t)\lambda-2} d\theta \le \frac{4p\varepsilon^{\lambda(2-t)}}{\lambda(2-t)} (1-r)^{-2p\lambda(2-t)-1}.$$

Next we have from inequality (3.10) of Theorem 3.2 applied with λt instead of λ, and (5.7), for $0 \le \rho \le r$, $r_0 \le r < 1$,

$$
\begin{aligned}
I_{\lambda t}(\rho, f) &= \frac{1}{2\pi} \int_0^{2\pi} |f(\rho e^{i\theta})|^{\lambda t} d\theta \\
&\le M(r_0, f)^{\lambda t} + \frac{p}{2}[(\lambda t)^2 + 1] \int_{r_0}^r \frac{M(x,f)^{\lambda t}}{x} dx \\
&\le \varepsilon^{\lambda t}(1-r_0)^{-2p\lambda t} + \frac{p}{2}[(\lambda t)^2 + 1] \int_{r_0}^r \varepsilon^{\lambda t}(1-x)^{-2p\lambda t} \frac{dx}{x} \\
&< (1-r)^{1-2p\lambda t},
\end{aligned}
$$

if $r_0 > \frac{1}{2}$, ε is chosen so small that

$$\frac{p\varepsilon^{\lambda t}}{(2p\lambda t - 1)}[(\lambda t)^2 + 1] < 1$$

and then r is chosen sufficiently near 1. With this choice of r_0, ε, r, ρ we deduce that

$$
\begin{aligned}
\frac{1}{2\pi} \int_0^{2\pi} |\phi'(\rho e^{i\theta})| d\theta &\le A(p,\lambda)\varepsilon^{\frac{1}{2}\lambda(2-t)}(1-r)^{\frac{1}{2}[1-2p\lambda t - 2p\lambda(2-t)-1]} \\
&= A(p,\lambda)\varepsilon^{\frac{1}{2}\lambda(2-t)}(1-r)^{-2p\lambda},
\end{aligned}
$$

where ρ is some number such that $2r - 1 \le \rho \le r$. We choose $r = 1 - 1/n$ and apply (5.5). Then if n is sufficiently large

$$
\begin{aligned}
|(n+\mu)b_n| &\le \left(1 - \frac{2}{n}\right)^{-n-\mu} \frac{1}{2\pi} \int_0^{2\pi} |\phi'(\rho e^{i\theta})| d\theta \\
&\le A(p,\lambda)\varepsilon^{\frac{1}{2}\lambda(2-t)} n^{2p\lambda}.
\end{aligned}
$$

Since ε may be chosen as small as we please, we deduce

$$b_n = o(n^{2p\lambda-1}) \quad (n \to +\infty),$$

and this proves Theorem 5.5 when $\alpha = 0$.

5.4 The case $\alpha > 0$: the minor arc   When $\alpha > 0$ it follows from Theorems 2.10 and 2.7 that $f(z)$ has a unique radius of greatest growth $\arg z = \theta_0$ and that $|f(re^{i\theta})|$ is relatively small except when θ is near θ_0.

We shall deduce Theorem 5.5 from formula (5.5) in this case by obtaining an asymptotic expansion for $\phi'(\rho e^{i\theta})$ on a *major arc*

$$\{\theta : |\theta - \theta_0| < K(1 - \rho)\},$$

where K is a large positive constant, and showing that the complementary *minor arc*

$$\gamma = \{\theta : K(1 - \rho) \leq |\theta - \theta_0| \leq \pi\}$$

contributes relatively little to the integral in (5.5). It is this latter result which we prove first.

Lemma 5.3 *Suppose that $\phi(z)$ is defined as in Theorem 5.5, that $\alpha > 0$ and that θ_0 satisfies (2.59). Then given $\eta > 0$, we can choose $K > 0$ and a positive integer n_0 with the following property. If $n > n_0$, there exists ρ in the range $1 - 1/n \leq \rho \leq 1 - 1/(2n)$ such that*

$$\frac{1}{2\pi} \int_\gamma |\phi'(\rho e^{i\theta})| d\theta < \eta n^{2p\lambda},$$

where γ is the minor arc $\{\theta : K(1 - \rho) \leq |\theta - \theta_0| \leq \pi\}$.

We again define t to satisfy the inequalities (5.6) and have

$$\frac{1}{2\pi} \int_\gamma |\phi'(\rho e^{i\theta})| d\theta \leq \lambda \left(\frac{1}{2\pi} \int_\gamma |f'(\rho e^{i\theta})|^2 |f(\rho e^{i\theta})|^{(2-t)\lambda - 2} \right)^{\frac{1}{2}}$$

$$\times \left(\frac{1}{2\pi} \int_\gamma |f(\rho e^{i\theta})|^{\lambda t} d\theta \right)^{\frac{1}{2}}. \quad (5.8)$$

Since $f(z)$ is mean p–valent, $f(z)$ satisfies (3.11) with a suitable constant C and $\beta = 2p$. Thus we may apply Lemma 3.1 with $r = 1 - 1/(2n)$, $\beta = 2p$ and $\lambda(2 - t)$ instead of λ. This gives, for some ρ such that $1 - 1/n \leq \rho \leq 1 - 1/(2n)$,

$$\frac{1}{2\pi} \int_\gamma |f'(\rho e^{i\theta})|^2 |f(\rho e^{i\theta})|^{(2-t)\lambda - 2} d\theta \leq \frac{4pC^{\lambda(2-t)}}{\lambda(2 - t)} (2n)^{2p\lambda(2-t)+1}. \quad (5.9)$$

To estimate the second factor on the right–hand side of (5.8) we note that the number θ_0 of (2.59) certainly has the properties of θ_0 in Theorem 2.6. We may thus apply Theorem 2.7. Hence if K is chosen sufficiently large, depending on ε, we have for $r_0 < \rho < 1$

$$|f(\rho e^{i\theta})| < \frac{1}{(1 - \rho)^\varepsilon |\theta - \theta_0|^{2p - \varepsilon}} \quad (K(1 - \rho) \leq |\theta - \theta_0| \leq \pi).$$

Writing $|\theta - \theta_0| = x$ we deduce

$$\frac{1}{2\pi} \int_\gamma |f(\rho e^{i\theta})|^{\lambda t} d\theta \leq \frac{2(1-\rho)^{-\varepsilon\lambda t}}{2\pi} \int_{K(1-\rho)}^\pi x^{-\lambda t(2p-\varepsilon)} dx.$$

We choose ε so small that $\lambda t(2p - \varepsilon) > 1$, which is possible by (5.6). Then

$$\begin{aligned}
\frac{1}{2\pi} \int_\gamma |f(\rho e^{i\theta})|^{\lambda t} d\theta &\leq \frac{(1-\rho)^{-\varepsilon\lambda t}}{\pi} \int_{K(1-\rho)}^\infty x^{-\lambda t(2p-\varepsilon)} dx \\
&= \frac{K^{1-\lambda t(2p-\varepsilon)}(1-\rho)^{1-\lambda t(2p-\varepsilon)-\varepsilon\lambda t}}{\pi[\lambda t(2p-\varepsilon)-1]} \\
&\leq \frac{K^{1-\lambda t(2p-\varepsilon)}}{\pi[\lambda t(2p-\varepsilon)-1]}(2n)^{2p\lambda t-1}.
\end{aligned} \tag{5.10}$$

We note that when λ, t, C, η and ε have been fixed, K may then be chosen so large that

$$\frac{\lambda^2 4pC^{\lambda(2-t)}}{\lambda(2-t)} \frac{K^{1-\lambda t(2p-\varepsilon)}}{\pi[\lambda t(2p-\varepsilon)-1]} 2^{4p\lambda} < \eta^2.$$

With this choice of K Lemma 5.3 follows from (5.8), (5.9) and (5.10).

5.5 The major arc Our next aim is to find an asymptotic formula for $f(z)$, $f'(z)$ and $\phi(z)$ on the major arc

$$\Gamma = \Gamma_n : |\theta - \theta_0| \leq K(1-\rho) \tag{5.11}$$

of $|z| = \rho$. Thus Γ is complementary to γ. For this purpose we use the asymptotic formulae (2.59) and (2.60) of Theorem 2.10. We define $r_n = 1 - 1/n$. Let $\rho = \rho_n$ be the number of Lemma 5.3 and write

$$z_n = \rho_n e^{i\theta_0}, \ n \geq 1$$

$$\alpha_n = (1 - \rho_n)^{2p} f(z_n), \tag{5.12}$$

and

$$f_n(z) = \frac{\alpha_n}{(1 - ze^{-i\theta_0})^{2p}}. \tag{5.13}$$

We deduce from (2.59), (2.60) that

$$\alpha_n \sim n^{-2p} f(r_n e^{i\theta_0}) \tag{5.14}$$

and

$$|\alpha_n| \to \alpha \text{ as } n \to \infty. \tag{5.15}$$

Using (2.60) we prove

Lemma 5.4 *We have*

$$f(z)/f_n(z) \quad \to \quad 1, \tag{5.16}$$

$$f'(z)/f'_n(z) \quad \to \quad 1 \tag{5.17}$$

uniformly as $n \to \infty$*, while* $z \in \Gamma_n$*.*

We write $z = \rho_n e^{i\theta}$ and recall that by (2.60) of Theorem 2.10 we have

$$\frac{f'(z)}{f(z)} = \frac{2p + o(1)}{e^{i\theta_0} - z},$$

uniformly for z on Γ_n as $n \to \infty$. Integrating along Γ_n from $z = z_n$ to $\rho_n e^{i\theta}$, we obtain

$$\begin{aligned}
\log \frac{f(\rho_n e^{i\theta})}{f(z_n)} &= 2p \log \left(\frac{e^{i\theta_0} - z_n}{e^{i\theta_0} - \rho_n e^{i\theta}} \right) + o(1) \\
&= 2p \log \left(\frac{1 - \rho_n}{1 - \rho_n e^{i(\theta - \theta_0)}} \right) + o(1),
\end{aligned}$$

since $|e^{i\theta_0} - z| \geq 1 - \rho_n \geq 1/(2n)$ and the length of Γ_n is at most $2K(1 - \rho_n) \leq 2K/n$, by (5.11). Thus

$$\frac{f(\rho_n e^{i\theta}) \left\{ 1 - \rho_n e^{i(\theta - \theta_0)} \right\}^{2p}}{f(z_n)(1 - \rho_n)^{2p}} \to 1.$$

Writing $z = \rho_n e^{i\theta}$ and recalling (5.12) and (5.13) we obtain (5.16).

Next we deduce from (5.13) and (2.60) that

$$\frac{f'(z)}{f(z)} \frac{f_n(z)}{f'_n(z)} = \frac{f'(z)}{f(z)} \frac{e^{i\theta_0} - z}{2p} \to 1$$

as $z \to e^{i\theta_0}$ on the union of the arcs Γ_n. Multiplying this by (5.16) we obtain (5.17). This proves Lemma 5.4. An alternative approach generalising the method of Lemmas 1.2 and 1.3 is provided for c.m.p–valent functions by Theorem 5.1 (see Example 5.5).

5.6 Proof of Theorem 5.5 We now conclude the proof of Theorem 5.5 when $\alpha > 0$, by proving the stronger

Theorem 5.6 *Suppose that* $\phi(z) = z^\mu \sum_{n=-\infty}^{+\infty} b_n z^n$ *satisfies the hypotheses of Theorem 5.5, that* $\alpha > 0$ *and that* θ_0 *satisfies* (2.59)*. Then*

$$nb_n \sim \frac{\phi \left[\left(1 - \frac{1}{n} \right) e^{i\theta_0} \right] e^{-i(n+\mu)\theta_0}}{\Gamma(2p\lambda)} \quad (n \to +\infty).$$

We have by (5.14), (5.15)

$$\left| \phi \left[\left(1 - \frac{1}{n} \right) e^{i\theta_0} \right] \right| = \left| f \left[\left(1 - \frac{1}{n} \right) e^{i\theta_0} \right] \right|^\lambda \sim \alpha^\lambda n^{2p\lambda} \ (n \to +\infty).$$

Thus Theorem 5.5 follows at once from Theorem 5.6. The latter result is significantly stronger, since it gives information about $\arg b_n$ as well as $|b_n|$.

We suppose, as we may without loss of generality, that $\theta_0 = 0$, since otherwise we may consider $f(ze^{i\theta_0})$, $\phi(ze^{i\theta_0})$ instead of $f(z)$, $\phi(z)$. Write

$$(1 - z)^{-2p\lambda} = \sum_{n=0}^{\infty} c_n z^n.$$

Then

$$\begin{aligned} c_n &= \frac{2p\lambda(2p\lambda + 1)\ldots(2p\lambda + n - 1)}{1.2\ldots n} \\ &= \frac{\Gamma(n + 2p\lambda)}{\Gamma(n + 1)\Gamma(2\lambda p)} \sim \frac{n^{2p\lambda - 1}}{\Gamma(2p\lambda)} \ (n \to +\infty). \end{aligned} \tag{5.18}$$

We also set $\phi_n(z) = f_n(z)^\lambda$, so that

$$\phi_n(z) = \alpha_n^\lambda \sum_{m=0}^{\infty} c_m z^m,$$

and hence

$$nc_n\alpha_n^\lambda = \frac{1}{2\pi i} \int_{|z|=\rho} \frac{\phi_n'(z)dz}{z^n} = \frac{\rho^{1-n}}{2\pi} \int_{-\pi}^{\pi} \phi_n'(\rho e^{i\theta})e^{-i(n-1)\theta}d\theta \ (0 < \rho < 1).$$

Finally we recall (5.5):

$$(n + \mu)b_n = \frac{\rho^{1-n-\mu}}{2\pi} \int_{-\pi}^{+\pi} \phi'(\rho e^{i\theta})e^{-i(n+\mu-1)\theta}d\theta \ (1 - 2\delta < \rho < 1).$$

Thus

$$2\pi\rho^{n-1}[(n + \mu)\rho^\mu b_n - nc_n\alpha_n^\lambda]$$

$$= \int_{-\pi}^{\pi} [\phi'(\rho e^{i\theta})e^{-i\mu\theta} - \phi_n'(\rho e^{i\theta})]e^{-i(n-1)\theta}d\theta. \tag{5.19}$$

We now suppose that $\eta > 0$ and choose K so large that for $n > n_0$ we can find $\rho = \rho_n$ in the range $1 - \frac{1}{n} \le \rho \le 1 - \frac{1}{2n}$ such that

$$\int_{\gamma} |\phi'(\rho e^{i\theta})|d\theta < \eta n^{2p\lambda},$$

where γ denotes the arc $K(1-\rho) \le |\theta| \le \pi$. This is possible by Lemma 5.3.

Next we note that

$$|\phi'_n(z)| = |2p\lambda\alpha_n^\lambda||1-z|^{-2p\lambda-1}.$$

Also for $z = \rho e^{i\theta}$, where $\frac{1}{2} \le \rho < 1$, $K(1-\rho) \le |\theta| \le \pi$, we have

$$\begin{aligned}
|1-z|^2 &= 1 - 2\rho\cos\theta + \rho^2 \\
&= (1-\rho)^2 + 4\rho\sin^2\frac{\theta}{2} \ge \frac{2\theta^2}{\pi^2},
\end{aligned}$$

while $|2p\lambda\alpha_n| < C_1$. Here $C_1, C_2, \ldots$ denote constants independent of K and n. Thus

$$\int_\gamma |\phi'_n(\rho e^{i\theta})|d\theta \le C_2 \int_\gamma |\theta|^{-1-2p\lambda}d\theta = 2C_2 \left[\frac{-\theta^{-2p\lambda}}{2p\lambda}\right]_{K(1-\rho)}^\pi.$$

Thus if $\rho = \rho_n$ we deduce that

$$\int_\gamma |\phi'_n(\rho e^{i\theta})|d\theta \le \frac{C_2[K(1-\rho)]^{-2p\lambda}}{p\lambda} \le \frac{C_2(\frac{1}{2}K)^{-2p\lambda}}{p\lambda}n^{2p\lambda},$$

since $(1-\rho) \ge 1/(2n)$. Hence if K is sufficiently large we obtain

$$\int_\gamma |\phi'_n(\rho e^{i\theta})|d\theta < \eta n^{2p\lambda}.$$

Thus

$$\left|\int_\gamma \left[\phi'(\rho e^{i\theta})e^{-i\mu\theta} - \phi'_n(\rho e^{i\theta})\right]e^{-i(n-1)\theta}d\theta\right|$$

$$\le \int_\gamma \left[|\phi'(\rho e^{i\theta})| + |\phi'_n(\rho e^{i\theta})|\right] d\theta < 2\eta n^{2p\lambda}, \qquad (5.20)$$

if K is sufficiently large, depending on η.

Having fixed K we now apply Lemma 5.4 on the complementary arc $\gamma' = \Gamma_n : 0 \le |\theta| \le K(1-\rho)$. Since $\phi(z) = f(z)^\lambda$, $\phi'(z) = \lambda f'(z)f(z)^{\lambda-1}$, and similarly $\phi'_n(z) = \lambda f'_n(z)f_n(z)^{\lambda-1}$, we deduce from Lemma 5.4 that, uniformly on γ',

$$\phi'(\rho e^{i\theta})e^{-i\mu\theta} \sim \phi'(\rho e^{i\theta}) \sim \phi'_n(\rho e^{i\theta}),$$

where $\rho = \rho_n$. Thus

$$\phi'(\rho e^{i\theta})e^{-i\mu\theta} - \phi'_n(\rho e^{i\theta}) = o\{\phi'_n(\rho e^{i\theta})\} = o(1-\rho)^{-1-2p\lambda} = o(n^{2p\lambda+1})$$

as $n \to \infty$. Hence

$$\left| \int_{\gamma'} [\phi'(\rho e^{i\theta}) e^{-i\mu\theta} - \phi'_n(\rho e^{i\theta})] e^{-i(n-1)\theta} d\theta \right|$$

$$\leq \int_{\gamma'} |\phi'(\rho e^{i\theta}) e^{-i\mu\theta} - \phi'_n(\rho e^{i\theta})| d\theta$$

$$= o(n^{2p\lambda+1}) 2K(1-\rho) = o(n^{2p\lambda}). \tag{5.21}$$

Since γ, γ' together make up the whole range $-\pi \leq \theta \leq \pi$, we deduce from (5.19), (5.20) and (5.21) that

$$|2\pi\rho^{n-1}[(n+\mu)\rho^\mu b_n - nc_n\alpha_n^\lambda]| < [2\eta + o(1)]n^{2p\lambda}$$

for large n. Also

$$\rho^{n-1} \geq \left(1 - \frac{1}{n}\right)^n \geq \frac{1}{4} \quad (n \geq 2),$$

η is arbitrarily small, and $\rho^\mu \to 1$ as $n \to +\infty$. Finally, using (5.14) and (5.18) we have

$$b_n = \frac{nc_n\alpha_n^\lambda}{(n+\mu)\rho^\mu} + o(n^{2p\lambda-1})$$

$$\sim \frac{n^{2p\lambda-1}\alpha_n^\lambda}{\Gamma(2p\lambda)} \sim \frac{[f(r_n)]^\lambda}{n\Gamma(2p\lambda)} = \frac{\phi(r_n)}{n\Gamma(2p\lambda)} \quad (n \to +\infty),$$

where $r_n = 1 - 1/n$. This proves Theorem 5.6.

5.7 Applications: the case $\lambda = 1$ The case $\lambda = 1$ of Theorem 5.5 is of most interest, since it refers directly to the coefficients of mean p–valent functions. If

$$f(z) = \sum_0^\infty a_n z^n$$

is mean p–valent in $|z| < 1$ and $p > \frac{1}{4}$, then

$$\frac{|a_n|}{n^{2p-1}} \to \frac{\alpha}{\Gamma(2p)} \quad (n \to \infty),$$

where α is the constant of Theorem 2.10. Whenever we can obtain bounds for α, we obtain correspondingly sharp bounds for the asymptotic growth of the coefficients. Thus we have for instance

Theorem 5.7 *Suppose that p is a positive integer and that*

$$f(z) = z^p + \sum_{n=p+1}^{\infty} a_n z^n$$

is mean p–valent in $|z| < 1$. Then

$$\lim_{n \to \infty} \frac{|a_n|}{n^{2p-1}} = \frac{\alpha}{(2p-1)!},$$

where $\alpha < 1$, except when $f(z) = z^p(1 - z e^{i\theta})^{-2p}$.

In fact the limiting relation holds by Theorem 5.5 and

$$\alpha = \lim_{r \to 1-} (1 - r)^{2p} M(r, f).$$

Also, by Theorem 2.11, $\alpha < 1$ except for the functions $z^p(1 - z\, e^{i\theta})^{2p}$. The proof was suggested in Example 2.11. For univalent functions the result is contained in Theorem 1.12. However the conclusion of Theorem 5.7 now extends to real p, such that $p > \frac{1}{4}$.

We have similarly

Theorem 5.8 *Suppose that $f(z) = \sum_{n=0}^{\infty} a_n z^n$ is c.m.p–valent and $f(z) \neq 0$ in $|z| < 1$, where $p > \frac{1}{4}$. Then*

$$\lim_{n \to \infty} \frac{|a_n|}{n^{2p-1}} = \frac{\alpha}{\Gamma(2p)},$$

where $\alpha \leq |a_0| 4^p$ with equality only for the functions

$$a_0 \left(\frac{1 + z e^{i\theta}}{1 - z e^{i\theta}} \right)^{2p}.$$

The limiting relation is again a consequence of Theorem 5.5. The inequality for α follows from the last statement of Theorem 5.1.

5.8 Functions with k–fold symmetry   Suppose now that k is a positive integer and that

$$f_k(z) = a_k z^k + a_{2k} z^{2k} + \cdots + a_{nk} z^{nk} + \cdots$$

is circumferentially or areally mean p–valent in $|z| < 1$. Then

$$\phi(z) = f_k(z^{1/k}) = a_k z + a_{2k} z^2 + \cdots$$

is mean (p/k)–valent there and conversely. For to each root z_0 of the equation $\phi(z) = w$ there correspond exactly k roots of the equation $f_k(z) = w$, given by $z^k = z_0$. Thus for every w

$$n[w, f_k(z)] = kn[w, \phi(z)],$$

and so for every positive R

$$p[R, \phi] = \frac{1}{k} p[R, f_k].\tag{5.22}$$

More generally if v is real and

$$f_k(z) = z^v + a_{k+v}z^{k+v} + a_{2k+v}z^{2k+v} + \cdots\tag{5.23}$$

in $0 < |z| < 1$ so that $|f(z)|$ is one-valued, then if

$$\phi(z) = f_k(z^{1/k})$$

the equation (5.22) still holds. To see this we note that each arc Γ of a level curve $|\phi(z)| = R$, which lies in $\theta_0 \le \arg z \le \theta_0 + 2\pi$, $0 < |z| < 1$, and on which $\arg \phi(z)$ increases by δ, corresponds by $z = w^k$ to k level curves γ_v, one of which lies in each sector $(\theta_0 + 2v\pi)/k < \arg w \le (\theta_0 + 2(v+1)\pi)/k$ and on each of which $\arg f_k(w)$ also increases by δ. The γ_v contribute a total of $k\delta/(2\pi)$ to $p[R, f_k]$, while Γ contributes $\delta/(2\pi)$ to $p[R, \phi]$ and this yields (5.22). Thus, if $f_k(z)$ is mean p–valent in $0 < |z| < 1$, $\phi(z)$ is mean p/k–valent and we may apply Theorem 5.5 with p/k instead of p, $\lambda = 1$ and $\phi(z)$ instead of f. Also

$$
\begin{aligned}
\alpha &= \lim_{r \to 1}(1 - r)^{2p/k} M(r, \phi)\\
&= \lim_{r \to 1}(1 - r^k)^{2p/k} M(r^k, \phi)\\
&= \lim_{r \to 1}(1 - r^k)^{2p/k} M(r, f_k)\\
&= k^{2p/k} \lim_{r \to 1}(1 - r)^{2p/k} M(r, f_k)
\end{aligned}\tag{5.24}
$$

and $b_n = a_{nk+v}$. We deduce

Theorem 5.9 *If $f_k(z)$, given by (5.23), is mean p–valent in $|z| < 1$, where $1 \le k < 4p$, then*

$$\lim_{n \to \infty}\frac{|a_{nk+v}|}{n^{2p/k-1}} = \frac{\alpha}{\Gamma(2p/k)},$$

where α is given by (5.24).

We note the special case $p = v$. In this case $\phi(z)$ satisfies the hypotheses of Theorem 2.11 with p/k instead of p. In particular $\alpha \le 1$. We deduce

Theorem 5.10 *Suppose that*

$$f_k(z) = z^p + a_{p+k}z^{p+k} + a_{p+2k}z^{p+2k} + \cdots$$

is mean p–valent in $|z| < 1$ *and that* $1 \le k < 4p$. *Then*

$$\lim_{n \to \infty} \frac{|a_{p+nk}|}{n^{2p/k-1}} = \frac{\alpha}{\Gamma(2p/k)},$$

and $\alpha \le 1$, *with equality only for the mean p–valent functions*

$$f_k(z) = z^p(1 - z^k e^{i\theta})^{-2p/k}.$$

In fact by Theorem 2.11 we have $\alpha \le 1$ with equality only when

$$\phi(z) = z^{p/k}(1 - ze^{i\theta})^{-2p/k}$$

so that

$$f_k(z) = z^p(1 - z^k e^{i\theta})^{-2p/k}.$$

We note the special case of an odd univalent function

$$f_2(z) = z + a_3 z^3 + a_5 z^5 + \cdots.$$

Taking $p = 1$, $k = 2$, we deduce from Theorem 5.10 that

$$\lim_{n \to \infty} |a_{2n+1}| < 1$$

in this case, except when $f_2(z) = z + z^3 e^{i\theta} + z^5 e^{2i\theta} + \cdots$. In particular,

$$|a_{2n+1}| \le 1 \quad (n > n_0(f)).$$

Nevertheless, for any fixed $n \ge 2$ an odd univalent function $f_2(z)$ can be found[†] for which $|a_{2n+1}| > 1$. We remark finally that if $f_k(z)$ is c.m.p–valent or univalent and given by (5.23) with $v = 1$, then

$$f(z) = \left\{ f_k(z^{1/k}) \right\}^k = \phi(z)^k = z \left(1 + \sum_{n=1}^{\infty} a_{kn+1} z^n \right)^k$$

is also c.m.p–valent or univalent respectively and conversely. For c.m.p–valent functions the result follows from Lemma 5.1 (a). For univalent functions we recall the argument for the proof of Theorem 1.1 for the case $k = 2$. For a.m.p–valent functions $f(z)$ we can still, using Lemma 5.1 (b), assert that $f_k(z) = f(z^k)^{1/k}$ is a.m.p–valent but the argument for the converse fails. That is why, in order to prove Theorems 5.9 and 5.10, we found it convenient to work with $\phi(z)$ directly instead of $\phi(z)^k = f(z)$.

[†] Schaeffer and Spencer [1943] (see also Duren [1983, p. 107]). Earlier Fekete and Szegö [1933] showed that the exact upper bound for $|a_5|$ in this case is $1/2 + e^{-2/3} = 1.01\ldots$ We shall prove this result in Chapter 7.

5.9 Some further results Some inequalities for the class $\mathfrak{F}$ of functions

$$f(z) = z + a_2 z^2 \cdots$$

circumferentially mean 1–valent in $|z| < 1$, which go beyond Theorem 5.3, have been proved by Jenkins [1958]. He obtained the exact upper bound for $|f(r)|$, when $0 < r < 1$ and either $|a_2|$ or $|f(-\rho)|$ is given, where ρ is fixed and $0 < \rho < 1$. In either case the extremals are univalent functions with real coefficients. It follows from Theorem 1.10 that for these extremals $|a_n| \le n$ and so

$$
\begin{aligned}
|f(\rho)| + |f(-\rho)| = f(\rho) - f(-\rho) &= 2(\rho + a_3\rho^3 + a_5\rho^5 \cdots) \\
&\le 2\sum_{n=0}^{\infty}(2n+1)\rho^{2n+1},
\end{aligned}
$$

$$|f(\rho)| + |f(-\rho)| \le \frac{\rho}{(1+\rho)^2} + \frac{\rho}{(1-\rho)^2}.$$

This latter inequality remains valid for all $f(z) \in \mathfrak{F}$, and hence $|a_3| \le 3$ for $f(z) \in \mathfrak{F}$. An unpublished example of Spencer shows this to be false in general for a.m. 1–valent functions, though $|a_2| \le 2$ remains true. We shall prove $|a_3| \le 3$ for univalent $f(z)$ by Löwner's original method in the next chapter.

Jenkins' results also imply that if $|a_2|$ is given and $f(z) \in \mathfrak{F}$ then

$$\alpha = \lim_{r\to 1-}(1-r)^2 M(r,f) \le \psi(|a_2|) = 4b^{-2}\exp(2 - 4b^{-1}),$$

where $b = 2 - (2 - |a_2|)^{\frac{1}{2}}$, and this inequality is sharp for $0 \le |a_2| \le 2$.

If $f(z) = z^p + a_{p+1}z^{p+1} + \cdots$ is c.m.p–valent in $|z| < 1$, then $[f(z)]^{1/p} \in \mathfrak{F}$, and from this Jenkins deduced the sharp inequality $|a_{p+2}| \le p(2p+1)$ in this case. It also follows that the number α in Theorem 5.10 satisfies the inequality

$$\alpha \le \left\{ \psi\left(\frac{k|a_{p+k}|}{p}\right) \right\}^{p/k},$$

which is again sharp for given $|a_{p+k}|$. These results are described in Jenkins [1958].

Regularity theorems for the means $I_\lambda(r,f)$ and their derivatives will be found in Hayman [1955] at least for the case of c.m.p–valent functions. (See Examples 5.2 and 5.4 below.)

Finally mention should be made of the following conjecture by A. W. Goodman [1948].

Suppose that p is a positive integer and that

$$f(z) = \sum_1^{\infty} b_n z^n$$

is p-valent in $|z| < 1$, i.e. that the equation $f(z) = w$ never has more than p roots in $|z| < 1$. Then for $n > p$

$$|b_n| \leq \sum_{k=1}^{p} \frac{2k(n+p)!}{(p+k)!(p-k)!(n-p-1)!(n^2-k^2)} |b_k|.$$

Goodman showed that equality is possible for arbitrary values of $|b_1|$ to $|b_p|$, with $f(z)$ a polynomial of degree p in $z/(1-z)^2$. The case $p = 1$ is de Branges' Theorem. For $p > 1$ only a few general results are known. If

$$a_1 = a_2 = \cdots = a_{p-1} = 0$$

the conjecture holds for $n = p+1$ [Spencer 1941b] and $n = p+2$ [Jenkins 1958]. The simplest general open case is

$$|b_3| \leq 5|b_1| + 4|b_2|$$

for 2-valent functions. Goodman and Robertson [1951] proved the conjecture for functions typically real of order $p^{\dagger}$. Livingston proved it for functions close to convex of order p for $n = p + 1$ in [1965] and for general n if $a_1 = a_2 = \cdots = a_{p-2} = 0$ [1969]. These results are all sharp. Now that de Branges' Theorem is known, Goodman's conjecture provides a fascinating challenge.

Examples

5.1   If $0 \leq R < 1$ let A be the cut annulus

$$A : R < |w| < 1, \quad |\arg w| < \pi$$

and let $f(z)$ map $|z| < 1$ uniformly onto A. Show that, if $q > 0$, the function

$$f_q(z) = f(z)^q$$

is c.m.p-valent in $|z| < 1$ and it is a.m.p-valent for $p = q(1-R^{2q})$, but for no smaller value of p. By comparing $f_q(z)$, $f_q(z)^n$ show that ηp cannot be replaced by any smaller number in Lemma 5.1 (a), and that $\eta_0 p$ cannot be replaced by any smaller number in Lemma 5.1 (b), (choose $R = 0$ or R close to 1).

---

† For definitions of these classes of functions we refer to the quoted papers or the interesting historical account by Goodman [1979]

5.2   If k is a positive integer prove that $f(z) \in S$ if and only if $f_k(z) = f(z^k)^{1/k} \in S$, and $f_k(z)$ is of the form (5.23) with $v = 1$.

5.3   If $f(z) = (1 - z)^{-2p}$, where $p > 0$, prove that, if $p\lambda > \frac{1}{2}$,

$$I_\lambda(r, f) \sim \frac{\Gamma(\lambda p - \frac{1}{2})}{2\sqrt{\pi}\Gamma(\lambda p)}(1 - r)^{1 - 2p\lambda}$$

as $r \to 1$. (Use Example 1.2 at the end of Chapter 1).

5.4   Prove that, with the hypotheses of Theorem 5.5, and if $p\lambda > \frac{1}{2}$,

$$(1 - r)^{2p\lambda - 1}I_\lambda(r, f) \to \frac{\Gamma(\lambda p - \frac{1}{2})}{2\sqrt{\pi}\Gamma(\lambda p)}\alpha^\lambda,$$

as $r \to 1$.

(Harder). Show also that, if $p\lambda = \frac{1}{2}$,

$$\frac{I_\lambda(r, f)}{\log \frac{1}{1-r}} \to \frac{\alpha^\lambda}{\pi}, \ r \to 1.$$

(Apply Theorem 3.1 with $f(z)^{\lambda/2}$ instead of $f(z)$ and $\lambda = 2$, considering separately the cases $\alpha = 0$, $\alpha > 0$.)

Corresponding results for the derivatives can also be obtained in suitable cases. We confine ourselves to the case of $I_2(r)$, where complete results are possible.

5.5   Prove that with the hypotheses of Theorem 5.5, we have, if $p\lambda > \frac{1}{4}$ and $q = 1, 2, \ldots$

$$(1 - r)^{4p\lambda + 2q - 1}I_2(r, \phi^{(q)}) \to \alpha^{2\lambda}\frac{\Gamma(2p\lambda + q)\Gamma(2p\lambda + q - \frac{1}{2})}{2\sqrt{\pi}\Gamma(2p\lambda)^2}.$$

Extend this result to $p > 0$, for $q = 1$ and hence for $q \geq 1$ by using Lemma 1.1 and Theorem 3.1. (If $q = 1$, $\alpha > 0$ and $kp\lambda \leq \frac{1}{4}$, mimic the argument for Lemma 1.4.)

5.6   If $f(z)$ is regular and c.m.p–valent and $f(z) \neq 0$ in $1 - 2\delta < |z| < 1$, prove that

$$\left|\frac{f'(z)}{f(z)}\right| \leq \frac{4p\delta}{(1 - |z|)(|z| + 2\delta - 1)}, \ 1 - \delta < |z| < 1,$$

and deduce that $((1 - r)/(r + 2\delta - 1))^{2p}|f(pe^{i\theta})|$ for fixed θ and $((1 - r)/(r + 2\delta - 1))^{2p}M(r, f)$ are decreasing functions of θ, for $1 - \delta \leq r < 1$. (Apply Theorem 5.1 to $f(z_0 + \delta\zeta)$, where $z_0 = (1 - \delta)e^{i\theta}$). Hence show how to extend the proof of Theorem 1.12 to c.m.p–valent functions.

6

Differences of successive coefficients

6.0 Introduction In this chapter we consider a function

$$f(z) = \sum a_n z^n. \tag{6.1}$$

We suppose that $p > \frac{1}{4}$, since for $p \leq \frac{1}{4}$ we cannot improve on Theorem 3.5. We also suppose that either

(i) $f(z)$ is regular and areally mean p–valent in $\Delta : |z| < 1$, in which case n goes through the integers from 0 to ∞ in (6.1),

or

(ii) if p is not an integer we also allow the possibility, as in Theorem 2.11, that $f(z)/z^p$ is regular in Δ and that $f(z)$ is mean p–valent in Δ, cut along a radius $z = \rho e^{i\theta}$, $0 \leq \rho < 1$. In this case $n = m + p$ in (6.1) where m goes through the integers from 0 to ∞.

With these hypotheses we shall obtain estimates for

$$\big||a_{n+1}| - |a_n|\big|.$$

We note that if $f(z)$ is a pth power of a Koebe function

$$f(z) = \frac{z^p}{(1-z)^{2p}},$$

where $p \neq \frac{1}{2}$ and $n = p + m$, we have

$$
\begin{aligned}
a_{n+1} - a_n &= \frac{\Gamma(2p+m)}{\Gamma(2p-1)\Gamma(m+2)} \\
&\sim \frac{m^{2p-2}}{\Gamma(2p-1)} \sim \frac{n^{2p-2}}{\Gamma(2p-1)}, \quad \text{as } n \to \infty
\end{aligned}
$$

while if

$$f(z) = \frac{z^p}{(1-z^2)^p},$$

165

and $n = p + m$, where m is even, then $a_{n+1} = 0$, and

$$a_n = |a_{n+1} - a_n| = \frac{\Gamma(p + \frac{1}{2}m)}{\Gamma(p)\Gamma(1 + \frac{1}{2}m)} \sim \frac{n^{p-1}}{2^{p-1}\Gamma(p)}.$$

This suggests the conjecture that

$$\big||a_{n+1}| - |a_n|\big| = O(n^{2p-2}), \quad p \geq 1 \tag{6.2}$$

$$\big||a_{n+1}| - |a_n|\big| = O(n^{p-1}), \quad \tfrac{1}{2} \leq p < 1 \tag{6.3}$$

and, if $p < \frac{1}{2}$,

$$\big||a_{n+1}| - |a_n|\big| = o(n^{-\frac{1}{2}}). \tag{6.4}$$

In fact the examples of Section 3.4 show that we cannot in any case expect more than (6.4).

We shall in this chapter prove (6.2). It is not known whether (6.3) is true. We prove the best known result, namely

$$\big||a_{n+1}| - |a_n|\big| = O(n^{2p-2\sqrt{p}}), \quad \tfrac{1}{4} < p \leq 1 \tag{6.5}$$

while (6.4) holds for $p < \frac{1}{4}$. For $p = \frac{1}{4}$ Theorem 3.5 yields

$$\big||a_{n+1}| - |a_n|\big| = O(n^{-\frac{1}{2}} \log n)$$

and we are unable to sharpen this.

The above results are due to Hayman [1963] when $p = 1$ and to Lucas [1969] for other values of p.

Suppose that

$$f_2(z) = z + c_3 z^3 + \cdots$$

is univalent and odd in Δ. Then (cf. Section 3.7)

$$f(z) = f_2(z^{\frac{1}{2}}) = z^{\frac{1}{2}}(1 + c_3 z + c_5 z^2 + \cdots)$$

is mean $\frac{1}{2}$-valent and so we obtain

$$\big||c_{2n+1}| - |c_{2n-1}|\big| = O(n^{1-\sqrt{2}})$$

in this case. This estimate due to Lucas [1969] is the best that is known but El Hosh [1984] has shown that if f is close to convex, and in particular starlike, we have

$$\big||c_{2n+1}| - |c_{2n-1}|\big| = O(n^{-\frac{1}{2}}). \tag{6.6}$$

As we saw above this order of magnitude is attained for the 4–symmetric function

$$f_2(z) = \frac{z}{(1 - z^4)^{\frac{1}{2}}}, \tag{6.7}$$

which is starlike and univalent. Shen [1992] has shown that (6.6) also holds if $f \in S$ and

$$M(r, f_2) = O(1-r)^{-\frac{1}{2}}.$$

He obtains a slightly weaker conclusion if, for some θ_0

$$\liminf_{r \to 1}(1-r)^{\frac{1}{2}}|f(re^{i\theta_0})| > 0.$$

Both these conditions are satisfied by the function (6.7). If

$$\lim_{r \to 1}(1-r)M(r, f_2) = \alpha > 0$$

Shen proves, using Baernstein's technique (cf. Section 3.5.3), that

$$\left|\,|c_{2n+1}| - |c_{2n-1}|\,\right| = O(n^{-\frac{1}{2} - \frac{1}{320}}).$$

These results show that functions $f_2(z)$ failing to satisfy (6.6) are likely to have $M(r, f_2)$ oscillating around the order $(1-r)^{-\frac{1}{2}}$.

The proof of (6.2) and (6.5) will occupy Sections 6.1–6.6. In Section 6.7 we consider k–symmetric univalent functions, in 6.8 asymptotic behaviour, in 6.9 starlike univalent functions, and in 6.10 the results of Shen.

6.1 The basic formalism   We suppose from now on that f is mean p–valent in Δ and of the form (6.1). We define

$$\mu = \sum_{\nu \leq p} |a_\nu|.$$

Next we choose a fixed n such that $n \geq 6$ and write

$$r_n = 1 - \tfrac{1}{n}, \text{ and } M_1 = M(r_n, f) = \max_{|z|=r_n} |f(z)|. \tag{6.8}$$

We also choose $z_1 = r_n e^{i\theta_0}$ so that $|f(z_1)| = M_1$. Then we define

$$M_\nu = 2^{1-\nu} M_1, \quad 1 \leq \nu < \infty. \tag{6.9}$$

Next we write

$$E_\nu = E_\nu(n)$$

$$= \{z \mid 1 - \tfrac{3}{n} < |z| < 1 - \tfrac{2}{n}, \text{ and } M_{\nu+1} < |f(z)| \leq M_\nu\}. \tag{6.10}$$

We note that, by (6.1), we have

$$(n-1)a_{n-1} - nz_1 a_n = (n-1)(a_{n-1} - e^{i\theta_0} a_n)$$

$$= \frac{1}{2\pi i} \int_{|z|=\rho} \frac{(z-z_1)f'(z)dz}{z^n}, \quad 1 - \tfrac{3}{n} \leq \rho \leq 1 - \tfrac{2}{n}.$$

Thus

$$(n-1)|a_{n-1} - e^{i\theta_0}a_n| \leq \frac{1}{2\pi\rho^n}\int_0^{2\pi}|\rho e^{i\theta} - z_1||f'(\rho e^{i\theta})|\rho d\theta. \qquad (6.11)$$

Also, since $n \geq 6$,

$$\rho^{-n} \leq \left(1 - \frac{3}{n}\right)^{-n} \leq \left(1 - \frac{1}{2}\right)^{-6} = 64 \quad \text{and} \quad \frac{n}{n-1} \leq \frac{6}{5}.$$

We integrate (6.11) from $\rho = 1 - 3/n$ to $\rho = 1 - 2/n$. This yields

$$|a_{n-1} - e^{i\theta_0}a_n| \leq \frac{32n}{(n-1)\pi}\int_0^{2\pi}d\theta\int_{1-3/n}^{1-2/n}|\rho e^{i\theta} - z_1||f'(\rho e^{i\theta})|\rho d\rho d\theta$$

$$< \frac{40}{\pi}\sum_{v=1}^{\infty}\int\int_{E_v}|z - z_1||f'(z)|\rho d\rho d\theta$$

$$< \frac{40}{\pi}\sum_{v=1}^{\infty}\left(\int\int_{E_v}\left|\frac{f'(z)}{f(z)}\right|^2\rho d\rho d\theta\right)^{\frac{1}{2}}\left(\int\int_{E_v}|z - z_1|^2|f(z)|^2\rho d\rho d\theta\right)^{\frac{1}{2}},$$

by Schwarz's inequality.

We note that, by (6.10) and since f is mean p–valent in E_v,

$$\int\int_{E_v}\left|\frac{f'(z)}{f(z)}\right|^2\rho d\rho d\theta \leq \left(\frac{1}{M_{v+1}}\right)^2\int\int_{E_v}|f'(z)|^2\rho d\rho d\theta$$

$$\leq \pi p\frac{M_v^2}{M_{v+1}^2} = 4\pi p.$$

Thus we obtain

$$|a_{n-1} - e^{i\theta_0}a_n|$$

$$< 80\left(\frac{p}{\pi}\right)^{\frac{1}{2}}\left\{\sum_{v=1}^{\infty}\int\int_{E_v}\rho d\rho\int|z - z_1|^2|f(\rho e^{i\theta})|^2 d\theta\right\}^{\frac{1}{2}}. \qquad (6.12)$$

In E_v we have $|f(z)| \leq M_v$ and so

$$|f(z)|^2 \leq \frac{2M_v^2|f(z)|^2}{M_v^2 + |f(z)|^2}.$$

Thus (6.12) yields finally

$$\big||a_n| - |a_{n-1}|\big| \leq |a_{n-1} - e^{i\theta_0}a_n|$$

$$\leq 80\sqrt{\left(\frac{2p}{\pi}\right)}\sum_{v=1}^{\infty}\left\{\int_{1-\frac{3}{n}}^{1-\frac{2}{n}}|z - z_1|^2\frac{M_v^2|f(z)|^2}{M_v^2 + |f(z)|^2}\rho d\rho d\theta\right\}^{\frac{1}{2}}. \qquad (6.13)$$

6.2 An application of Green's formula   We need to transform the right-hand side of (6.13) further.

Lemma 6.1   *Suppose that $v(z) = v(x, y)$ is twice continuously differentiable in $0 < |z| \le \rho$, and that v remains continuous at $z = 0$ and $t \, \partial v(te^{i\theta})/\partial t \, \log(1/t) \to 0$ as $t \to 0$. If*

$$\nabla^2 = \left(\frac{\partial}{\partial x}\right)^2 + \left(\frac{\partial}{\partial y}\right)^2,$$

we have

$$\int_0^{2\pi} v(\rho e^{i\theta})d\theta = 2\pi v(0) + \int_0^{\rho} dt \int_0^{2\pi} \nabla^2 v(te^{i\theta})t \log\frac{\rho}{t}d\theta \qquad (6.14)$$

and so, for $\frac{1}{2} \le \rho < 1$,

$$\begin{aligned}
\int_0^{2\pi} v(\rho e^{i\theta})d\theta &= \int_0^{2\pi} v(\tfrac{1}{2}e^{i\theta})d\theta \\
&+ \left\{\int_0^{\frac{1}{2}} t(\log 2\rho)dt + \int_{\frac{1}{2}}^{\rho} t\log\frac{\rho}{t}dt\right\} \int_0^{2\pi} \nabla^2 v(te^{i\theta})d\theta.
\end{aligned}$$

$$(6.15)$$

We write

$$I(t) = \int_0^{2\pi} v(te^{i\theta})d\theta,$$

and apply Green's formula, Lemma 4.2, to v in the annulus $\varepsilon < t < s$. This yields

$$sI'(s) - \varepsilon I'(\varepsilon) = \int_\varepsilon^s t \, dt \int_0^{2\pi} \nabla^2 v(te^{i\theta})d\theta, \quad 0 < \varepsilon < s < \rho.$$

We divide by s and integrate both sides from $s = \varepsilon$ to $s = \rho$. We obtain

$$\begin{aligned}
I(s) - I(\varepsilon) - \varepsilon I'(\varepsilon)\log\frac{\rho}{\varepsilon} &= \int_\varepsilon^\rho \frac{ds}{s} \int_\varepsilon^s t \, dt \int_0^{2\pi} \nabla^2 v(te^{i\theta})d\theta \\
&= \int_{t=\varepsilon}^\rho t \, dt \int_{s=t}^\rho \frac{ds}{s} \int_0^{2\pi} \nabla^2 v(te^{i\theta})d\theta \\
&= \int_{t=\varepsilon}^\rho t\log\frac{\rho}{t} \int_0^{2\pi} \nabla^2 v(te^{i\theta})d\theta.
\end{aligned}$$

We let ε tend to zero, and note that by hypothesis

$$\varepsilon I'(\varepsilon) \log \frac{\rho}{\varepsilon} \to 0, \quad I(\varepsilon) \to v(0)$$

in this case and now (6.14) follows. We deduce (6.15) by subtraction.

We shall apply Lemma 6.1 to the right-hand side of (6.13). To do so we need

Lemma 6.2   *If $f(z)$ is regular in $0 < |z| < 1$, $|f(z)| = R$ and*

$$G(R) = \frac{M^2 R^2}{M^2 + R^2}, \tag{6.16}$$

then

$$\nabla^2 G(R) = \frac{4M^4(M^2 - R^2)}{(M^2 + R^2)^3} |f'(z)|^2. \tag{6.17}$$

If

$$v(z) = |z - z_1|^2 G(R), \tag{6.18}$$

then

$$\nabla^2 v(z) \le 4G(R) + |z - z_1|^2 \nabla^2 G(R) + \frac{8|z - z_1| M^4 R}{(M^2 + R^2)^2} |f'(z)|. \tag{6.19}$$

Also if $f(z)$ is given by (6.1) subject to the conditions (i) and (ii) of the introduction, $v(z)$ and $G(R)$ satisfy the hypotheses of Lemma 6.1.

We write $R = |f(z)| = e^u$, so that u is harmonic in $0 < |z| < 1$, except where $R = 0$. Also

$$\frac{\partial}{\partial x} G(R) = \frac{\partial}{\partial x} G(e^u) = e^u \frac{\partial u}{\partial x} G'(e^u), \tag{6.20}$$

$$\left(\frac{\partial}{\partial x} \right)^2 G(R) = [e^{2u} G''(e^u) + e^u G'(e^u)] \left(\frac{\partial u}{\partial x} \right)^2 + e^u G'(e^u) \frac{\partial^2 u}{\partial x^2}.$$

Differentiating similarly with respect to y and adding we have

$$\nabla^2 G(R) = [R^2 G''(R) + R G'(R)] \left\{ \left(\frac{\partial u}{\partial x} \right)^2 + \left(\frac{\partial u}{\partial y} \right)^2 \right\}$$

except at points where $R = 0$. Also

$$\frac{\partial u}{\partial x} - i \frac{\partial u}{\partial y} = \frac{d}{dz} \log f(z) = \frac{f'(z)}{f(z)}, \tag{6.21}$$

so that

$$\left(\frac{\partial u}{\partial x}\right)^2 + \left(\frac{\partial u}{\partial y}\right)^2 = \frac{|f'(z)|^2}{R^2}.$$

Thus

$$\nabla^2 G(R) = \left[G''(R) + \frac{G'(R)}{R}\right]|f'(z)|^2,$$

except perhaps where $R = 0$. Using (6.16) we obtain (6.17) when $R \neq 0$ and the result remains true by continuity at points where $R = 0$. Again (6.20), (6.21) yield for $|z| = \rho$ in case (ii)

$$\left|\left(\frac{\partial}{\partial x} - i\frac{\partial}{\partial y}\right) G(R)\right| = |f'(z)|G'(R)$$

$$= O(\rho^{p-1})\rho^p = O(\rho^{2p-1}), \text{ as } \rho \to 0$$

and $G(R) = O(\rho^{2p})$. Thus $G(R)$ satisfies the hypotheses of Lemma 6.1. Also

$$\frac{\partial}{\partial x}v(z) = |z - z_1|^2 \left|\frac{\partial}{\partial x}G(R)\right| + 2(x - x_1)G(R) = O(\rho^{2p-1})$$

and

$$\frac{\partial}{\partial y}v(z) = O(\rho^{2p-1})$$

similarly. Thus $v(\rho e^{i\theta}) \to 0$ and

$$\rho \log \rho \frac{\partial}{\partial \rho}v(\rho e^{i\theta}) = O(\rho^{2p})\log\frac{1}{\rho} \to 0$$

as $\rho \to 0$ in case (ii), and in case (i) all the partials of v are continuous at $z = 0$. Thus $v(z)$ satisfies the hypotheses of Lemma 6.1.

It remains to verify (6.19). We set $X = |z - z_1|^2$, $Y = G(R)$ and note that

$$\nabla^2(XY) = Y\nabla^2 X + X\nabla^2 Y + 2\left(\frac{\partial X}{\partial x}\frac{\partial Y}{\partial x} + \frac{\partial X}{\partial y}\frac{\partial Y}{\partial y}\right)$$

$$= 4G(R) + |z - z_1|^2\nabla^2 G(R) + 2\left(\frac{\partial X}{\partial x}\frac{\partial Y}{\partial x} + \frac{\partial X}{\partial y}\frac{\partial Y}{\partial y}\right). \tag{6.22}$$

By Schwarz's inequality

$$\left(\frac{\partial X}{\partial x}\frac{\partial Y}{\partial x} + \frac{\partial X}{\partial y}\frac{\partial Y}{\partial y}\right)^2 \leq \left[\left(\frac{\partial X}{\partial x}\right)^2 + \left(\frac{\partial X}{\partial y}\right)^2\right]\left[\left(\frac{\partial Y}{\partial x}\right)^2 + \left(\frac{\partial Y}{\partial y}\right)^2\right]$$

$$= 4|z - z_1|^2 G'(R)^2|f'(z)|^2$$

by (6.20) and (6.21). Substituting in (6.22) and noting that

$$G'(R) = \frac{2M^4R}{(M^2 + R^2)^2},$$

we deduce (6.19).

6.3 Estimates for the first term in (6.19) In order to prove (6.2) and (6.5) we apply (6.13) and estimate the terms on the right-hand side. For each ρ, such that

$$1 - \frac{3}{n} \leq \rho \leq 1 - \frac{2}{n},$$

we apply Lemma 6.1 with $v(z)$ given by (6.18) and $M = M_v$. We obtain three integrals corresponding to the three terms on the right-hand side of (6.19). We denote by C a positive constant depending on p only, not necessarily the same each time, and by $C_1, C_2, \ldots$ particular constants C.

Lemma 6.3 *Suppose that $f(z)$ is as in (6.1), mean p–valent in $0 < |z| < 1$ and normalized so that*

$$\mu = \sum_{v \leq p} |a_v| = 1; \tag{6.23}$$

further that $G(R)$ is given by (6.16) with $R = |f(z)|$, $z = re^{i\theta}$. Then we have for $0 < r < 1$

$$\int_0^{2\pi} G(|f(re^{i\theta})|)d\theta \leq C_1 \min\left\{M^{(4p-1)/(2p)}, (1-r)^{1-4p}\right\}, \quad p > \frac{1}{4} \tag{6.24}$$

and

$$\int_0^\rho (1-r)r dr \int_0^{2\pi} G(|f(re^{i\theta})|)d\theta$$

$$\leq C_2 \min\left\{M^{(4p-3)/(2p)}, (1-\rho)^{3-4p}\right\}, \quad p > \frac{3}{4}. \tag{6.25}$$

We apply Lemmas 6.1, 6.2 with $v(z) = G(R)$ and use (6.17). Suppose first that $M < 1$. Then

$$G(R) < M^2 < M^{2-1/(2p)} < (1-r)^{1-4p}.$$

Thus (6.24) holds with $C_1 = 2\pi$. Suppose next that $M > 1$ and $r \leq \frac{1}{2}$. Then $|f(re^{i\theta})| < C$ by Theorems 2.3 and 2.11 so that $G(|f(re^{i\theta})|) < C$ and (6.24) follows again. Suppose next that $\frac{1}{2} < r < 1$, and

$$1 \leq M \leq (1-r)^{-2p}.$$

Let F_ν be the set of points in $\frac{1}{2} \leq |z| \leq r$, where $M2^{-\nu} < |f(z)| \leq M2^{1-\nu}$. Then Theorems 2.3 and 2.11 yield

$$|f(z)| < C_3(1 - |z|)^{-2p}. \tag{6.26}$$

Thus in F_ν we have

$$M2^{-\nu} < C_3(1 - |z|)^{-2p}$$

i.e.

$$|z| > 1 - C_4 M^{-1/(2p)} 2^{\nu/(2p)}, \quad \text{where } C_4 = C_3^{1/(2p)}.$$

Since also $t \log(r/t) < r - t < 1 - t$, we have for $z = te^{i\theta}$ in F_ν

$$\log \frac{r}{t} < \frac{1-t}{t} < 2(1-t) < 2C_4(2^\nu/M)^{1/2p}.$$

Also, if $R = |f(z)|$, we have

$$\frac{M^4(M^2 - R^2)}{(M^2 + R^2)^3} < 1.$$

Thus

$$\int \int_{F_\nu} \frac{4M^4(M^2 - R^2)}{(M^2 + R^2)^3} |f'(z)|^2 \log \frac{r}{t} t \, dt \, d\theta$$

$$< \; 8C_4 \int \int_{F_\nu} |f'(z)|^2 \left(\frac{2^\nu}{M}\right)^{1/(2p)} t \, dt \, d\theta$$

$$< \; 8C_4 \left(\frac{2^\nu}{M}\right)^{1/(2p)} p\pi (M2^{1-\nu})^2$$

$$= \; C_5(M2^{-\nu})^{2 - 1/(2p)}, \tag{6.27}$$

since $f(z)$ is mean p-valent in F_ν. Summing from $\nu = 1$ to ∞ we obtain, writing $z = te^{i\theta}$, $R = |f(z)|$

$$\int_{\frac{1}{2}}^{r} \int_0^{2\pi} \frac{4M^4(M^2 - R^2)}{(M^2 + R^2)^3} |f'(z)|^2 t \log \frac{r}{t} \, dt \, d\theta < C_6 M^{2 - 1/(2p)}, \quad p > \frac{1}{4}. \tag{6.28}$$

To estimate the integral over $|z| < \frac{1}{2}$, we note that by (6.26)

$$|f(z)| < C_7, \quad \text{for } |z| < \frac{3}{4}$$

so that Cauchy's inequality yields in case (i)

$$|f'(z)| < 4C_7, \quad |z| < \frac{1}{2}.$$

In case (ii) we have by Schwarz's Lemma

$$\left| \frac{f(z)}{z^p} \right| < \left(\frac{4}{3} \right)^p C_7, \quad |z| < \frac{3}{4}.$$

This yields

$$\left| \frac{f'(z)}{z^p} - p \frac{f(z)}{z^{p+1}} \right| < \frac{4^{p+1}}{3^p} C_7, \quad |z| < \frac{1}{2}$$

i.e.

$$|f'(z)| < C_8 |z|^{p-1}, \quad |z| < \frac{1}{2}.$$

Hence we have in all cases in $|z| < \frac{1}{2}$

$$|f'(z)| < C_9 |z|^\gamma, \quad \text{where } \gamma = \inf(0, p-1).$$

Thus

$$\int_0^{2\pi} \int_0^{\frac{1}{2}} |f'(z)|^2 t \log \frac{r}{t} dt d\theta < 2\pi C_9^2 \int_0^{\frac{1}{2}} t^{2\gamma+1} \log \frac{1}{t} dt < C_{10}.$$

On combining this with (6.28) we obtain

$$\int \int_{|z|<r} \frac{M^4(M^2 - R^2)}{(M^2 + R^2)^2} |f'(z)|^2 t \log \frac{r}{t} dt d\theta < C_{11} M^{2-1/(2p)}, \quad p > \frac{1}{4}.$$

On applying Lemma 6.1, with $v(z) = G(|f(z)|)$ and using (6.14) and $v(0) < |f(0)|^2 \le 1$ we obtain (6.24) if $p > \frac{1}{4}$ and $M \le (1-r)^{-2p}$. If $M > (1-r)^{-2p}$, (6.26) yields for $|z| \le r$

$$R \le M(r, f) < C_3(1-r)^{-2p} = C_3 M_0$$

say, and

$$G(R) = \frac{M^2 R^2}{(M^2 + R^2)} < R^2 \le \frac{M_0^2 R^2}{M_0^2 + R^2}(1 + C_3^2) = (1 + C_3^2)G_0(R)$$

say. Applying (6.24) with M_0 instead of M we obtain

$$\begin{aligned} \int_0^{2\pi} G(R) d\theta &\le (1 + C_3^2) \int_0^{2\pi} G_0(R) d\theta \\ &< C_{12} M_0^{2-1/(2p)} = C_{12}(1-r)^{1-4p}. \end{aligned}$$

This concludes the proof of (6.24).

To obtain (6.25) we integrate (6.24). Suppose first that $M \le 1$. Then

$$G(R) < M^2 \le M^{(4p-3)/(2p)},$$

so that (6.25) holds with $C_2 = \pi/3$. Next if $M \geq (1 - \rho)^{-2p}$, we deduce from (6.24) that

$$\int_0^\rho (1 - r)dr \int_0^{2\pi} G(|f(re^{i\theta})|)d\theta$$

$$< C_1 \int_0^\rho (1 - r)^{2-4p}dr < \frac{C_1}{4p - 3}(1 - \rho)^{3-4p} \quad (6.29)$$

as required. Finally if $1 < M < (1 - \rho)^{-2p}$, we define ρ_0 by

$$M = (1 - \rho_0)^{-2p}, \text{ so that } 0 < \rho_0 < \rho.$$

Then (6.24) and (6.29) applied with ρ instead of ρ_0 yield

$$\int_0^\rho (1 - r)dr \int_0^{2\pi} G(|f(re^{i\theta})|)d\theta = \int_0^{\rho_0} + \int_{\rho_0}^\rho$$

$$\leq \frac{C_1}{4p - 3}(1 - \rho_0)^{3-4p} + C_1 \int_{\rho_0}^\rho M^{(4p-1)/(2p)}(1 - r)dr$$

$$< C_1 \left\{ \frac{1}{4p - 3}(1 - \rho_0)^{3-4p} + \frac{1}{2}(1 - \rho_0)^2 M^{(4p-1)/(2p)} \right\} = C_2 M^{(4p-3)/(2p)}.$$

This proves (6.25) and concludes the proof of Lemma 6.3. We leave the case $\frac{1}{4} < p \leq \frac{3}{4}$ as an exercise.

Examples

6.1   Prove that, if $M \geq 1$,

$$\int_0^\rho (1 - r)rdr \int_0^{2\pi} G(|f(re^{i\theta})|)d\theta$$

$$< \begin{cases} C_{13} \min\{\log(eM), \log \frac{1}{1-\rho}\}, & p = \frac{3}{4} \\ C_{13}, & \frac{1}{4} < p < \frac{3}{4}. \end{cases}$$

6.2   Prove that if $M < 1, p > 0$

$$\int_0^\rho (1 - r)rdr \int_0^{2\pi} G(|f(re^{i\theta})|)d\theta < \frac{\pi}{3}M^2.$$

6.4 A 2–point estimate   In order to deal with the second and third terms on the right-hand side of (6.19) we need a two point estimate for mean p–valent functions, which has independent interest.

Theorem 6.1   *Suppose that $f(z)$ satisfies the hypotheses (i) or (ii) of Section 6.0 and is normalized so that (6.23) holds. If $z_1 = \rho_1 e^{i\theta_1}$, $z_2 = \rho_2 e^{i\theta_2}$ are two points in $|z| < 1$, such that*

$$|f(z_1)| \leq |f(z_2)| \tag{6.30}$$

and a, b are positive numbers, then we have

$$|f(z_1)|^{(a^2+2ab)/2p}|f(z_2)|^{b^2/2p}$$

$$< C_0(p, a, b)\frac{1}{(1 - \rho_1)^{a^2}}\frac{1}{(1 - \rho_2)^{b^2}}\frac{1}{|z_1 - z_2|^{2ab}}, \tag{6.31}$$

where the constant $C_0(p, a, b)$ depends only on a, b and p.

The result is due to Hayman [1963] when $a = b$, and in the general case to Lucas [1968]. To prove Theorem 6.1 we shall use the case $k = 2$ of Theorem 2.4. In the proof $C_{14}, C_{15}, \ldots$ will denote constants depending on p, a and b. Suppose first that

$$|f(z_2)| = |f(z_1)| = R, \text{ and that } \rho_1 \leq \rho_2. \tag{6.32}$$

We define

$$\eta = |e^{i\theta_2} - e^{i\theta_1}|, \tag{6.33}$$

and first prove (6.31) with η instead of $|z_2 - z_1|$. We may also assume that

$$\eta \geq 3^{p+2}(1 - \rho_1). \tag{6.34}$$

For if this is false we have by (6.26)

$$|f(z_2)|^{(a^2)/2p}|f(z_1)|^{(2ab+b^2)/(2p)} = R^{(a+b)^2/2p} \leq M(\rho_1, f)^{(a+b)^2/2p}$$

$$\leq C_3^{(a+b)^2/2p}(1 - \rho_1)^{-(a+b)^2} \leq \frac{C_{14}}{(1 - \rho_1)^{a^2}(1 - \rho_2)^{b^2}\eta^{2ab}}, \tag{6.35}$$

which is the desired conclusion.

We suppose therefore from now on that (6.34) holds, so that $\rho_2 \geq \rho_1 > \frac{1}{2}$, since $\eta \leq 2$. We choose r_0 to be the smallest number such that

$$\frac{1}{2} \leq 1 - \frac{\eta}{4} \leq r_0 \leq 1 - 3^{-(p+2)}\eta \leq \rho_1, \tag{6.36}$$

and

$$f(z) \neq 0 \quad \text{for} \quad r_0 - \tfrac{1}{2}(1 - r_0) < |z| < r_0 + \tfrac{1}{2}(1 - r_0). \quad (6.37)$$

We start by showing that (6.36) and (6.37) are compatible.

We recall (see Section 2.3) that, since $f(z)$ is mean p–valent, $f(z)$ has q zeros in $0 < |z| < 1$, where $q \leq p$. Thus at least one of the $q + 1$ annuli

$$\frac{\eta}{8}3^{-\nu} < (1 - |z|) < \frac{\eta}{8}3^{1-\nu}, \quad \nu = 0 \text{ to } q$$

is free from zeros of $f(z)$. We choose the smallest associated value of ν and define

$$r_0 = 1 - \frac{1}{4}\eta 3^{-\nu}.$$

Thus (6.37) holds. Also

$$1 - r_0 \geq \frac{\eta}{4}3^{-q} \geq 3^{-(p+2)}\eta \geq 1 - \rho_1$$

by (6.34), while $r_0 \geq 1 - \frac{\eta}{4} \geq \frac{1}{2}$. This yields (6.36).

We now apply Theorem 2.4 to the discs

$$\Delta_n : \{|z - r_0 e^{i\theta_n}| < 1 - r_0\}, \quad n = 1, 2.$$

These discs are disjoint since, by (6.36),

$$r_0|e^{i\theta_2} - e^{i\theta_1}| = r_0\eta \geq \tfrac{1}{2}\eta \geq 2(1 - r_0).$$

Also by (6.37) $f(z) \neq 0$ in $|z - r_0 e^{i\theta_j}| < \tfrac{1}{2}(1 - r_0)$ and $f(z)$ is mean p–valent in $\Delta_1 \bigcup \Delta_2$. We take z_1, z_2, $r_0 e^{i\theta_1}$, $r_0 e^{i\theta_2}$, $(1 - r_0)$, instead of z'_1, z'_2, z_1, z_2, r_n in Theorem 2.4. Then

$$\delta_n = \frac{1 - \rho_n}{1 - r_0}, \quad (6.38)$$

and

$$R_2 = |f(z_2)| = |f(z_1)| = R$$

by (6.32) while

$$2^p|f(r_0 e^{i\theta_j})| \leq 2^p M(r_0, f) = R_1 \quad (6.39)$$

say. Now Theorem 2.4 yields, if $R_2 > eR_1$,

$$\left(\log \frac{A_0}{\delta_1}\right)^{-1} + \left(\log \frac{A_0}{\delta_2}\right)^{-1} < \frac{2p}{\log(R_2/R_1) - 1}. \quad (6.40)$$

We write

$$\log \frac{A_0}{\delta_1} = \frac{X}{a^2}, \quad \log \frac{A_0}{\delta_2} = \frac{Y}{b^2}$$

and note that, by Schwarz's inequality,

$$(X+Y)\left(\frac{a^2}{X}+\frac{b^2}{Y}\right) = a^2\frac{Y}{X}+b^2\frac{X}{Y}+a^2+b^2$$

$$\geq 2ab\left(\frac{Y}{X}\cdot\frac{X}{Y}\right)^{\frac{1}{2}}+a^2+b^2 = (a+b)^2,$$

with equality when $Y/X = b/a$, i.e. $(A_0/\delta_2)^b = (A_0/\delta_1)^a$. Now (6.40) yields

$$\frac{(a+b)^2}{X+Y} < \frac{2p}{\log(R_2/eR_1)}.$$

Taking exponentials and using (6.26), (6.36), (6.38) and (6.39) we obtain

$$\begin{aligned}
R^{(a+b)^2/2p} &< (eR_1)^{(a+b)^2/2p}\left(\frac{A_0}{\delta_1}\right)^{a^2}\left(\frac{A_0}{\delta_2}\right)^{b^2}\\
&< C_{15}(1-r_0)^{-(a+b)^2}\left(\frac{1-r_0}{1-\rho_1}\right)^{a^2}\left(\frac{1-r_0}{1-\rho_2}\right)^{b^2}\\
&= C_{15}(1-r_0)^{-2ab}(1-\rho_1)^{-a^2}(1-\rho_2)^{-b^2}\\
&< C_{16}\eta^{-2ab}(1-\rho_1)^{-a^2}(1-\rho_2)^{-b^2}.
\end{aligned} \tag{6.41}$$

We have assumed that $R > eR_1$. If $R \leq eR_1$ we have

$$\begin{aligned}
R^{(a+b)^2/2p} &\leq C_{17}(1-r_0)^{-(a+b)^2}\\
&\leq C_{17}(1-r_0)^{-2ab}(1-\rho_1)^{-a^2}(1-\rho^2)^{-b^2}
\end{aligned}$$

since $r_0 \leq \rho_1 \leq \rho_2$ by (6.32) and (6.36), so that (6.41) always holds.

6.4.1 The case $|f(z_1)| < |f(z_2)|$ To complete the proof of Theorem 6.1, we now suppose that $|f(z_1)| < |f(z_2)|$. We recall (6.26), so that

$$|f(\rho e^{i\theta})| < C_3(1-\rho)^{-2p}, \ 0 \leq \rho < 1, \ 0 \leq \theta \leq 2\pi. \tag{6.42}$$

Since $f(z) = z^p(1-z)^{-2p}$ satisfies our hypotheses, $C_3 \geq 1$. Suppose now that

$$|f(z_1)| \leq C_3(2 \times 3^{p+1})^{2p}.$$

Then (6.31) is a consequence of

$$|f(z_2)| < C_3(1-\rho_2)^{-2p}.$$

So we suppose from now on that

$$2^p|f(z_1)| = R_1, \ |f(z_2)| = R_2$$

where

$$R_2 > 2^{-p}R_1 > C_3(2 \times 3^{p+1})^{2p} > 1. \tag{6.43}$$

Then $|f(0)| < R_1$ by (6.23) and so we can find $z_1' = \rho_1' e^{i\theta_2}$, such that

$$2^p|f(z_1')| = R_1 \text{ and } 1 - \tfrac{1}{2}3^{-(p+1)} < \rho_1' < \rho_2$$

by (6.42) and (6.43). We choose the smallest such ρ_1', and define

$$\delta = 1 - \rho_1', \text{ so that } 3^{p+1}\delta < \tfrac{1}{2}. \tag{6.44}$$

We note that, since f has q zeros in $0 < |z| < 1$ where $q \le p$, at least one of the annuli

$$\tfrac{1}{2}3^v\delta < (1 - |z|) < \tfrac{1}{2}3^{v+1}\delta, \ v = 0, 1, \ldots, q$$

is free from zeros of $f(z)$. We choose the smallest such v, set

$$z_0 = (1 - 3^v\delta)e^{i\theta_2},$$

and apply Theorem 2.4 to the single disc $|z - z_0| < 3^v\delta$, with z_0, z_2 instead of z_1, z_1'. Then

$$2^p|f(z_0)| \le R_1, \ |f(z_2)| = R_2, \ \delta_1 = \frac{1 - \rho_2}{3^v\delta}$$

and $f(z) \ne 0$ for $|z - z_0| < \tfrac{1}{2}3^v\delta$. Thus the hypotheses of Theorem 2.4 are satisfied and we obtain, using (6.44), either $R_2 < eR_1$ or

$$R_2 < eR_1\left(\frac{A_0}{\delta_1}\right)^{2p} < C_{18}R_1\left(\frac{1 - \rho_1'}{1 - \rho_2}\right)^{2p}, \tag{6.45}$$

so that (6.45) holds in any case. We next apply (6.41) with z_1' instead of z_2. This is legitimate since $|f(z_1')| = 2^pR_1 = |f(z_1)|$ and we have proved (6.41) in this case. We obtain

$$R_1^{(a+b)^2/2p} < C_{16}\frac{1}{(1 - \rho_1)^{a^2}(1 - \rho_1')^{b^2}\eta^{2ab}}.$$

Combining this with (6.45) we obtain

$$\begin{aligned} R_1^{(a^2+2ab)/(2p)}R_2^{b^2/(2p)} &= R_1^{(a+b)^2/(2p)}(R_2/R_1)^{b^2/(2p)} \\ &< C_{19}(1/(1 - \rho_1)^{a^2}(1 - \rho_2)^{b^2}\eta^{2ab}). \end{aligned} \tag{6.46}$$

To complete the proof of Theorem 6.1 we need to replace η by $|z_2 - z_1|$ in (6.46). Suppose first that $|\rho_2 - \rho_1| \le \eta$. Then

$$|z_2 - z_1| \le |\rho_1 e^{i\theta_1} - \rho_2 e^{i\theta_1}| + |\rho_2 e^{i\theta_2} - \rho_2 e^{i\theta_1}| \le |\rho_2 - \rho_1| + \eta < 2\eta$$

so that (6.31) follows. Suppose next that

$$\eta < |\rho_2 - \rho_1| < \max(1 - \rho_2, 1 - \rho_1).$$

If $\rho_2 \leq \rho_1$, we have $\eta < (1 - \rho_2)$, and $|z_2 - z_1| < 2(1 - \rho_2)$. Then (6.42) yields

$$\begin{aligned}
R_1^{(a^2+2ab)/(2p)} &\leq R_2^{(a^2+2ab)/(2p)} < C_3^{(a^2+2ab)/(2p)}(1 - \rho_2)^{-(a^2+2ab)} \\
&< C_3^{(a^2+2ab)/(2p)} 2^{2ab}(1 - \rho_1)^{-a^2}|z_2 - z_1|^{-2ab}.
\end{aligned}$$

If $\rho_1 < \rho_2$, we have $|z_2 - z_1| < 2(1 - \rho_1)$. Hence

$$\begin{aligned}
R_1^{(a^2+2ab)/(2p)} &< C_3^{(a^2+2ab)/(2p)}(1 - \rho_1)^{-a^2-2ab} \\
&< C_3^{(a^2+2ab)/(2p)} 2^{2ab}(1 - \rho_1)^{-a^2}|z_2 - z_1|^{-2ab}. \quad (6.47)
\end{aligned}$$

So (6.47) holds in both cases. Also, again by (6.42),

$$R_2^{b^2/(2p)} < C_3^{b^2/(2p)}(1 - \rho_2)^{-b^2}.$$

On combining this with (6.47) we obtain (6.31) and the proof of Theorem 6.1 is complete.

6.5 Statement of the basic theorem   We now state our result on coefficient differences.

Theorem 6.2   *Suppose that $f(z)$, given by (6.1), is mean p–valent in the sense* (i) *or* (ii) *of the introduction. Then for $n \geq 1$*

$$||a_{n+1}| - |a_n|| \leq \begin{cases} C_{20}\,\mu\,n^{2p-2}, & \text{if } p \geq 1 \\ C_{20}\,\mu\,n^{2p-2}\sqrt{p}, & \text{if } \frac{1}{4} < p < 1 \end{cases}$$

where $\mu = \sum_{v \leq p} |a_v|$, and the constant C_{20} depends only on p.

We continue to assume that $\mu = 1$, since otherwise we may consider $f(z)/\mu$ instead of $f(z)$.

In this section we deal with the second and third term on the right-hand side of (6.19) by means of Theorem 6.1. We shall put our results together to prove Theorem 6.2 in the next section. We define

$$s(p) = \begin{cases} 4p - 3, & p \geq 1 \\ (2\sqrt{p} - 1)^2, & \frac{1}{4} < p < 1 \end{cases} \quad (6.48)$$

and

$$\gamma(p) = s(p)/(4p). \quad (6.49)$$

Lemma 6.4 *We have*

$$\int\int |z - z_1|^2 |f'(z)|^2 (1-r) r dr d\theta < C_{21} n^{s(p)} 2^{-2\gamma\nu}. \tag{6.50}$$

Here the integral is taken over all points, $z = re^{i\theta}$ for which $|f(z)| < M_\nu$, and $0 \le r < r_n$.

Let F_k be the set of points in $|z| < r_n = 1 - 1/n$, for which

$$M_{k+1} \le |f(z)| < M_k.$$

We recall the definitions (6.8) and (6.9), and have $|f(z_1)| = M_1$ and $|f(z)| \ge M_{k+1} = M_1 2^{-k}$ in F_k. Thus Theorem 6.1 applied with z, z_1 instead of z_1, z_2 gives

$$M_{k+1}^{(a^2+2ab)/(2p)} M_1^{b^2/(2p)} < \frac{C_0 n^{b^2}}{(1-r)^{a^2}|z_1 - z|^{2ab}}. \tag{6.51}$$

Suppose first that $p \ge 1$. In this case we choose $a = b = 1$ and obtain in F_k

$$(1-r)|z_1 - z|^2 < C_0 n M_{k+1}^{-3/(2p)} M_1^{-1/(2p)} = C_0 n M_1^{-2/p} 2^{3k/(2p)}.$$

Also since f is mean p–valent and $|f| < M_k$ in F_k we have

$$\int\int_{F_k} |f'(z)|^2 r dr d\theta < \pi p M_k^2 = 4\pi p M_1^2 2^{-2k}. \tag{6.52}$$

Thus

$$\int\int_{F_k} |f'(z)|^2 |z - z_1|^2 (1-r) r dr d\theta < 4 C_0 n M_1^{-2/p} 2^{3k/(2p)} \pi p M_1^2 2^{-2k}$$

$$= 4 C_0 n \pi p M_1^{2-2/p} 2^{(3/(2p)-2)k} = 4 C_0 n \pi p M_1^{2-2/p} 2^{-2\gamma k}.$$

Summing from $k = \nu$ to ∞ we obtain

$$\int\int_{\substack{|z|<r_n \\ |f(z)|<M_\nu}} |f'(z)|^2 |z - z_1|^2 (1-r) r dr d\theta < 4 C_0 n \pi p M_1^{2-2/p} \frac{2^{-2\gamma\nu}}{1 - 2^{-2\gamma}}.$$

Also, by (6.26), $M_1 < C_3 n^{2p}$. Now (6.50) follows.

Suppose next that $\frac{1}{4} < p < 1$. We apply (6.51) with $a = 1$, $b = 2\sqrt{p}-1$. This yields in F_k

$$M_1^2 2^{-k((4\sqrt{p}-1)/(2p))} = M_{k+1}^{(4\sqrt{p}-1)/(2p)} M_1^{(2\sqrt{p}-1)^2/(2p)} < \frac{C_0 n^{s(p)}}{(1-r)|z-z_1|^{4\sqrt{p}-2}},$$

so that

$$(1-r)|z-z_1|^2 \le 4(1-r)|z-z_1|^{4\sqrt{p}-2} < 4 C_0 n^{s(p)} 2^{k(4\sqrt{p}-1)/(2p)} M_1^{-2},$$

since $0 < 4 - 4\sqrt{p} < 2$, and $|z - z_1| < 2$. Using this and (6.52) we obtain

$$\int\int_{F_k} |f'(z)|^2 |z - z_1|^2 (1 - r) r dr d\theta < 16\pi p C_0 n^{s(p)} 2^{-k(2\sqrt{p}-1)^2/(2p)}$$
$$= 16\pi p C_0 n^{s(p)} 2^{-2\gamma k}.$$

Summing from $k = v$ to ∞ we obtain (6.50) also in this case. This completes the proof of Lemma 6.4.

It remains to deal with the third term on the right-hand side of (6.19).

Lemma 6.5 *If $z = re^{i\theta}$, $|f(z)| = R$ and $M = M_v$ we have*

$$I_v = \int\int_{|z|<r_n} |z - z_1| |f'(z)| \frac{M^4 R}{(M^2 + R^2)^2} (1 - r) r dr d\theta$$
$$< C_{22} n^{s(p)} 2^{-\gamma v}. \tag{6.53}$$

We have by Schwarz's inequality

$$I_v < \left(\int\int_{|z|<r_n} |z - z_1|^2 |f'(z)|^2 (1 - r) r dr d\theta \right)^{\frac{1}{2}}$$
$$\times \left(\int\int_{|z|<r_n} \frac{M^8 R^2}{(M^2 + R^2)^4} (1 - r) r dr d\theta \right)^{\frac{1}{2}}. \tag{6.54}$$

For the first integral we apply Lemma 6.4 with $v = 1$ and obtain

$$\int\int_{|z|<r_n} |z - z_1|^2 |f'(z)|^2 (1 - r) r dr d\theta < C_{21} n^{s(p)}. \tag{6.55}$$

To estimate the second integral we note that

$$\frac{M^8 R^2}{(M^2 + R^2)^4} \leq \frac{M^2 R^2}{(M^2 + R^2)} = G(R),$$

and apply Lemma 6.3 and Examples 6.1 and 6.2. Suppose first that $M \geq 1$. If $p > \frac{3}{4}$, we note that $s(p) \geq 4p - 3$. Then by (6.26)

$$M^{(4p-3)/(2p)} \leq M^{s(p)/(2p)} = (M_1 2^{1-v})^{s(p)/(2p)} < C_{23} n^{s(p)} 2^{-2\gamma v}. \tag{6.56}$$

If $p \leq \frac{3}{4}$ we use Example 6.1. Then

$$\max\{1, \log(eM)\} < C_{24} M^{s(p)/(2p)} < C_{24} C_{23} n^{s(p)} 2^{-2\gamma v}.$$

If $M < 1$ we use Example 6.2. Again $M^2 < M^{s(p)/(2p)} < C_{23} n^{s(p)} 2^{-2\gamma v}$. Thus in all cases

$$\int\int_{|z|<r_n} \frac{M^8 R^2}{(M^2 + R^2)^4} (1 - r) r dr d\theta < C_{25} n^{s(p)} 2^{-2\gamma v}.$$

On combining this with (6.54) and (6.55) we obtain (6.53).

6.6 Proof of Theorem 6.2 We write $z = re^{i\theta}$, $R = |f(z)|$ and

$$v_\nu(z) = |z - z_1|^2 G_\nu(R),$$

where

$$G_\nu(R) = \frac{M_\nu^2 R^2}{M_\nu^2 + R^2}.$$

Suppose that $1 - 3/n < \rho < 1 - 2/n$. Then, by (6.15) of Lemma 6.1,

$$\int_0^{2\pi} v_\nu(\rho e^{i\theta})d\theta - \int_0^{2\pi} v_\nu \left(\tfrac{1}{2}e^{i\theta}\right) d\theta$$

$$= \int_0^{\frac{1}{2}} \left(\log \frac{\rho}{\max\left(t, \frac{1}{2}\right)}\right) t\,dt \int_0^{2\pi} \nabla^2 v_\nu(te^{i\theta})d\theta.$$

We note that if $\frac{1}{2} < t < \rho$

$$\log \frac{\rho}{\max\left(t, \frac{1}{2}\right)} = \log \frac{\rho}{t} < \frac{\rho - t}{t} < 2(1 - t)$$

and since $\log(2\rho) < \log 2 < 1$, this inequality remains valid for $t < \frac{1}{2}$. We deduce from (6.17) that $0 < \nabla^2 G_\nu(R) < 4|f'(z)|^2$, if $|f(z)| < M_\nu$, and $\nabla^2 G_\nu(R) \leq 0$ otherwise.

We define $\chi_\nu(z) = 1$, where $|f(z)| < M_\nu$, and $\chi_\nu(z) = 0$ elsewhere. Now (6.19) yields

$$\int_0^{2\pi} v_\nu(\rho e^{i\theta})d\theta - \int_0^{2\pi} v_\nu \left(\tfrac{1}{2}e^{i\theta}\right) d\theta$$

$$\leq 2 \int_0^\rho (1 - r)r\,dr \int_0^{2\pi} \left\{ 4G_\nu(R) + 4|z - z_1|^2 \chi_\nu(z)|f'(z)|^2 \right.$$

$$\left. + \frac{8|z - z_1|M_\nu^4 R}{(M_\nu^2 + R^2)^2}|f'(z)| \right\} d\theta$$

$$= I_{\nu,1} + I_{\nu,2} + I_{\nu,3} \tag{6.57}$$

say. By Lemma 6.3, (6.25) and Example 6.1 we have, if $M_\nu \geq 1$,

$$\int_0^\rho (1 - r)r\,dr \int_0^{2\pi} G_\nu(R)d\theta < \begin{cases} C_2 M_\nu^{(4p-3)/(2p)}, & p > \frac{3}{4} \\ C_{13} \log(eM_\nu), & p \leq \frac{3}{4} \end{cases}.$$

Using (6.56) and noting that $s(p) > 0$, we deduce that

$$I_{v,1} < C_{26}M_v^{s(p)/(2p)} < C_{26}C_{23}n^{s(p)}2^{-\gamma v}.$$

If $M_v < 1$ we have $G_v(R) < M_v^2$ so that

$$
\begin{aligned}
I_{v,1} \quad &< \quad 16M_v^2\pi \int_0^1 (1-t)t\,dt = \frac{8M_v^2\pi}{3} \\
&< \quad \frac{8\pi}{3}M_v^{s(p)/(2p)} < \frac{8\pi}{3}C_{23}n^{s(p)}2^{-\gamma v}.
\end{aligned}
$$

Thus in all cases

$$I_{v,1} < C_{27}n^{s(p)}2^{-\gamma v}.$$

Next Lemma 6.4 shows that

$$I_{v,2} = 8\int\int \chi_v(z)|z - z_1|^2|f'(z)|^2r(1-r)dr d\theta < 8C_{21}n^{s(p)}2^{-\gamma v}.$$

Finally Lemma 6.5 shows that

$$I_{v,3} < 16C_{22}n^{s(p)}2^{-\gamma v}.$$

Now (6.57) yields

$$\int_0^{2\pi} v_v(\rho e^{i\theta})d\theta < \int_0^{2\pi} v_v\left(\tfrac{1}{2}e^{i\theta}\right)d\theta + C_{28}n^{s(p)}2^{-\gamma v}.$$

Again for $|z| = \frac{1}{2}$ we have by (6.26) and (6.56) if $M_v \geq 1$

$$
\begin{aligned}
v_v(z) < 4R^2 \quad &< \quad 4C_3^2(16)^p < 4C_3^2(16)^p M_v^{s(p)/(2p)} \\
&< \quad (16)^{p+1}C_3^2 C_{23}n^{s(p)}2^{-2\gamma v},
\end{aligned}
$$

and if $M_v < 1$

$$v_v(z) < 4M_v^2 < 4M_v^{s(p)/(2p)} < 4C_{23}n^{s(p)}2^{-2\gamma v}.$$

Thus we obtain finally

$$\int_0^{2\pi} v_v(\rho e^{i\theta})d\theta < C_{29}n^{s(p)}2^{-\gamma v}, \quad 1 - \frac{3}{n} < \rho < 1 - \frac{2}{n}.$$

On substituting this in (6.13) we have

$$||a_n| - |a_{n+1}|| \leq 80\sqrt{\left(\frac{2p}{\pi}C_{29}\right)}\sum_{v=1}^{\infty}\left(n^{s(p)-1}2^{-\gamma v}\right)^{\frac{1}{2}} < C_{20}n^{\frac{1}{2}s(p)-\frac{1}{2}}.$$

This proves Theorem 6.2.

6.7 Coefficient differences of k–symmetric functions   We apply Theorem 6.2 to k–symmetric functions. Suppose e.g. that $f(z)$ is mean p–valent in $\Delta : |z| < 1$ and has a power series development

$$f(z) = z^p + a_{p+k}z^{p+k} + a_{p+2k}z^{p+2k} + \cdots \tag{6.58}$$

Here $p > 0$, k is a positive integer and we assume as in (ii) of the introduction that, if p is not an integer, $f(z)/z^p$ is regular in Δ and that $f(z)$ is mean p–valent in Δ cut along a radius $z = re^{i\theta}$, $0 \leq r < 1$. We make a substitution $\zeta = z^k$ and note that

$$F(\zeta) = f(\zeta^{1/k}) = \zeta^{p/k}(1 + a_{p+k}\zeta + a_{p+2k}\zeta^2 + \cdots)$$

is mean (p/k)–valent in $|\zeta| < 1$ in the above sense. (cf. Example 2.8, or Section 5.8.) In fact a sector

$$S : \theta_1 < \arg z < \theta_1 + \frac{2\pi}{k}, \ 0 < |z| < 1$$

corresponds (1,1) conformally to

$$S' : k\theta_1 < \arg \zeta < k\theta_1 + 2\pi.$$

Since

$$F(\zeta e^{i\lambda}) = e^{ip/(\lambda k)} F(\zeta)$$

we deduce that $p(R, S', F) = p(R, S, f)$ is independent of θ_1. On applying this conclusion in turn with $\theta_1 + 2\pi v/k$ instead of θ_1, where $v = 0, 1, 2, \ldots, k - 1$, we deduce that

$$p(R, S', F) = p(R, \Delta, f)/k.$$

Thus, since f is mean p–valent,

$$\int_0^R p(t, S', F)t\,dt = \frac{1}{k} \int_0^R p(t, S, f)t\,dt \leq \frac{p}{k}R^2, \ 0 < R < \infty$$

so that F is mean (p/k)–valent. We may now apply Theorem 6.2 to $F(\zeta)$ and deduce Lucas' [1969]

Theorem 6.3   *If $f(z)$ is mean p–valent in $|z| < 1$ with the expansion (6.58) then*

$$|a_{p+nk} - a_{p+(n-1)k}| < \begin{cases} C_{30} \, n^{2p/k-2}, & \text{if } p \geq k \\ C_{30} \, n^{2p/k-2(p/k)^{\frac{1}{2}}}, & \text{if } \frac{k}{4} < p < k. \end{cases}$$

If $p = k = 1$ we obtain the right order of magnitude [Hayman 1963]

$$\left| |a_{n+1}| - |a_n| \right| < C_{30} \qquad (6.59)$$

for the coefficient differences of mean univalent and in particular univalent functions. If $p = 1$, $k = 2$ we obtain the bound

$$\left| |a_{2n+1}| - |a_{2n-1}| \right| < C_{30}\, n^{1-\sqrt 2} = C_{30}\, n^{-.4142\ldots} \qquad (6.60)$$

for the coefficient differences of odd univalent functions. If $p = 1$, $k = 3$, we obtain

$$\left| |a_{3n+1}| - |a_{3n-2}| \right| < C_{30} n^{2/3-2/\sqrt 3} = C_{30} n^{-.4880\ldots}. \qquad (6.61)$$

If $k = 4$ we have from Theorem 3.7

$$\left| |a_{4n+1}| - |a_{4n-3}| \right| < A n^{-\frac{1}{2}}$$

if f is univalent, and from Theorem 3.5

$$\left| |a_{4n+1}| - |a_{4n-3}| \right| < A n^{-\frac{1}{2}} \log(n + 1),$$

if f is mean univalent. The example of Section 3.4 shows that in the latter case we cannot at least replace $n^{-\frac{1}{2}} \log(n + 1)$ by anything smaller than $o(n^{-\frac{1}{2}})$. The results (6.60) and (6.61) are the best that are known even for univalent functions, and no other proof than that of Lucas exists as far as I know. However in the univalent case Grinspan [1976] has obtained the best known bounds

$$-2.97 < |a_{n+1}| - |a_n| < 3.61.$$

6.8 Asymptotic behaviour   It is natural to ask what functions attain the maximal growth in Theorem 6.2, at least when $p \geq 1$, so that the conclusion of Theorem 6.2 is sharp. If $\alpha > 0$, $p = 1$ in Theorem 2.10 or 2.11 and f is circumferentially mean p–valent I proved [1963] that

$$|a_{n+1}| - |a_n| \to \alpha \qquad (6.62)$$

and in fact

$$|a_{n+1} - e^{i\theta_0} a_n| \to \alpha \qquad (6.63)$$

as $n \to \infty$, where $\arg z = \theta_0$ is the radius of greatest growth and $p \geq 1$. The argument extends to a.m.p–valent functions, now that Eke's Theorems 2.10 and 2.11 are known.

Also if $p > 1$ and $\alpha = 0$, (6.62) and (6.63) still hold. It is merely necessary to modify the estimates of Lemmas 6.3, 6.4 and 6.5 to the case when (6.26) is replaced by

$$M(r, f) = o(1 - r)^{-2p} \text{ as } r \to 1.$$

This conclusion breaks down when $p = 1$ as the examples

$$f_\theta(z) = \frac{z}{1 - 2z \cos \theta + z^2} = \sum_0^\infty z^n \frac{\sin n\theta}{\sin \theta} \quad (6.64)$$

show. However Eke [1967b] showed that the only mean univalent functions for which $|a_{n+1}| - |a_n|$ does not tend to zero are the functions with positive α, for which (6.62) holds, and functions with maximal growth on two rays for which $|a_{n+1}| - |a_n|$ oscillates. Finally Hamilton proved [1982] if $f \in \mathfrak{S}$, and [1984] if f is mean univalent and $f'(0) = 1$, that

$$\limsup_{n\to\infty} ||a_{n+1}| - |a_n|| < 1$$

unless there is a real ϕ such that

$$e^{i\phi} f(ze^{-i\phi}) = \frac{z}{(1 - z)^2} \text{ or } f_\theta(z),$$

where $f_\theta(z)$ is given by (6.64).

Examples

6.3   If $f_\theta(z)$ is given by (6.64) prove that

$$||a_{n+1}| - |a_n|| \le 1.$$

Show also that, if θ is a rational multiple of π, equality holds infinitely often while, if θ/π is irrational, strict inequality holds for all n, but

$$\limsup_{n\to\infty}(a_{n+1} - a_n) = 1$$

and

$$\liminf_{n\to\infty}(a_{n+1} - a_n) = -1.$$

Eke's Theorems are refinements of the regularity theorems of Chapter 2 and their proofs would take us too far. However some other results can be proved fairly simply.

Examples

6.4   Suppose that $f(z)$, given by (6.1), is mean p–valent in the sense
of the introduction, where $p > 1$, and that $\alpha > 0$ in Theorem
2.10 or 2.11 respectively and $\arg z = \theta_0$ is a radius of greatest
growth. Show that

$$\left|a_{n+1} - a_n e^{i\theta_0}\right| = \left\{\frac{\alpha}{\Gamma(2p-1)} + o(1)\right\} n^{2p-2}, \text{ as } n \to \infty.$$

(Apply the Cauchy integral formula in the form

$$a_{n+1} - e^{i\theta_0} a_n = \int_{|z|=r_n} \frac{(z - e^{i\theta_0})f(z)dz}{z^{n+1}}$$

to $f(z)$ and $f_n(z) = \alpha_n(1 - ze^{-i\theta_0})^{-2p}$, where $r_n = 1 - 1/n$, and
$\alpha_n = n^{-2p}f(r_n e^{i\theta_0})$, and subtract. Use Theorem 2.10 on the
major arc $|\theta - \theta_0| < K/n$, where K is a large constant, and
Theorem 2.7 on the minor arc.)

6.5   Extend the conclusion of the previous example to the case $p > \frac{3}{4}$,
by using the formalism of Section 6.1, and in particular (6.12),
on the minor arc. (Deduce from Theorem 2.7 that

$$\int_{E_v} |z - z_1|^2 |f(\rho e^{i\theta})|^2 \rho d\rho d\theta = O\left\{n^{4p-4}2^{v[3/(2p-\varepsilon)-2]}\right\}.)$$

6.6   If $p > 1$ and $\alpha = 0$, show that

$$|a_{n+1}| - |a_n| = o(n^{2p-2}) \text{ as } n \to \infty.$$

(In this case $I_v = o\left\{n^{4p-3}2^{-\gamma v}\right\}$ in (6.57) for $j = 1, 2, 3$. Where
does the argument break down for $p = 1$?)

6.9 Starlike functions   Leung [1978] has shown that for starlike functions the behaviour of the functions (6.64) is extremal. We prove his
elegant

Theorem 6.4   *If*

$$f(z) = z + \sum_2^\infty a_n z^n \in \mathfrak{S}$$

and f is starlike then for $n = 1, 2, \ldots$,

$$\left||a_{n+1}| - |a_n|\right| \leq 1. \tag{6.65}$$

We shall see in the next chapter that this conclusion fails for general f in $\mathfrak{S}$.

We need a lemma of MacGregor [1969]:

Lemma 6.6 *Suppose that*

$$\phi(z) = 1 + \sum_{1}^{\infty} c_n z^n$$

and

$$\psi(z) = \sum_{1}^{\infty} \lambda_n c_n z^n$$

are regular in $\Delta : |z| < |$, *that* $\lambda_n \geq 0$ *and that for* $z \in \Delta$

$$\Re\phi(z) \geq 0,$$
$$\Re\psi(z) \leq M.$$

Then

$$\sum_{n=1}^{\infty} \lambda_n |c_n|^2 \leq 2M. \tag{6.66}$$

We write $c_n = x_n + i y_n$ and for $0 \leq r < 1, 0 \leq \theta \leq 2\pi$,

$$u(r, \theta) = \Re\phi(re^{i\theta}) = 1 + \sum_{n=1}^{\infty} (x_n \cos n\theta - y_n \sin n\theta) r^n$$

and

$$v(r, \theta) = \Re\psi(re^{i\theta}) = \sum_{1}^{\infty} \lambda_n (x_n \cos n\theta - y_n \sin n\theta) r^n.$$

Then since $u(r, \theta) \geq 0$ and $v(r, \theta) \leq M$ we have

$$2\pi M = \int_0^{2\pi} M u(r, \theta) d\theta \geq \int_0^{2\pi} u(r, \theta) v(r, \theta) d\theta$$

$$= \pi \sum_{n=1}^{\infty} \lambda_n (x_n^2 + y_n^2) r^{2n}.$$

Letting r tend to 1 we obtain (6.66). We deduce

Lemma 6.7 *If* $\phi(z)$ *is as in Lemma 6.6 and* $n > 0$, *there exists* ζ, *such that* $|\zeta| = 1$ *and*

$$\sum_{k=1}^{n} \frac{1}{k} \left| c_k - \zeta^k \right|^2 \leq \sum_{1}^{n} \frac{1}{k}.$$

We choose $\lambda_k = 1/k$ for $k \le n$, $\lambda_k = 0$ for $k > n$ in Lemma 6.6 so that

$$\psi(z) = \sum_{k=1}^{n} \frac{1}{k} c_k z^k,$$

$$M = \sup_{|z|=1} \Re\psi(z).$$

Then, by Lemma 6.6,

$$\sum_{k=1}^{n} \frac{1}{k} |c_k - \zeta^k|^2 = \sum_{1}^{n} \left\{ \frac{1}{k} |c_k|^2 - 2\Re\psi(\zeta) + \sum_{k=1}^{n} \frac{1}{k} \right\}$$

$$\le 2M - 2\Re\psi(\zeta) + \sum_{k=1}^{n} \frac{1}{k}.$$

Choosing ζ so that $\Re\psi(\zeta) = M$, we obtain the desired result.

We can now prove Theorem 6.4. If $f(z)$ is the function of that theorem we write

$$\phi(z) = \frac{zf'(z)}{f(z)} = 1 + \sum_{k=1}^{\infty} c_k z^k$$

and recall from (1.15) that $\Re\phi(z) \ge 0$. Let ζ be the point whose existence is asserted in Lemma 6.7. We have

$$\log \frac{f(z)}{z} = \int_0^z \frac{\phi(t) - 1}{t} dt = \sum_{k=1}^{\infty} \frac{c_k}{k} z^k.$$

Thus

$$\log \left\{ (1 - \zeta z) \frac{f(z)}{z} \right\} = \sum_{k=1}^{\infty} \alpha_k z^k, \text{ where } \alpha_k = \frac{1}{k}(c_k - \zeta^k).$$

On the other hand

$$(1 - \zeta z) \frac{f(z)}{z} = \sum_{k=0}^{\infty} \beta_k z^k, \text{ where } \beta_k = a_{k+1} - \zeta a_k.$$

Thus

$$\sum_{k=0}^{\infty} \beta_k z^k = \exp \left(\sum_{k=1}^{\infty} \alpha_k z^k \right).$$

It now follows from a theorem of Milin [1971] that

$$|\beta_n|^2 \le \exp \left\{ \sum_{k=1}^{n} \frac{k^2 |\alpha_k|^2 - 1}{k} \right\}. \tag{6.67}$$

We defer the proof to Chapter 8. The inequality (6.67) is a special case of Theorem 8.4, with $A_k = \alpha_k$, $D_k = \beta_k$, $\lambda = 1$. Using Lemma 6.7 we obtain

$$|\beta_n| = |\alpha_n - \zeta a_n| \le 1,$$

and this yields Theorem 6.4.

We note that the functions (6.64) are starlike and yield equality in Theorem 6.4 if $\sin n\theta = 0$, i.e. $\theta = k\pi/n$, where k is a positive integer. Equality also holds for all n, when $f(z) = z/(1-z)^2$.

6.10 The theorems of Dawei Shen We conclude the chapter by proving some of the results of Shen [1992] referred to in the introduction. We start with

Theorem 6.5 *If*

$$f(z) = z + c_3 z^3 + c_5 z^5 + \cdots \in \mathfrak{S} \tag{6.68}$$

and

$$\alpha = \lim_{r \to \infty} (1 - r) M(r, f) > 0$$

then

$$\big||c_{2n+1}| - |c_{2n-1}|\big| = O(n^{\beta-1}) \text{ as } n \to \infty,$$

if $\beta > \beta_0$, and $\beta_0 = .490\ldots$ is the constant of Theorem 3.6.

We write

$$F(z) = f(z^{\frac{1}{2}}) = z^{\frac{1}{2}}(1 + c_3 z + c_5 z^2 + \cdots), \tag{6.69}$$

so that $F(z)$ is mean $\frac{1}{2}$-valent in $|z| < 1$ in the sense of the introduction. Also $F(z)$ has maximal growth and so a unique radius of greatest growth $\arg z = \theta_0$ by Theorem 2.11. Again (6.11), with $n = m + \frac{1}{2}$, $z_1 = \left(1 - 2/(2m+1)\right)e^{i\theta_0}$ and $\rho = (2m-1)/(2m+1)$ yields

$$\left(m - \frac{1}{2}\right)\left|c_{2m-1} - e^{i\theta_0} c_{2m+1}\right| \le \frac{1}{2\pi\rho^{m-3/2}} \int_0^{2\pi} |e^{i\theta} - e^{i\theta_0}||F'(\rho e^{i\theta})|d\theta. \tag{6.70}$$

We choose a positive number ε, and note that, by Theorems 2.7, 2.11

$$
\begin{aligned}
|F(\rho e^{i\theta})| &\le \frac{C_0}{(1-\rho)^\varepsilon |\theta - \theta_0|^{1-\varepsilon}} \\
&< \frac{\pi C_0}{(1-\rho)^\varepsilon |\theta - \theta_0|}, \quad \text{if } \rho_0 < \rho < 1 \text{ and } 0 < |\theta - \theta_0| < \pi,
\end{aligned}
$$

where C_0 is a constant. Also $|e^{i\theta} - e^{i\theta_0}| < |\theta - \theta_0|$. Thus

$$
\begin{aligned}
\int_{\theta_0-\pi}^{\theta_0+\pi} |e^{i\theta} - e^{i\theta_0}| \, |F'(\rho e^{i\theta})| d\theta &\leq \frac{\pi C_0}{(1-\rho)^\varepsilon} \int_{\theta_0-\pi}^{\theta_0+\pi} \frac{|F'(\rho e^{i\theta})|}{|F(\rho e^{i\theta})|} d\theta \\
&= \frac{\pi C_0}{2(1-\rho)^\varepsilon} \rho^{-\frac{1}{2}} \int_{\theta_0-\pi}^{\theta_0+\pi} \left| \frac{f'(\rho^{\frac{1}{2}} e^{\frac{1}{2}i\theta})}{f(\rho^{\frac{1}{2}} e^{\frac{1}{2}i\theta})} \right| d\theta \\
&< C_1 \rho^{-\frac{1}{2}} (1-\rho)^{-\beta-\varepsilon}
\end{aligned}
$$

by Theorem 3.6. Here C_0, C_1 are constants depending on β, ε and we can choose $\beta + \varepsilon$ as close as we please to β_0. Now (6.70) yields Theorem 6.5.

6.10.1 Some weaker growth conditions enable Shen [1992] to obtain (6.6) or something close to it.

Theorem 6.6 *Suppose that $f(z)$ is as in (6.68). If*

$$M(r,f) = O(1-r)^{-\frac{1}{2}}, \tag{6.71}$$

we have

$$|c_n| = O(n^{-\frac{1}{2}}) \tag{6.72}$$

and hence

$$\left| |c_{2n+1}| - |c_{2n-1}| \right| = O(n^{-\frac{1}{2}}).$$

On the other hand if there is a radius $\arg z = \theta_0$ such that

$$\liminf_{r \to 1} (1-r)^{\frac{1}{2}} |f(re^{i\theta_0})| > 0, \tag{6.73}$$

we have

$$\left| |c_{2n+1}| - |c_{2n-1}| \right| = O\{n^{-\frac{1}{2}} \log n\}, \text{ as } n \to \infty. \tag{6.74}$$

The conclusion (6.72) is an immediate consequence of Baernstein's Theorem 3.7. Shen's deduction of (6.74) from (6.73) lies deeper. In Theorem 6.6 the hypothesis $f(z) \in \mathfrak{S}$ was only used to obtain (6.72). For the rest of the proof we assume only that $f(z)$ is mean univalent. Following Shen we base our argument on

Lemma 6.8 *Suppose that $F(z)$ given by (6.1) is mean p–valent in $|z| < 1$ in the sense of Section 6.0 and that*

$$\liminf_{\rho \to 1} (1-\rho)^p |F(\rho e^{i\theta_0})| = \alpha > 0, \tag{6.75}$$

where $0 \leq \theta_0 \leq 2\pi$. Then there is a constant K_0, independent of $z = \rho e^{i\theta}$ such that

$$|F(z)| < \frac{K_0}{(1-\rho)^p |\theta - \theta_0|^{2p}} \quad 0 < \rho < 1, \ |\theta - \theta_0| \leq \pi. \tag{6.76}$$

We denote by $K_0, K_1, \ldots$, positive constants independent of z. By (6.75) there exists ρ_0 such that $0 \leq \rho_0 < 1$, and

$$|F(\rho_1 e^{i\theta_0})| \geq \frac{K_1}{(1-\rho_1)^p}, \quad \text{if } \rho_0 < \rho_1 < 1. \tag{6.77}$$

Hence if $R \geq K_2 = |F(\rho_0 e^{i\theta_0})|$, there exists ρ_1, such that $\rho_0 \leq \rho_1 < 1$ and

$$|f(\rho_1 e^{i\theta_0})| = R. \tag{6.78}$$

Suppose that $|F(z)| = R$. If $R < K_2$, (6.76) holds. If $R \geq K_2$ we choose ρ_1 to satisfy (6.78) and apply Theorem 6.1 with $z_1 = \rho_1 e^{i\theta_0}$, z, instead of z_1, z_2, $|F(z_1)| = |F(z)| = R$, and $a = b = 1$. This yields

$$R^{\frac{2}{p}} < \frac{C_0}{(1-\rho_1)(1-\rho)|z - z_1|^2}. \tag{6.79}$$

We have, if $|\theta - \theta_0| \leq \frac{\pi}{2}$,

$$\begin{aligned} |z - z_1|^2 &= \rho_1^2 + \rho^2 - 2\rho\rho_1 \cos(\theta - \theta_0) \\ &= (\rho_1 \cos(\theta - \theta_0) - \rho)^2 + \rho_1^2 \sin^2(\theta - \theta_0) \\ &\geq \frac{4\rho_1^2(\theta - \theta_0)^2}{\pi^2}, \end{aligned}$$

while if $\frac{\pi}{2} \leq |\theta - \theta_0| \leq \pi$, we have $|z - z_1|^2 \geq \rho_1^2$. Thus in all cases

$$|z - z_1|^2 \geq \rho_1^2 \frac{(\theta - \theta_0)^2}{\pi^2} \geq \rho_0^2 \frac{(\theta - \theta_0)^2}{\pi^2}.$$

Also by (6.77) we have

$$\frac{1}{1-\rho_1} \leq \left(\frac{R}{K_1}\right)^{1/p}.$$

Thus (6.79) yields

$$R^{1/p} < \frac{K_0^{1/p}}{(1-\rho)(\theta - \theta_0)^2},$$

which is (6.76).

We can now prove Theorem 6.6. We start with (6.11), applied to $F(z) = f(z^{\frac{1}{2}})$, with $n + \frac{1}{2}$ instead of n, so that $z_1 = \{(2n - 1)/(2n + 1)\}e^{i\theta_0}$,

$1 - 3/\left(n + \tfrac{1}{2}\right) \le \rho \le 1 - 2/\left(n + \tfrac{1}{2}\right)$. This yields

$$|c_{2n-1} - e^{i\theta_0}c_{2n+1}| \le K \int_0^{2\pi} d\theta \int_{1-\frac{6}{2n+1}}^{1-\frac{4}{2n+1}} |\rho e^{i\theta} - z_1||F'(z)|\rho d\rho d\theta$$

$$\le K \left(\int\int \left|\frac{F'(z)}{F(z)}\right|^2 \rho d\rho d\theta \right)^{\frac{1}{2}} \left(\int\int |z - z_1|^2|F(z)|^2 \right)^{\frac{1}{2}}. \qquad (6.80)$$

By Lemma 3.2 we have

$$\int\int \left|\frac{F'(z)}{F(z)}\right|^2 \rho d\rho d\theta < K \log n. \qquad (6.81)$$

To the second integral we apply Lemmas 6.1, 6.2 with $G(R) = R^2$. Letting M tend to ∞ in (6.19) or directly we obtain for $v(z) = |z - z_1|^2 R^2$,

$$\nabla^2 v(z) \le 4R^2 + 4|z - z_1|^2|f'(z)|^2 + 8|z - z_1|R|f'(z)|. \qquad (6.82)$$

Also Lemma 6.1 gives for $\tfrac{1}{2} \le \rho < 1$

$$\int_0^{2\pi} v(\rho e^{i\theta})d\theta \le \int_0^{2\pi} v\left(\tfrac{1}{2}e^{i\theta}\right) d\theta + K \int_0^{2\pi} d\phi \int_0^\rho \nabla^2 v(te^{i\phi})(1-t)t dt. \qquad (6.83)$$

We estimate the integral on the right-hand side by means of (6.76). We have from Theorems 3.2 and 2.3 that

$$\int_0^{2\pi} |F(te^{i\theta})|^2 dt < K \int_0^t (1 - \xi)^{-2}d\xi < \frac{K}{1-t},$$

so that

$$\int_0^{2\pi} d\phi \int_0^\rho |F(te^{i\phi})|^2(1 - t)t dt < K. \qquad (6.84)$$

Next we have

$$\int\int_{|z|<\rho} (1 - |z|)|z - z_1|^2|F'(z)|^2|dz|^2 = \sum_{v=0}^N I_v(z)$$

where

$$I_v(z) = \int\int_{E_v} (1 - |z|)|z - z_1|^2|F'(z)|^2|dz|^2.$$

Here if $0 \le v < N$, E_v is the subset of $|z| < \rho$ where $M_{v+1} < |F(z)| < M_v$ and M_v is defined in (6.8). We choose N to be the first integer such that $M_N < 1$, and define E_N to be the set where $|F(z)| < M_N$. We recall that by Lemma 6.8, with $p = \tfrac{1}{2}$,

$$|F(te^{i\theta})| < \frac{K_0}{(1-t)^{\frac{1}{2}}|\theta - \theta_0|},$$

where θ_0 is as in (6.73). Hence if $|\theta-\theta_0| > (1-t)$ we have, with $z_1 = \rho_1 e^{i\theta_0}$, $a = te^{i\theta}$, $0 \le t < \rho_1$,

$$|z - z_1| \le |t - \rho_1| + |\rho_1 e^{i\theta} - \rho e^{i\theta_0}|$$

$$\le |\rho_1 - t| + |\theta - \theta_0| < (1 - t) + |\theta - \theta_0| < 2|\theta - \theta_0|.$$

Thus in this case we have in E_ν, if $0 \le \nu < N$,

$$|z - z_1|^2(1 - |z|) < 4(\theta - \theta_0)^2(1 - t) < \frac{4K_0^2}{M_{\nu+1}^2} = \frac{16K_0^2}{M_\nu^2}.$$

On the other hand if $|\theta - \theta_0| \le (1 - t)$ we have

$$|z - z_1| < |\theta - \theta_0| + (1 - t) < 2(1 - t).$$

Thus in this case

$$|z - z_1|^2(1 - t)M_\nu^2 < K(1 - t)^3 M_1^2 < K.$$

Thus for $0 \le \nu < N$ we have in F_ν

$$|z - z_1|^2(1 - t) < \frac{K}{M_\nu^2},$$

so that

$$\int\int_{F_\nu} |z - z_1|^2(1 - t)|F'(z)|^2|dz|^2 < \frac{K}{M_\nu^2} M_\nu^2 = K.$$

Also in F_N we have $|F(z)| < 1$ so that

$$\int\int_{F_N} |z - z_1|^2(1 - t)|F'(z)|^2|dz|^2 \le 4\int_{F_N} |F'(z)|^2|dz|^2 < K.$$

Thus, for $n \ge 6$,

$$\int\int_{|z|<\rho} (1 - |z|)|z - z_1|^2|F'(z)||dz|^2 < K(N + 1) < K\log n, \qquad (6.85)$$

since $M_{N-1} = 2^{2-N}M_1 \ge 1$, so that

$$2^{N+1} \le 8M_1 < Kn.$$

Finally by (6.84) and (6.85) we have

$$\int\int_{|z|<\rho} |z - z_1||F'(z)||F(z)|(1 - |z|)|dz|^2$$

$$\le \left\{ \int\int |z - z_1|^2 |1 - |z|| |F'(z)|^2|dz|^2 \int\int (1 - |z|)|F(z)|^2|dz|^2 \right\}^{\frac{1}{2}}$$

$$\le K(\log n)^{\frac{1}{2}}.$$

On combining this with (6.82) to (6.85) and noting that $v\left(\frac{1}{2}e^{i\theta}\right) \le K$ we obtain

$$\int_0^{2\pi} v(\rho e^{i\theta})d\theta \le K \log n, \quad 1 - \frac{3}{n} \le \rho \le 1 - \frac{2}{n}, \quad n \ge 6.$$

Now (6.80) and (6.81) yield

$$|c_{2n-1} - e^{i\theta_0}c_{2n+1}| \le Kn^{-\frac{1}{2}} \log n.$$

This is (6.74) and completes the proof of Theorem 6.6.

Examples

6.7   If $F(z)$ satisfies the hypotheses of Lemma 6.8 prove that,

$$\begin{aligned}
\big||a_{n+1}| - |a_n|\big| &= O(n^{p-1}), & \text{as } n \to \infty \quad \text{if } \tfrac{1}{2} < p < 1; \\
\big||a_{n+1}| - |a_n|\big| &= o(n^{-\frac{1}{2}}), & \text{as } n \to \infty, \quad \text{if } p < \tfrac{1}{2}.
\end{aligned}$$

(If $p < \frac{1}{2}$, choose λ so that $2 < \lambda < 1/p$, and use instead of (6.12)

$$\big||a_{n-1}| - |a_n|\big|$$

$$< A(p) \sum_{\nu=1}^{\infty} \left\{ \iint_{E_\nu} \frac{|f'(z)|^2|dz|^2}{1 + |f(z)|^\lambda} \iint_{E_\nu} |z - z_1|^2 \left(|f(z)|^\lambda + 1\right) |dz|^2 \right\}.)$$

6.8   If $f(z)$ is mean p–valent in $|z| < 1$ and satisfies

$$\liminf_{r \to 1}(1 - r)^\alpha |f(re^{i\theta_0})| > 0,$$

where $0 < \alpha < 2p$, prove that

$$|f(\rho e^{i\theta})| < \frac{K}{(1 - \rho)^{2p-\alpha}|\theta - \theta_0|^{2\alpha}}, \quad 0 < |\theta - \theta_0| < \pi;$$

(apply Theorem 6.1 with $a = \alpha$, $b = 2p - \alpha$).

7

The Löwner theory

7.0 Introduction We shall in this chapter give a deeper theory due to Löwner [1923], which enables us to obtain sharp bounds for the class $\mathfrak{S}$ of functions

$$f(z) = z + a_2 z^2 + \cdots$$

univalent in $|z| < 1$. These results do not seem to be accessible by the methods of Chapter 1. We shall say that the class $\mathfrak{S}'$ is dense in $\mathfrak{S}$, if $\mathfrak{S}'$ is a subclass of $\mathfrak{S}$ and if every function $f(z)$ in $\mathfrak{S}$ can be approximated by a sequence of functions $f_n(z)$ in $\mathfrak{S}'$ so that $f_n(z) \to f(z)$ uniformly on every compact subset of $|z| < 1$ as $n \to \infty$. It will then follow that pth derivatives at an arbitrary point in $|z| < 1$, and in particular the coefficients of $f_n(z)$, approach those of $f(z)$ as $n \to \infty$. Thus bounds obtained for the class $\mathfrak{S}'$ will remain true for the wider class $\mathfrak{S}$. Löwner's basic result can now be stated as follows:

Theorem 7.1 *Let $\kappa(t)$ be measurable and complex valued for $0 \leq t < \infty$ and satisfy $|\kappa(t)| \leq 1$. Then if $|z| < 1$, there exists a unique function $w = f(z, t)$ absolutely continuous in t for $0 \leq t < \infty$, and satisfying for almost all t Löwner's differential equation*

$$\frac{\partial w}{\partial t} = -w \frac{1 + \kappa(t)w}{1 - \kappa(t)w}, \tag{7.1}$$

with the initial condition $f(z, 0) = z$. Also

$$g(z, t) = e^t f(z, t) \in \mathfrak{S}, \quad 0 \leq t < \infty. \tag{7.2}$$

Finally there is a dense subclass $\mathfrak{S}_1$ of $\mathfrak{S}$ such that if $f(z) \in \mathfrak{S}_1$, there exists $\kappa(t)$ continuous and with $|\kappa(t)| = 1$, such that the associated function

197

g(z, t) satisfies

$$g(z, t) \to f(z) \quad \text{as } t \to \infty, \qquad (7.3)$$

uniformly on compact subsets of $|z| < 1$.

We shall see that the functions $g(z)$ in $\mathfrak{S}_1$ map the unit disc $|z| < 1$ onto the complement of a sectionally analytic slit. We recall that if $f(z)$ is the limit of a sequence $f_n(z)$ of univalent functions then $f(z)$ is either constant or univalent (C. A. p. 231). If $f_n(z) \in \mathfrak{S}_1$, $f'(0) = \lim f_n'(0) = 1$ so that $f(z) \in \mathfrak{S}$. Thus if the limit $f(z)$ in (7.3) exists $f(z) \in \mathfrak{S}$.

In the first part of this chapter we shall prove Theorem 7.1 by (a) constructing the class $\mathfrak{S}_1$ of mappings $f(z)$, which is dense in $\mathfrak{S}$; showing (b) that if $f(z) \in \mathfrak{S}_1$ then there exists $g(z, t)$ satisfying (7.1) in (7.3) for a suitable continuous $\kappa(t)$; and (c) given $\kappa(t)$ as in Theorem 7.1, the functions $g(z, t)$ given by (7.2) belong to $\mathfrak{S}$. The proof of Theorem 7.1 is rather long and not easy. It is however fully justified by its many beautiful applications. In the last part of this chapter we shall use $\mathfrak{S}_1$ to obtain the exact bounds for $|a_3|$ and all the coefficients of the inverse function $z = f^{-1}(w)$ as well as the arguments of $f(z)/z$ and $f'(z)$ and the radii of convexity and starlikeness. In the next chapter we use $\mathfrak{S}_1$ to prove de Branges' Theorem $|a_n| \le n$ for all n.

Löwner Theory is a very powerful tool for finding exact bounds for functionals, but the method makes it difficult to discuss the form that the extremals take. Sometimes other techniques, which are outside the scope of this book, are more effective for this. (See e.g. Duren [1983] and Jenkins [1958].)

7.1 Boundary behaviour in conformal mapping   In this section we prove two preliminary results:

Lemma 7.1 *Suppose that* $w = \phi(z)$ *maps a domain* D_1 *(1,1) conformally onto a domain* D_2 *lying in* $|w| < 1$. *For any point* z_0 *let* $l(R)$ *be the total length of the image in* D_2 *of that part of the circumference* $|z - z_0| = R$ *which lies in* D_1. *Then if* $R_1 > 0$, $k > 1$, *there exists* R *such that* $R_1 < R < kR_1$ *and* $l(R) \le \pi(2/\log k)^{\frac{1}{2}}$. *In particular, there exists a sequence* R_n, *decreasing to zero as* $n \to \infty$, *such that* $l(R_n) \to 0$ *as* $n \to \infty$.

We consider the mapping $z = \phi^{-1}(w)$ of D_2 onto D_1 and put

$$\psi(w) = \phi^{-1}(w) - z_0.$$

Then the level curves γ_R in D_2 corresponding to $|z - z_0| = R$ are the level curves $|\psi(w)| = R$. Let $l(R)$ be their total length. Then since $\psi(w)$ is univalent, we may apply Theorem 2.1 with $p(R) \leq 1$, $A \leq \pi$. This gives

$$\int_{R_1}^{kR_1} \frac{l(R)^2 dR}{R} \leq \int_0^\infty \frac{l(R)^2 dR}{Rp(R)} \leq 2\pi A \leq 2\pi^2.$$

If l is the lower bound of $l(R)$ in $R_1 < R < kR_1$, we deduce

$$l^2 \log k \;\leq\; 2\pi^2,$$

$$l \;\leq\; \pi \left(\frac{2}{\log k} \right)^{\frac{1}{2}},$$

as required.

If we now define R_n' inductively by $R_0' = 1$, $R_{n+1}' = e^{-n}R_n'$. Then it follows that there exists R_n such that $R_{n+1}' < R_n < R_n'$ and

$$l(R_n) \leq \pi\sqrt{\left(\frac{2}{n} \right)}.$$

This completes the proof of Lemma 7.1.

We shall also need

Lemma 7.2 *Let γ be a simple arc which lies in $|z| < 1$ but approaches $|z| = 1$ at both ends, and suppose that the length l of γ is less than 1. Let D be the set of all those points P in $|z| < 1$ such that any curve joining the origin to P in $|z| < 1$ meets γ. Then the diameter of D is at most l.[†]*

We may parametrize the curve γ by $z = \alpha(t)$ $(a < t < b)$. Let t_n $(n = 0, 1, 2, \ldots)$ be an increasing sequence of numbers such that

$$a < t_n < b \qquad (n \geq 1),$$
$$t_n \to b \qquad (n \to \infty).$$

Then

$$\sum_{r=1}^n |\alpha(t_r) - \alpha(t_{r-1})| \leq l,$$

and so $\sum_{r=1}^\infty |\alpha(t_r) - \alpha(t_{r-1})| \leq l$, $\sum_{r=1}^\infty [\alpha(t_r) - \alpha(t_{r-1})]$ converges.

Thus $\alpha(t_r)$ approaches a definite limit $\alpha(b)$ as $r \to \infty$ and $\alpha(b)$ is clearly independent of the sequence t_r. Thus $\alpha(t) \to \alpha(b)$ as $t \to b-$ and similarly $\alpha(t)$ approaches a finite limit $\alpha(a)$ as $t \to a+$.

[†] The diameter of a point set E is the upper bound of distances $|z_1 - z_2|$ of pairs of points z_1, z_2 in E.

We write $z_1 = \alpha(a)$, $z_2 = \alpha(b)$, $z_3 = \frac{1}{2}(z_1 + z_2)$. Then $|z_1| = |z_2| = 1$ by hypothesis. If z is any point on γ, then we have

$$|z - z_1| + |z - z_2| \le l,$$

since γ has length l and so z lies inside an ellipse of centre z_3 and major axis l. Hence z lies inside the circle C of centre z_3 and radius $\frac{1}{2}l$. This circle has diameter l, and since C contains z_1 on $|z| = 1$, C cannot contain the origin. Any point P in $|z| < 1$ but outside C can clearly by joined to the origin by a curve lying in $|z| < 1$ and outside C, and so P lies outside D. Thus D lies inside C and so has diameter at most l.

7.2 Transformations   Consider

$$f(z) = \beta(z + a_2 z^2 + \cdots), \qquad (7.4)$$

where $\beta > 0$, $f(z)$ is univalent in $|z| < 1$ and $|f(z)| < 1$. Such a function $f(z)$ will be called a *transformation*. The transformation $w = f(z)$ maps $|z| < 1$ onto a domain D in $|w| < 1$. We denote by $S = S_f$ the set of all points of $|w| < 1$ not in D, and by $d = d_f$ the diameter of S_f. We ignore the trivial case when S_f is null and $f(z) = z$.

We shall say that two points z, w on $|z| = 1$ and the frontier of D respectively *correspond* by the transformation $w = f(z)$, if there exists a sequence z_n, such that

$$|z_n| < 1 \quad (n \ge 1)$$

and

$$z_n \to z, \ f(z_n) \to w \ (n \to \infty).$$

Let $B = B_f$ be the set of all points of $|z| = 1$, which correspond to points of S. We note that points of $|z| = 1$ not in B can correspond only to points on $|w| = 1$. We write $\delta = \delta_f$ for the diameter of B.

Our aim is to study the limiting behaviour of transformations when δ or d is small. In this case D approximates to $|w| < 1$ and $f(z)$ approximates to z. Our first aim is to show that, if either of δ or d is small, then so is the other, and in this case β is nearly equal to 1.

Lemma 7.3   *If $f(z)$ is a transformation given by (7.4), then we have with the above notation $1 - d \le \beta \le 1$.*

The inequality $\beta \le 1$ follows from Schwarz's Lemma. To prove $\beta \ge 1 - d$, we may suppose that $d < 1$. It follows from Lemma 5.2 that if S meets $|w| = r$ for some $r < 1$, then S meets $|w| = \rho$ for $r < \rho < 1$. Hence S has at least one limit point $e^{i\theta}$ on $|w| = 1$. Thus S lies entirely in $|w - e^{i\theta}| \leq d$, and so in $|w| \geq 1 - d$. Thus D contains the disc $|w| < 1 - d$.

Hence the inverse function $z = f^{-1}(w)$ maps $|w| < 1 - d$ onto a subdomain of $|z| < 1$ and

$$f^{-1}[(1-d)w] = \frac{(1-d)w}{\beta} + \cdots$$

satisfies the conditions of Schwarz's Lemma. Thus $1 - d \leq \beta$ as required and Lemma 7.3 is proved.

Lemma 7.4 *We have with the above notation*

$$\delta \leq 4\pi [\log(2/d)]^{-\frac{1}{2}}, \quad d \leq 4\pi [\log(2/\delta)]^{-\frac{1}{2}}.$$

To prove the first inequality, we may assume that $d < 2e^{-4\pi^2}$. For since $\delta \leq 2$, the inequality is trivial otherwise.

Let w_0 be a limit point of S_f on $|w| = 1$. Then S_f lies entirely in $|w - w_0| \leq d$, and so for $R > d$ that arc, c_R say, of $|w - w_0| = R$ which lies in $|w| < 1$, lies also in D. Let γ_R be the image of c_R by $z = f^{-1}(w)$, and let $l(R)$ be the length of γ_R. Then by Lemma 7.1 with $R_1 = d$, $k = d^{-1}$, we can choose R_0 satisfying

$$d < R_0 < 1, \quad l(R_0) \leq \pi \left(\frac{1}{2} \log \frac{1}{d} \right)^{-\frac{1}{2}} \leq \pi \left(\frac{1}{4} \log \frac{2}{d} \right)^{-\frac{1}{2}} < 1.$$

Next since c_{R_0} separates $w = 0$ from S_f, it follows that γ_{R_0} separates $z = 0$ from B_f. Hence by Lemma 7.2

$$\delta_f \leq l(R_0) \leq 2\pi \left(\log \frac{2}{d} \right)^{-\frac{1}{2}},$$

as required. The proof of the second inequality of Lemma 7.4 is similar, and so the lemma is proved.

7.2.1 We shall also need the following:

Lemma 7.5 *Suppose that $\psi(z) = u + iv$ is regular in $|z| < 1$, that $u(z)$ has constant sign there and $v(0) = 0$. Suppose further that*

$$u(re^{i\phi}) \to 0 \quad (r \to 1),$$

uniformly for $\delta \leq |\phi - \phi_0| \leq \pi$. *Then we have*

$$\psi(z) = \psi(0)\left[\frac{e^{i\phi_0} + z}{e^{i\phi_0} - z} + \varepsilon(z)\right],$$

where $|\varepsilon(z)| < 5\delta|e^{i\phi_0} - z|^{-2}$ *for* $|e^{i\phi_0} - z| > 2\delta$, $|z| < 1$.

We may without loss of generality assume that $\phi_0 = 0$. Then we have by Poisson's formula for $|z| < r < 1$,

$$\psi(z) = \frac{1}{2\pi}\int_{-\pi}^{+\pi} u(re^{i\theta})\frac{re^{i\phi} + z}{re^{i\phi} - z}d\phi + iC,$$

where C is a constant in $|z| < r$, and since $\psi(0)$ is real, $C = 0$. Thus we have for a fixed z,

$$\psi(z) = \frac{1}{2\pi}\int_{-\delta}^{\delta} u(re^{i\phi})\frac{re^{i\phi} + z}{re^{i\phi} - z}d\phi + o(1) \quad (r \to 1). \tag{7.5}$$

Writing $\psi(0) = \alpha$ we obtain

$$\frac{1}{2\pi}\int_{-\delta}^{\delta} u(re^{i\phi})d\phi \to \alpha \quad (r \to 1). \tag{7.6}$$

Also

$$\frac{re^{i\phi} + z}{re^{i\phi} - z} \longrightarrow \frac{e^{i\phi} + z}{e^{i\phi} - z},$$

and if $|z - 1| \geq 2\delta$, $|\phi| \leq \delta$ we have $|e^{i\phi} - 1| \leq \delta$ and so

$$\left|\frac{e^{i\phi} + z}{e^{i\phi} - z} - \frac{1 + z}{1 - z}\right| = \left|\frac{2z(e^{i\phi} - 1)}{(e^{i\phi} - z)(1 - z)}\right|$$

$$\leq \frac{2\delta}{|1 - z|[|1 - z| - \delta]} \leq \frac{4\delta}{|1 - z|^2}.$$

Thus if $|z - 1| > 2\delta$, $|\phi| < \delta$, and r is sufficiently near to 1, then

$$\left|\frac{re^{i\phi} + z}{re^{i\phi} - z} - \frac{1 + z}{1 - z}\right| < \frac{5\delta}{|1 - z|^2}.$$

Hence we deduce from (7.5), as $r \to 1$,

$$\left|\psi(z) - \frac{1+z}{1-z}\frac{1}{2\pi}\int_{-\delta}^{\delta} u(re^{i\phi})d\phi\right| \leq \frac{5\delta}{|1-z|^2}\frac{1}{2\pi}\int_{-\delta}^{\delta} |u(re^{i\phi})|d\phi + o(1).$$

Using (7.6) and the fact that u has constant sign in $|z| < 1$ we deduce

$$\left|\psi(z) - \alpha\frac{1+z}{1-z}\right| \leq \frac{5\delta|\alpha|}{|1-z|^2},$$

if $|1 - z| \geq 2\delta$, and this is Lemma 7.5.

7.3 Structure of infinitesimal transformations We are now able to describe infinitesimal transformations, i.e. those for which $\delta(f)$ and $d(f)$ are small.

Lemma 7.6 *Let $f_n(z) = \beta_n(z + \cdots)$ be a sequence of transformations of $|z| < 1$, let $d(f_n)$, $\delta(f_n)$ be defined as in §7.2 and suppose that either $d(f_n)$ or $\delta(f_n) \to 0$ as $n \to \infty$. Then*

$$f_n(z) \to z \quad (n \to \infty), \tag{7.7}$$

uniformly in $|z| < 1$ and in particular $\beta_n \to 1$.

Further, if z_n is a point on $|z| = 1$ which corresponds to a point w_n in $|w| < 1$ by $w = f_n(z)$, and if $0 < r < 1$, then

$$f_n(z) - z \sim (\log \beta_n) z \frac{z_n + z}{z_n - z} \quad (n \to \infty), \tag{7.8}$$

uniformly in $|z| \leq r$.

It follows from Lemma 7.4 that if $\delta(f_n) \to 0$ then $d(f_n) \to 0$ and conversely. It then follows from Lemma 7.3 that $\beta_n \to 1$. Write

$$\psi_n(z) = \log \frac{f_n(z)}{z} = u_n(z) + iv_n(z).$$

Since $f_n(z)$ vanishes only at $z = 0$, $\psi_n(z)$ is regular in $|z| < 1$, and since $f_n(z)$ satisfies the hypotheses of Schwarz's Lemma, $u_n(z) \leq 0$ in $|z| < 1$. Let $z_n = e^{i\phi_n}$ correspond to w_n, where $|w_n| < 1$, by $w = f_n(z)$. Then if $|e^{i\phi} - e^{i\phi_n}| > \delta(f_n)$, the point $z = e^{i\phi}$ corresponds only to points w on $|w| = 1$, and given $\delta > 0$, this will be true for $\delta \leq |\phi - \phi_n| \leq \pi$ if $n > n_0(\delta)$, since $\delta(f_n) \to 0$. In this case

$$u_n(re^{i\phi}) \to 0 \quad (r \to 1),$$

uniformly for $\delta \leq |\phi - \phi_n| \leq \pi$ and a fixed $n > n_0(\delta)$.

We now apply Lemma 7.5 and obtain

$$\psi_n(z) = \log \beta_n \left[\frac{z_n + z}{z_n - z} + \varepsilon_n(z) \right], \tag{7.9}$$

where $|\varepsilon_n(z)| < 5\delta|z - z_n|^{-2}$ for $|z - z_n| > 2\delta$. Thus if η is fixed and positive we may choose n_1 so large that

$$|\psi_n(z)| \leq \eta$$

if $n > n_1$ and $|z - z_n| \geq \eta$, and then

$$|f_n(z) - z| = |z| \left| e^{\psi_n(z)} - 1 \right| \leq e^\eta - 1 < 2\eta \tag{7.10}$$

if $\eta < \frac{1}{2}$. We also suppose n so large that the end–points of the arc $|z - z_n| = \eta$ on $|z| = 1$ correspond to points on $|w| = 1$ only. By (7.10) we have

$$|f_n(z) - z| \leq 2\eta \quad \text{and so} \quad |f_n(z) - z_n| \leq 3\eta$$

on this arc. The values which $w = f_n(z)$ assumes for $|z - z_n| < \eta$ form a domain Δ in $|w| < 1$, which is separated from $w = 0$ by the image of the arc $|z - z_n| = \eta$ by $w = f_n(z)$. Hence Δ also lies in $|w - z_n| < 3\eta$. Thus we have finally

$$|f_n(z) - z_n| \leq 3\eta \quad \text{in} \quad |z - z_n| < \eta,$$

and so

$$|f_n(z) - z| < 4\eta \quad \text{for} \quad |z_n - z| \leq \eta.$$

This, together with (7.10), gives (7.7) since η is arbitrary.

Next (7.9) shows that if r is fixed, $0 < r < 1$, then $\psi_n(z) \to 0$ ($n \to \infty$), uniformly for $|z| \leq r$, and so we have as $n \to \infty$ uniformly for $|z| \leq r$,

$$f_n(z) - z = z[e^{\psi_n(z)} - 1] \sim z\psi_n(z) \sim z \log \beta_n \frac{z_n + z}{z_n - z}$$

by (7.9), which is (7.8). This proves Lemma 7.6.

7.4 The class $\mathfrak{S}_1$ Let BP be an analytic Jordan arc lying in $|z| \leq r$ and with one end-point $P = re^{i\theta}$. Let $P\infty$ be the ray

$$z = \rho e^{i\theta}, \ r \leq \rho < \infty. \tag{7.11}$$

We shall call the Jordan arc $\gamma : BP\infty$ a *slit*.

The set of points G consisting of all points w not on γ will be a simply connected domain. In fact if Q_1, Q_2 lie in G near different points of γ we can pass from Q_1 to Q_2 in G by a curve near γ, which if necessary will go round the tip of γ and along the other side. Thus any two points of G near γ can be joined in G, and any point in G can be joined to some point near γ, for instance, by a straight–line segment to the tip of γ. Thus G is connected. Further, the complement of G consists of γ and so is closed and connected. Thus[†] G is a simply connected domain containing $w = 0$, and so by Riemann's mapping theorem[‡] there exists a unique function

$$w = f(z) = \beta \left(z + a_2 z^2 + \cdots \right), \tag{7.12}$$

[†] C. A. p. 139
[‡] C. A. p. 230

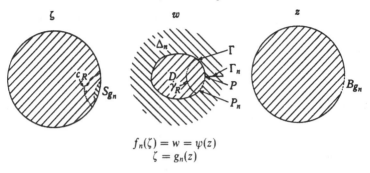

$$f_n(\zeta) = w = \psi(z)$$
$$\zeta = g_n(z)$$

Fig. 6.

mapping $|z| < 1$ (1,1) conformally onto G so that $f(0) = 0$ and $f'(0) = \beta > 0$. Clearly $f(z)/\beta \in \mathfrak{S}$. We have further

Lemma 7.7 *The functions* $f(z)/\beta$, *where* $f(z)$ *is constructed as in* (7.12), *form a dense subclass* $\mathfrak{S}_1$ *of* $\mathfrak{S}$.

Suppose that $f(z) \in \mathfrak{S}$ and $0 < \rho < 1$. Then $f(\rho z)$ maps $|z| < 1$ onto the interior of an analytic curve, namely, the image of $|z| = \rho$ by $f(z)$. Also $\frac{1}{\rho}f(\rho z) \in \mathfrak{S}$, $\frac{1}{\rho}f(\rho z) \to f(z)$ as $\rho \to 1$, uniformly for $|z| \le r$, when $0 < r < 1$.

It is thus sufficient to show that the functions $\frac{1}{\rho}f(\rho z)$ can be approximated by functions in $\mathfrak{S}_1$. Next if M is large

$$w = \psi(z) = \frac{1}{\rho M}f(\rho z) = \frac{1}{M}z + \cdots$$

maps $|z| < 1$ onto the interior D of a closed analytic curve Γ lying entirely in $|w| < 1$.

Let $P = re^{i\theta}$ be a point of largest modulus on Γ. We obtain the arc PP_n of Γ by going along the curve Γ in an anticlockwise sense from P until the whole of Γ apart from an arc of diameter $1/n$ has been described. Then ∞PP_n is a slit Γ_n.

Let Δ_n be the complement of Γ_n and let

$$f_n(z) = \beta_n \left(z + a_2 z^2 + \cdots \right)$$

map $|z| < 1$ onto Δ_n. To prove Lemma 7.7 it remains to show that, if $t < 1$,

$$f_n(\zeta) \to \psi(\zeta), \text{ as } n \to \infty, \text{ uniformly for } |\zeta| \le t. \tag{7.13}$$

For then $f_n(z)/\beta_n$ approximates $\rho^{-1}f(\rho z)$ for a fixed ρ, which in turn approximates $f(z)$.

We consider

$$g_n(z) = f_n^{-1}\{\psi(z)\}.$$

We verify that $g_n(0) = 0$, $g_n'(0) > 0$, so that $g_n(z)$ is a transformation. Let S_{g_n} be defined as in §7.2. Then S_{g_n} consists of all those points in $|\zeta| < 1$ which correspond to points outside D by $w = f_n(\zeta)$.

Choose now δ so small that the circle of centre P and radius R meets D in a single arc γ_R for $0 < R < \delta$, lies in $|w| < 1$, and that the origin $w = 0$ lies outside this circle. Then if $1/n < R$, γ_R corresponds to a single arc c_R in $|\zeta| < 1$ by $\zeta = f_n^{-1}(w)$ and all points of S_{g_n} are separated from $\zeta = 0$ by c_R. By Lemma 7.1 we may choose R so that the length $l(R)$ of c_R satisfies

$$l(R) \leq \pi \left[\frac{1}{2}\log(n\delta)\right]^{-\frac{1}{2}},$$

and so by Lemma 7.2 the diameter $d(g_n)$ of S_{g_n} satisfies

$$d(g_n) \leq \pi \left[\frac{1}{2}\log(n\delta)\right]^{-\frac{1}{2}}.$$

It now follows from Lemma 7.6 that

$$g_n(z) \to z \quad (n \to \infty),$$

uniformly in $|z| < 1$. We note that $w = f_n(\zeta)$ maps $|\zeta| < 1 - d(g_n)$ into D and so into $|w| < 1$. Given $0 < t < 1$, we now choose $\rho = \frac{1}{2}(1 + t)$ and suppose that $n > n_0$ so that $d(g_n) < \frac{1}{2}(1 - \rho) = \frac{1}{4}(1 - t)$. Then for $|\zeta| \leq \rho$ we have by Cauchy's inequality

$$|f_n'(\zeta)| < \frac{1}{1 - d(g_n) - |\zeta|} \leq \frac{2}{1 - \rho}.$$

Hence if $|\zeta_1| \leq \rho$, $|\zeta_2| \leq \rho$, $n > n_0$ we obtain

$$|f_n(\zeta_1) - f_n(\zeta_2)| = \left|\int_{\zeta_1}^{\zeta_2} f_n'(\zeta)d\zeta\right| \leq \frac{2}{1 - \rho}|\zeta_1 - \zeta_2|.$$

Suppose now that $|\zeta| \leq t$. Then if $n > n_1$ say, we have $|g_n(\zeta)| \leq \rho$. Now

$$|f_n(\zeta) - \psi(\zeta)| = |f_n(\zeta) - f_n\{g_n(\zeta)\}| \leq \frac{2}{1 - \rho}|\zeta - g_n(\zeta)| \to 0, \text{ as } n \to \infty.$$

This is (7.13) and the proof of Lemma 7.7 is complete.

7.5 Continuity properties Following Löwner we now investigate the class $\mathfrak{S}_1$ and show that the functions $f(z)$ in (7.12) can be obtained by a series of successive infinitesimal transformations from $w = z$.

Let γ be a slit given by $w = \alpha(t)$ $(0 \leq t \leq \infty)$. We denote by $\gamma_{t't''}$ the arc $t' \leq t \leq t''$ of γ and by γ_t the arc $\gamma_{t\infty}$. Let $G(t)$ consist of the complement of γ_t. As t increases from 0 to ∞, $G(t)$ expands from $G = G(0)$ to the whole plane. We denote by

$$w = g_t(z) = \beta(t)\left(z + a_2(t)z^2 + \cdots\right) \quad (\beta(t) > 0),$$

the function which maps $|z| < 1$ onto $G(t)$ and proceed to show that $g_t(z)$ varies continuously with t, as t increases, from $g_0(z) = f(z)$. We have first

Lemma 7.8 *If $w = g_t(z)$ is defined as above, then the inverse function $z = g_t^{-1}(w)$ remains continuous at $w = \alpha(t)$. Thus as $w = g_t(z) \to \alpha(t)$ in any manner from $G(t)$, z approaches a point $\lambda(t)$ such that $|\lambda(t)| = 1$.*

Choose δ so small that the circle $|w - \alpha(t)| = R$ meets γ_t in exactly one point for $0 < R < \delta$. The choice is possible since γ_t has a continuous tangent at $\alpha(t)$. Then for $0 < R < \delta$ the circle $|w - \alpha(t)| = R$ lies in $G(t)$ except for a single point. The image of this circle is an arc c_R lying, except for end–points, in $|z| < 1$. By Lemma 7.1 we can then choose a sequence R_n such that $l(R_n) \to 0$ and $R_n \to 0$, where $l(R_n)$ is the length of c_{R_n}. If $R_n < |\alpha(t)|$, then the disc $|w - \alpha(t)| < R_n$ cut along the arc γ_t corresponds to one of the domains into which c_{R_n} divides $|z| < 1$, namely, that one, Δ_n, which does not contain $z = 0$. By Lemma 7.2 the diameter of Δ_n is not greater than $l(R_n)$ and so tends to zero as $n \to \infty$. Since $\Delta_n \subset \Delta_{n-1}$ when $R_n < R_{n-1}$, it follows that Δ_n shrinks to a single point $\lambda(t)$ as $n \to \infty$.

Thus we may choose n so large that Δ_n lies in the disc $|z - \lambda(t)| < \varepsilon$. Then if $|w - \alpha(t)| < R_n$ and $w = g_t(z)$, we have $|z - \lambda(t)| < \varepsilon$ and this proves Lemma 7.8.

7.5.1 We next write

$$h\left(z, t', t''\right) = g_{t''}^{-1}[g_{t'}(z)] \quad \left(0 \leq t' < t'' < \infty\right).$$

We shall study the behaviour of $h(z, t', t'')$, as $t'' - t' \to 0$.

We note that $w = g_{t'}(z)$ maps $|z| < 1$ onto $G(t')$, that is the plane cut along $\gamma_{t'}$. Also $\zeta = g_{t''}^{-1}(w)$ maps $G(t'')$ onto $|\zeta| < 1$, and so $G(t')$ corresponds to $|\zeta| < 1$ except for the image of $\gamma_{t't''}$ by $\zeta = g_{t''}^{-1}(w)$. For $\gamma_{t't''}$ lies, except for one end–point $\alpha(t'')$, in $G(t'')$ but not in $G(t')$. By

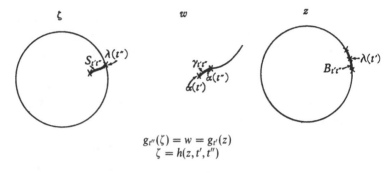

$$g_{t''}(\zeta) = w = g_{t'}(z)$$
$$\zeta = h(z, t', t'')$$

Fig. 7.

Lemma 7.8, $\zeta = g_{t''}^{-1}(w)$ is continuous also at $\alpha(t'')$ and so on the whole of $\gamma_{t't''}$ and maps $\gamma_{t't''}$ onto a Jordan arc $S_{t't''}$ in $|\zeta| \leq 1$.

Thus $\zeta = h(z, t', t'')$ maps $|z| < 1$ onto $|\zeta| < 1$ cut along $S_{t't''}$. We write $B_{t't''}$ for the set of boundary points on $|z| = 1$ which correspond to $S_{t't''}$ by this transformation. Other points of $|z| = 1$ correspond only to $|\zeta| = 1$. We note also that $z = \lambda(t')$ belongs to $B_{t't''}$ and corresponds to the tip of the cut $S_{t't''}$ and that the other end–point of $S_{t't''}$ on $|\zeta| = 1$ is $\zeta = \lambda(t'')$.

We now have

Lemma 7.9 *As $t'' - t' \to 0$, while either $t = t'$ or $t = t''$ remains fixed, both $S_{t't''}$ and $B_{t't''}$ approach the point $\lambda(t)$ in their respective planes. Also $\lambda(t)$ is continuous.*

Suppose first that $t'' \to t'$ while t' remains fixed. Then the arc $\gamma_{t't''}$ shrinks to the fixed point $\alpha(t')$ and so finally lies in a disc of centre $\alpha(t')$ and radius δ. If ε is sufficiently small it follows from Lemma 7.8 that we can choose δ depending on ε, such that if $w = g_{t'}(z)$ and $|w - \alpha(t')| < \delta$, then $|z - \lambda(t')| < \varepsilon$. Thus if $\gamma_{t't''}$ has diameter less than δ, $B_{t't''}$ has diameter less than 2ε. Thus as $t'' \to t'$, the diameter of $B_{t't''}$ tends to zero and hence so does that of $S_{t't''}$ by Lemma 7.4.

Similarly, if $t' \to t''$, while t'' remains fixed, $\gamma_{t't''}$ shrinks to the point $\alpha(t'')$. It then follows from the continuity of $g_{t''}^{-1}(w)$ at $w = \alpha(t'')$, that $S_{t't''}$, which corresponds to $\gamma_{t't''}$ by $\zeta = g_{t''}^{-1}(w)$ shrinks to the point $g_{t''}^{-1}[\alpha(t'')] = \lambda(t'')$, and so the diameter of $S_{t't''}$ tends to zero. Thus the diameter of $B_{t't''}$ tends to zero also by Lemma 7.4.

It now follows from Lemma 7.6 that in either case

$$h(z, t', t'') \to z \tag{7.14}$$

uniformly in $|z| < 1$ as $t'' - t' \to 0$ through any sequence of values. Thus if z lies on $B_{t't''}$ and ζ is a corresponding point on $S_{t't''}$, we have $|z - \zeta| < \varepsilon$ finally.

Also $S_{t't''}$ always contains $\lambda(t'')$ and $B_{t't''}$ contains $\lambda(t')$. Taking $z = \lambda(t')$ fixed, choose t'' so near t' that the diameter of $S_{t't''}$ is less than ε. Then $|\zeta - \lambda(t'')| < \varepsilon$ and $|\zeta - z| < \varepsilon$, and so

$$|\lambda(t'') - \lambda(t')| < 2\varepsilon.$$

Similarly, if t'' is fixed, choose $\zeta = \lambda(t'')$ and t' so near t'' that the diameter of $B_{t't''}$ is less than ε. Then $|\zeta - z| < \varepsilon$ and $|z - \lambda(t')| < \varepsilon$, and so $|\lambda(t'') - \lambda(t')| < 2\varepsilon$. Thus in either case $\lambda(t'') - \lambda(t') \to 0$, and if $t = t'$ or t'' and t is fixed $\lambda(t')$ and $\lambda(t'')$ approach $\lambda(t)$. Thus $\lambda(t)$ is continuous. Since the diameters of $S_{t't''}$, $B_{t't''}$ tend to zero and these sets contain $\lambda(t'')$ and $\lambda(t')$ respectively, both these sets approach $\lambda(t)$ as $t'' - t' \to 0$, while one of t', t'' remains fixed. This proves Lemma 7.9.

7.6 The differential equation We note that for $0 \le t' < t'' < \infty$,

$$h(z, t', t'') = \frac{\beta(t')}{\beta(t'')} z + \cdots$$

satisfies the conditions of Schwarz's Lemma in $|z| < 1$. Thus $\beta(t)$ is a strictly increasing function of t in $0 \le t < \infty$. Again by (7.14)

$$\frac{\beta(t')}{\beta(t'')} \to 1 \quad (t'' - t' \to 0),$$

while either t' or t'' remains fixed. Thus $\beta(t)$ is continuous. Hence

$$\tau = \log \frac{\beta(t)}{\beta(0)}$$

is a continuous strictly increasing function of t for $0 \le t < \infty$, and we may take τ for our parameter t, which has been left undetermined so far. We shall do so in what follows. With this normalization

$$g_t(z) = \beta e^t \left(z + a_2(t) z^2 + \cdots \right) \quad (0 \le t < \infty), \tag{7.15}$$

where $\beta = \beta(0) = f'(0)$. Also

$$h(z, t', t'') = e^{t' - t''} z + \cdots$$

We now prove

Lemma 7.10   *With the normalization (7.15) suppose that $t'' - t' \to 0$ while $t = t'$ or t'' remains fixed in the range $0 \le t' < t'' < \infty$. Then we have, uniformly for $|z| \le r$, when $0 < r < 1$,*

$$\frac{h(z, t', t'') - z}{t'' - t'} \longrightarrow -z\frac{1 + \kappa(t)z}{1 - \kappa(t)z},$$

where $\kappa(t) = \lambda(t)^{-1}$.

In fact Lemmas 7.6 and 7.9 show that, for $|z| \le r < 1$,

$$h(z, t', t'') - z \sim (t' - t'')z\frac{e^{i\phi} + z}{e^{i\phi} - z},$$

as $t'' - t' \to 0$, where $e^{i\phi}$ is a point of $B_{t't''}$ so that $e^{i\phi} \to \lambda(t)$. Now Lemma 7.10 follows.

We can now prove part (b) of our fundamental theorem.

Theorem 7.2   *If $f(z) \in \mathfrak{S}_1$ so that $f(z)$ is given by (7.12) with $\beta = 1$, if further $g_t(z)$ is defined by (7.15) and $f(z, t) = g_t^{-1}[f(z)]$, $0 \le t < \infty$, then $w = f(z, t)$ satisfies the differential equation (7.1) and $f(z, 0) = z$. Also $g(z, t) = e^t f(z, t)$ satisfies (7.2) and (7.3).*

The function $w = f(z)$ maps $|z| < 1$ onto the w–plane cut along the slit γ. Also $\zeta = g_t^{-1}(w)$ maps $|z| < 1$, the plane cut along the smaller slit γ_t onto $|\zeta| < 1$. Thus $\zeta = f(z, t)$ maps $|z| < 1$ (1,1) conformally onto a subset of $|\zeta| < 1$. Using also (7.15) we see that $e^t f(z, t) \in \mathfrak{S}$ so that (7.2) holds. Also, when t is large, γ_t is a ray given by (7.11) and

$$w = g_t(z) = \frac{4rz}{(1 + ze^{-i\theta})^2},$$

where $4r = e^t$. Suppose that $|z| \le \rho < 1$, so that $|f(z)| \le K$, where K is a constant. As $t \to \infty$ we see that $z \to 0$, $w \sim 4rz$ uniformly for $|w| \le K$. Thus

$$f(z, t) = g_t^{-1}\{f(z)\} \sim \frac{f(z)}{4r} = e^{-t}f(z)$$

as $t \to \infty$, uniformly for $|z| \le \rho$. This proves (7.3).

Next write $f(z, t')$ instead of z in Lemma 7.10. This is permissible since $|f(z, t')| < 1$ for $|z| < 1$. Also

$$h[f(z, t'), t', t''] = g_{t''}^{-1}\left\{g_{t'}\left(g_{t'}^{-1}[f(z)]\right)\right\} = g_{t''}^{-1}[f(z)] = f(z, t'').$$

Thus Lemma 7.10 gives

$$\frac{f(z, t'') - f(z, t')}{t'' - t'} \sim -f(z, t')\frac{1 + \kappa(t)f(z, t')}{1 - \kappa(t)f(z, t')} \qquad (7.16)$$

if $t'' - t' \to 0$, while t' or t'' remains fixed. If $t = t'$ is fixed this gives the required result for the right derivative. If t'' remains fixed the required result for the left derivative also follows. In fact by Lemma 7.9 and since $\kappa(t) = 1/\lambda(t)$

$$\kappa(t') \to \kappa(t'') \quad (t' \to t''),$$

while t'' is fixed, and by (7.14)

$$h(z, t', t'') - z \to 0 \quad (t' \to t''),$$

uniformly in $|z| < 1$. Writing $f(z, t')$ instead of z we obtain

$$f(z, t'') - f(z, t') \to 0 \quad (t' \to t''),$$

uniformly in $|z| < 1$, so that (7.16) gives, also if $t = t''$ is fixed,

$$\frac{f(z, t) - f(z, t')}{t - t'} \longrightarrow -f(z, t)\frac{1 + \kappa(t)f(z, t)}{1 - \kappa(t)f(z, t)}$$

as $t' \to t$ from below. This proves (7.1) and completes the proof of Theorem 7.2.

7.7 Completion of proof of Theorem 7.1   In order to complete the proof of Theorem 7.1 it remains to show that given $\kappa(t)$, as in Theorem 7.1, there exists a unique solution $w = f(z, t)$ of the differential equation (7.1), such that $f(z, 0) = z$, and further that $e^t f(z, t) \in \mathfrak{S}$.

It is convenient to put

$$W = U + iV = \log\left(\frac{1}{w}\right).$$

Then (7.1) becomes

$$\frac{\partial W}{\partial t} = \frac{1 + \kappa(t)e^{-W}}{1 - \kappa(t)e^{-W}} = \phi(t, W), \tag{7.17}$$

say. We note that $\Re\phi(t, W) > 0$ for $\Re W > 0$. Also if $U > \delta > 0$,

$$\left|\frac{\partial\phi}{\partial W}\right| = \frac{|2\kappa(t)e^{-W}|}{|1 - \kappa(t)e^{-W}|^2} \leq \frac{2}{(1 - e^{-\delta})^2} = K,$$

say, and

$$|\phi(t, W)| \leq \frac{2}{1 - e^{-\delta}} < K. \tag{7.18}$$

If we integrate along the straight line segment from W_1 to W_2 we deduce, if $\Re W_1 > \delta$, $\Re W_2 > \delta$,

$$|\phi(t, W_2) - \phi(t, W_1)| \leq \int_{W_1}^{W_2} \left|\frac{\partial \phi}{\partial W}\right| |dW| \leq K|W_2 - W_1|. \qquad (7.19)$$

Thus $\phi(t, W)$ is Lip in W in the half–plane $U > \delta$ uniformly with respect to t.

We now define a sequence of functions $F_n(t, \omega)$ ($n \geq 0$) as follows. We suppose that $\Re \omega > \delta > 0$ and set

$$\left.\begin{array}{l} F_0(t, \omega) \equiv \omega, \\ F_n(t, \omega) = \omega + \int_0^t \phi[\tau, F_{n-1}(\tau, \omega)]d\tau \quad (0 \leq t < \infty, \ n \geq 1). \end{array}\right\} \qquad (7.20)$$

Since $\kappa(t)$ is measurable so is $\phi(t, \omega)$ for fixed ω. Also $\phi(t, \omega)$ is bounded, so that the integral exists. We note first that $\Re F_n(t, \omega)$ increases with t for fixed ω and so remains greater than δ. For if this is true for n, then

$$\Re \phi[\tau, F_n(\tau, \omega)] > 0$$

and the result remains true for $n + 1$.

Next we have

$$|F_n(t, \omega) - F_{n-1}(t, \omega)| \leq \frac{K^n t^n}{n!} \quad (0 \leq t < \infty, \ n \geq 1).$$

In fact if $n = 1$ we have from (7.18)

$$F_1(t, \omega) - F_0(t, \omega) = \left|\int_0^t \phi(\tau, \omega)d\tau\right| \leq \int_0^t K d\tau = Kt.$$

Also if our result is true for n, then (7.19) gives

$$|F_{n+1}(t, \omega) - F_n(t, \omega)| = \left|\int_0^t \{\phi[\tau, F_n(\tau, \omega)] - \phi[\tau, F_{n-1}(\tau, \omega)]\}d\tau\right|$$

$$\leq K \int_0^t |F_n(\tau, \omega) - F_{n-1}(\tau, \omega)|d\tau \leq K \int_0^t \frac{K^n \tau^n}{n!}d\tau = \frac{K^{n+1}t^{n+1}}{(n+1)!}.$$

Thus the result is true for $n + 1$.

It follows that the sequence $F_n(t, \omega)$ converges uniformly for $\Re \omega > \delta$ and $0 \leq t \leq t_0$ to a limit function $F(t, \omega)$. Taking the limit in (7.20) we deduce

$$F(t, \omega) = \omega + \int_0^t \phi[\tau, F(\tau, \omega)]d\tau. \qquad (7.21)$$

Thus $\Re F(t, \omega) \geq \delta$, and so, by (7.18), the integrand in (7.21) is bounded by K. Hence $F(t, \omega)$ is absolutely continuous in any finite interval

$0 \leq t \leq t_0$, and satisfies (7.17) almost everywhere, and conversely an absolutely continuous function $W(t) = F(t, \omega)$, which satisfies (7.17) almost everywhere and $W(0) = \omega$, also satisfies (7.21).

Next we prove that $F(t, \omega)$ is analytic in ω. Since $F_n(t, \omega)$ converges to $F(t, \omega)$ uniformly for fixed t and $\Re\omega \geq \delta$, it is enough to show that $F_n(t, \omega)$ is analytic in ω for $n = 0, 1, 2, \ldots$ We prove by induction on n that $F_n(t, \omega)$ is analytic and further that, for $\Re\omega_1 \geq \delta$, $\Re\omega_2 \geq \delta$, $t \geq 0$, we have

$$|F_n(t, \omega_2) - F_n(t, \omega_1)| \leq e^{Kt}|\omega_2 - \omega_1|.$$

These results are clearly true for $n = 0$ by (7.20). Suppose that they hold for n. Then (7.19) and (7.20) yield

$$|F_{n+1}(t, \omega_2) - F_{n+1}(t, \omega_1)| \leq |\omega_2 - \omega_1| \left\{ 1 + \int_0^t Ke^{K\tau} d\tau \right\} = e^{Kt}|\omega_2 - \omega_1|.$$

Also $\{\phi[\tau, F_n(\tau, \omega_2)] - \phi[\tau, F_n(\tau, \omega_1)]\}/(\omega_2 - \omega_1)$ is bounded for $0 \leq \tau \leq t$ and tends to

$$\left\{ \frac{\partial}{\partial\omega} \phi[\tau, F_n(\tau, \omega)] \right\}_{\omega_1} = \left[\frac{\partial F_n(t, \omega)}{\partial\omega} \right]_{\omega_1} \left[\frac{\partial \phi(\tau, W)}{\partial W} \right]_{F_n(\tau, \omega_1)}$$

as $\omega_2 \to \omega_1$ through any sequence. Thus by Lebesgue's dominated convergence theorem (Titchmarsh [1939, p. 345])

$$\frac{F_{n+1}(t, \omega_2) - F_{n+1}(t, \omega_1)}{\omega_2 - \omega_1} = 1 + \int_0^t \frac{\phi[\tau, F_n(\tau, \omega_2)] - \phi[\tau, F_n(\tau, \omega_2)]}{\omega_2 - \omega_1} d\tau$$

$$\longrightarrow 1 + \int_0^t \left\{ \frac{\partial}{\partial\omega} \phi[\tau, F_n(\tau, \omega)] \right\}_{\omega_1} d\tau$$

as $\omega_2 \to \omega_1$ through any sequence and so generally. Thus $F_{n+1}(t, \omega)$ is analytic and the inductive step is proved.

Suppose next that $W(t)$ is any absolutely continuous solution of (7.17) which satisfies $\Re W(0) > \delta$ and so $\Re W(t) > \delta$ $(0 \leq t \leq t_0)$ and further that

$$W(t_1) = F(t_1, \omega)$$

for some pair (t_1, ω), such that $0 \leq t_1 \leq t_0$ and $\Re\omega > \delta$. Then $W(t) \equiv F(t, \omega)$ $(0 \leq t \leq t_0)$. In fact write

$$M = \sup_{0 \leq t \leq t_0} |W(t) - F(t, \omega)|.$$

Then (7.17) gives, for $0 \le t \le t_0$,

$$W(t) - W(t_1) = \int_{t_1}^{t} \phi[\tau, W(\tau)]d\tau,$$

$$F(t, \omega) - F(t_1, \omega) = \int_{t_1}^{t} \phi[\tau, F(t, \omega)]d\tau,$$

and so

$$|F(t, \omega) - W(t)| = \left| \int_{t_1}^{t} \{\phi[\tau, W(\tau)] - \phi[\tau, F(t, \omega)]\}d\tau \right|. \qquad (7.22)$$

We deduce by induction that

$$|F(t, \omega) - W(t)| \le \frac{MK^n|t - t_1|^n}{n!}$$

for every positive integer n. For this holds for $n = 0$ by hypothesis. If it is true for n, then (7.19) and (7.22) give

$$|F(t, \omega) - W(t)| \le K \left| \int_{t_1}^{t} \frac{MK^n(\tau - t_1)^n d\tau}{n!} \right| = \frac{MK^{n+1}|t - t_1|^{n+1}}{(n+1)!}$$

as required. Since n is arbitrary $W(t) \equiv F(t, \omega)$.

Taking $t_1 = 0$, we see that $F(t, \omega)$ is the unique solution of (7.17) such that $F(0, \omega) = \omega$. We also see that $F(t, \omega)$ is a univalent function of ω for $\Re\omega > 0$ and a fixed positive t. For if

$$F(t, \omega_1) = F(t, \omega_2),$$

the above uniqueness theorem shows that $F(0, \omega_1) = F(0, \omega_2)$, i.e. $\omega_1 = \omega_2$. Again, if $\omega_2 - \omega_1 = 2\pi i$, we easily see that

$$F(t, \omega_2) - F(t, \omega_1) = 2\pi i.$$

Thus $f(z, t)$, given by

$$-\log f(z, t) = F(t, -\log z) \quad (0 \le t \le t_0),$$

is for fixed t an analytic one–valued univalent function of z for $0 < |z| < 1$, and is the unique solution of the differential equation (7.1), which satisfies $f(z, 0) = z$. Also $|f(z, t)| < 1$ in $|z| < 1$, and so $f(z, t)$ remains regular at $z = 0$ and is clearly zero there. If we apply (7.1) to $g(t) = f(z, t)/z$, we obtain

$$\frac{\partial g(t)}{\partial t} = -g \frac{1 + z\kappa(t)g}{1 - z\kappa(t)g},$$

and putting $z = 0$, we deduce

$$\frac{\partial g}{\partial t} = -g, \quad g(t) = e^{-t}g(0) = e^{-t}.$$

Thus $f(z, t) = ze^{-t} + \cdots$ near $z = 0$ and so $e^t f(z, t) \in \mathfrak{S}$. This completes the proof of Theorem 7.1.

7.8 The third coefficient We proceed to give a number of applications of Theorem 7.1. Let $\mathfrak{S}$ denote as usual the class of functions

$$f(z) = z + a_2 z^2 + \cdots$$

which are univalent in $|z| < 1$. In this section we prove

Theorem 7.3 *If $f(z) \in \mathfrak{S}$, then $|a_3| \le 3$.*[†]

By Theorem 7.1, we may confine ourselves to the functions $e^{t_0} f(z, t_0)$, $0 < t_0 < \infty$, of that theorem, since they form a dense subclass of $\mathfrak{S}$. We write $\beta = e^{-t_0}$,

$$f(z, t_0) = \beta \left(z + \sum_{n=2}^{\infty} a_n z^n \right),$$

and proceed to develop Löwner's formulae for a_n in terms of $\kappa(t)$. It is convenient to work with the function $g_t(\zeta)$ defined by

$$g_t[f(z, t)] = f(z, t_0) \quad (0 \le t \le t_0).$$

We write $\zeta = f(z, t)$ and differentiate the above relation with respect to t. This gives

$$\frac{\partial g}{\partial \zeta} \frac{\partial \zeta}{\partial t} + \frac{\partial g}{\partial t} = 0.$$

Substituting in (7.1) we obtain

$$\frac{\partial g}{\partial t} = \frac{\partial g}{\partial \zeta} \zeta \frac{1 + \kappa \zeta}{1 - \kappa \zeta}. \tag{7.23}$$

We have $f(z, t) = e^{-t}(z + \cdots)$ near $z = 0$ and so

$$g_t(\zeta) = \beta e^t \left[\zeta + \sum_{n=2}^{\infty} c_n(t) \zeta^n \right]$$

[†] A rather more detailed discussion shows that $|a_3| < 3$ except when $f(z) = z(1 - ze^{i\theta})^{-2}$. A similar remark applies to Theorem 7.4. See Löwner [1923].

near $\zeta = 0$. We substitute in (7.23), equate coefficients of ζ^n, and obtain for $n \geq 2$

$$c_n'(t) + c_n(t) = nc_n(t) + 2\sum_{r=1}^{n-1}(n-r)c_{n-r}(t)[\kappa(t)]^r,$$

i.e.

$$c_n'(t) = (n-1)c_n(t) + 2\sum_{r=1}^{n-1}rc_r(t)[\kappa(t)]^{n-r} \quad (0 \leq t \leq t_0). \tag{7.24}$$

The term–by–term differentiation with respect to t may be justified by expressing the coefficients as contour integrals in terms of $g_t(\zeta)$. Also when $t = 0$, $f(z,0) \equiv z$, $g_0(z) \equiv f(z,t_0)$, $c_n(0) = a_n$. When $t = t_0$, $g_t(z) \equiv z$ and so $c_n(t_0) = 0$ $(n \geq 2)$. Finally, $c_1(t) \equiv 1$. Thus we have the boundary conditions

$$c_1(t) \equiv 1; \quad c_n(0) = a_n, \quad c_n(t_0) = 0 \quad (n \geq 2).$$

From this system we can successively determine the coefficients. We obtain in fact for $n \geq 2$

$$c_n(t) = -2e^{(n-1)t}\int_t^{t_0}\left\{\sum_{r=1}^{n-1}rc_r(\tau)[\kappa(\tau)]^{n-r}\right\}e^{-(n-1)\tau}d\tau \quad (0 \leq t \leq t_0).$$

This gives

$$c_2(t) = -2e^t\int_t^{t_0}\kappa(\tau)e^{-\tau}d\tau. \tag{7.25}$$

Also

$$c_3(t) = -2e^{2t}\left\{\int_t^{t_0}\kappa(\tau)^2e^{-2\tau}d\tau + 2\int_t^{t_0}\kappa(\tau)c_2(\tau)e^{-2\tau}d\tau\right\}$$

$$= -2e^{2t}\left\{\int_t^{t_0}[\kappa(\tau)]^2e^{-2\tau}d\tau - 2\left(\int_t^{t_0}\kappa(\tau)e^{-\tau}d\tau\right)^2\right\}, \tag{7.26}$$

since

$$\frac{d}{dt}\left(\int_t^{t_0}\kappa(\tau)e^{-\tau}d\tau\right)^2 = -2\kappa(t)e^{-t}\int_t^{t_0}\kappa(\tau)e^{-\tau}d\tau = \kappa(t)c_2(t)e^{-2t}.$$

We deduce at once that $|a_2| = |c_2(0)| \leq 2$. To find the upper bound for $|a_3| = |c_3(0)|$, we may without loss of generality suppose a_3 real and positive, since this may be achieved by considering $e^{-i\phi}f(ze^{i\phi})$ instead of $f(z)$. We write $\kappa(t) = e^{i\theta(t)}$ and consider

$$\Re a_3 = -2\int_0^{t_0} \left[2\cos^2\theta(t) - 1\right]e^{-2t}dt + 4\left(\int_0^{t_0}\cos\theta(t)e^{-t}dt\right)^2$$

$$-4\left(\int_0^{t_0}\sin\theta(t)e^{-t}dt\right)^2.$$

To obtain an upper bound for $\Re a_3$ we omit the third term on the right. The first term is

$$-4\int_0^{t_0}e^{-2t}\cos^2\theta(t)dt + 1 - e^{-2t_0},$$

and by Schwarz's inequality the second term is at most

$$4\int_0^{t_0}e^{-t}dt\int_0^{t_0}e^{-t}\cos^2\theta(t)dt < 4\int_0^{t_0}e^{-t}\cos^2\theta(t)dt.$$

Thus we obtain

$$|a_3| = \Re a_3 \leq 1 + 4\int_0^{t_0}\cos^2\theta(t)(e^{-t} - e^{-2t})dt$$

$$\leq 1 + 4\int_0^{\infty}(e^{-t} - e^{-2t})dt = 3.$$

This proves Theorem 7.3.

7.8.1 The Fekete–Szegö Theorem

While Theorem 7.3 is a special case of de Branges' Theorem, which will be proved in the next chapter, Löwner's technique can be refined to yield the following result which has interesting applications to coefficients of powers of univalent functions and in particular of odd univalent functions. (Examples 7.3 and 7.4)

Theorem 7.4 (Fekete–Szegö [1933].)   *If $f \in \mathfrak{S}$ and $0 < \alpha < 1$, then*

$$\left|a_3 - \alpha a_2^2\right| \leq 1 + 2e^{-2\alpha/(1-\alpha)}.$$

Equality holds for a function $f(z)$ in $\mathfrak{S}$ and real for real z.

We follow the argument given by Duren [1983, p. 104] and start with the following lemma essentially due to Valiron and Landau [see Landau 1929].

Lemma 7.11   *Let $\phi(t)$ be real–valued and continuous for $t \geq 0$ and satisfy $|\phi(t)| \leq e^{-t}$ and*

$$\int_0^\infty \phi(t)^2 dt = \left(\lambda + \frac{1}{2}\right) e^{-2\lambda}, \tag{7.27}$$

where $0 \leq \lambda < \infty$. Then

$$\left| \int_0^\infty \phi(t) dt \right| \leq (\lambda + 1) e^{-\lambda}. \tag{7.28}$$

Equality holds if and only if $\phi(t) = \pm\psi(t)$, where

$$\left. \begin{array}{rcll} \psi(t) & = & e^{-\lambda}, & 0 \leq t \leq \lambda \\ \psi(t) & = & e^{-t}, & \lambda < t < \infty. \end{array} \right\} \tag{7.29}$$

We note that

$$\int_0^\infty \phi(t)^2 dt \leq \int_0^\infty e^{-2t} dt = \frac{1}{2}.$$

Since $\left(\lambda + \frac{1}{2}\right) e^{-2\lambda}$ decreases from $\frac{1}{2}$ to 0 as λ increases from 0 to ∞, λ is uniquely determined by (7.27). We also note that if $\phi(t) = \pm\psi(t)$ equality holds in (7.27) and (7.28). Finally we observe that

$$F(t) = [\psi(t) - |\phi(t)|] \left[2e^{-\lambda} - \psi(t) - |\phi(t)| \right] \geq 0, \quad 0 \leq t < \infty.$$

For if $t \leq \lambda$ the right-hand side is $\left(e^{-\lambda} - |\phi(t)| \right)^2$ and if $t > \lambda$

$$|\phi(t)| \leq e^{-t} = \psi(t) \quad \text{and} \quad 2e^{-\lambda} - \psi(t) - |\phi(t)| \geq e^{-t} - \phi(t) \geq 0.$$

Hence

$$\begin{aligned} 0 \leq & \int_0^\infty F(t) dt \\ = & \; 2e^{-\lambda} \left\{ \int_0^\infty \psi(t) dt - \int_0^\infty |\phi(t)| dt \right\} - \int_0^\infty \psi(t)^2 dt + \int_0^\infty \phi(t)^2 dt \\ = & \; 2e^{-\lambda} \left\{ (\lambda + 1) e^{-\lambda} - \int_0^\infty |\phi(t)| dt \right\} \end{aligned}$$

by (7.27) and (7.29). Also equality holds if and only if $F(t) \equiv 0$, i.e. $\phi(t) \equiv \pm\psi(t)$. This proves Lemma 7.11.

We can now prove Theorem 7.4. We suppose without loss of generality that $\Delta = a_3 - \alpha a_2^2$ is real and positive. For if we replace $f(z)$ by $e^{i\phi} f(ze^{-i\phi})$, Δ is replaced by $e^{-2i\phi}\Delta$, and we may choose ϕ so that $\Delta e^{-2i\phi} > 0$. We write $\kappa(t) = \cos\theta(t) + i\sin\theta(t)$. Using (7.25), (7.26) with $t = 0$ we obtain

$$\Delta = \mathfrak{R}\left(a_3 - \alpha a_2^2\right)$$

$$= 2\int_0^{t_0} \left(1 - 2\cos^2\theta(t)\right)e^{-2t}dt + 4(1-\alpha)\left\{\int_0^{t_0} e^{-t}\cos\theta(t)dt\right\}^2$$

$$-4(1-\alpha)\left(\int_0^{t_0} e^{-t}\sin\theta(t)dt\right)^2$$

$$\leq 4(1-\alpha)\left(\int_0^\infty \phi(t)dt\right)^2 - 4\int_0^\infty \phi^2(t)dt + 1, \qquad (7.30)$$

where $\phi(t) = e^{-t}\cos\theta(t)$, $0 \leq t \leq t_0$, $\phi(t) = 0$, $t > t_0$. We apply Lemma 7.11 and deduce that

$$\mathfrak{R}(a_3 - \alpha a_2^2) \leq 4(1-\alpha)(\lambda+1)^2 e^{-2\lambda} - (4\lambda+2)e^{-2\lambda} + 1 \qquad (7.31)$$

where

$$\int_0^{t_0} \phi(t)^2 dt = \left(\lambda + \frac{1}{2}\right)e^{-2\lambda}.$$

The right-hand side of (7.31) attains its maximum $1 + 2\exp\{-2\alpha/(1-\alpha)\}$ when $\lambda = \alpha/(1-\alpha)$. Thus we obtain

$$\Delta = \mathfrak{R}\left(a_3 - \alpha a_2^2\right) \leq 2e^{-2\alpha/(1-\alpha)} + 1$$

as required.

To find an extremal we need to choose $\kappa(t) = e^{i\theta(t)}$ so that equality holds in (7.31) with $\lambda = \alpha/(1-\alpha)$. Setting $\kappa(t) = e^{i\theta(t)}$, and using Lemma 7.11, we see that we must have

$$\begin{aligned} \phi(t) &= e^{-t}\cos\theta(t) &= e^{-\lambda}, &\quad 0 < t < \lambda \\ \phi(t) & &= e^{-t}, &\quad t > \lambda. \end{aligned}$$

Thus $\theta(t)$ is given by

$$\cos\theta(t) = e^{t-\lambda}, \quad 0 \leq t \leq \lambda$$

and

$$\theta(t) = 0, \quad t > \lambda.$$

We may suppose that $-\frac{\pi}{2} < \theta(t) < \frac{\pi}{2}$ for $0 \leq t < \lambda$. To obtain equality in (7.30) we also choose the sign of $\theta(t)$ so that

$$\int_0^\infty e^{-t}\sin\theta(t)dt = \int_0^\lambda e^{-t}\sin\theta(t)dt = 0.$$

Clearly this can be done in many different ways. For instance we may choose $\theta(t) > 0$ for $0 < t < \tau$, $\theta(t) < 0$ for $\tau < t < \lambda$. Then

$$I(\tau) = \int_0^\tau e^{-t} \sqrt{\{1 - e^{2(t-\lambda)}\}} \, dt - \int_\tau^1 e^{-t} \sqrt{\{1 - e^{2(t-\lambda)}\}} \, dt$$

is an increasing function of τ and $I(0) < 0 < I(\lambda)$. Thus there exists a unique value of τ, such that $0 < \tau < \lambda$ and $I(\tau) = 0$. (See Duren [1983, p. 107].)

To find a function with real coefficients we proceed as follows. Let N be a large positive integer and define $\theta(t) = \theta_N(t)$ by

$$\left.\begin{array}{llll}\theta_N(t) & > & 0, & \frac{2\lambda}{2N} \leq t < \frac{(2m+1)\lambda}{2N}, \\[2mm] \theta_N(t) & < & 0, & \frac{(2m+1)\lambda}{2N} \leq t < \frac{(2m+2)\lambda}{2N},\end{array}\right\} \quad 0 \leq m \leq N - 1.$$

Then evidently

$$\int_0^\lambda e^{-t} \sin \theta_N(t) dt \to 0 \quad \text{as} \quad N \to \infty. \tag{7.32}$$

If $\kappa(t) = e^{i\theta_N(t)}$, let $w = f_N(z,t)$ be the corresponding solution of (7.1) for $0 \leq t \leq \lambda$. We write $\eta = \lambda/N$ and $w_m = f_N(z, m\eta)$,

$$\delta_m = w_{m+1} - w_m.$$

Then (7.1) shows that to a first approximation

$$\begin{aligned} \delta_m & \doteq -\frac{\eta}{2} w_m \left\{ \frac{1 + \kappa(m\eta)w_m}{1 - \kappa(m\eta)w_m} + \frac{1 + \overline{\kappa(m\eta)}w_m}{1 - \overline{\kappa(m\eta)}w_m} \right\} \\[2mm] & = -\eta w_m \frac{1 - w_m^2}{1 + w_m^2 - 2w_m \Re\kappa(m\eta)} \\[2mm] & = -\eta w_m \frac{1 - w_m^2}{1 + w_m^2 - 2w_m e^{m\eta - \lambda}}. \end{aligned}$$

Hence, as $N \to \infty$ and uniformly for $|z| \leq r$, $0 \leq t \leq \lambda$, where $r < 1$,

$$f_N(z,t) \to f(z,t),$$

where $w = f(z,t)$ is the solution of

$$\frac{\partial w}{\partial t} = -w \frac{1 - w^2}{1 + w^2 - 2we^{t-\lambda}}, \quad f(z,0) = z. \tag{7.33}$$

Also $e^t f_N(z,t) \in \mathfrak{S}$, so that for $z = 0$

$$\frac{\partial}{\partial z} f(z,t) = \lim \frac{\partial}{\partial z} f_N(z,t) = e^{-t}.$$

Hence $g(z,t) = e^t f(z,t)$ is not constant and so belongs to $\mathfrak{S}$. Clearly $g(z,t)$ is real for real z and so has real coefficients.

Once $f(z,\lambda)$ has been obtained we apply (7.1) with $\kappa(t) = 1$ for $t > \lambda$, i.e.

$$\frac{\partial f}{\partial t} = -f\frac{1+f}{1-f},$$

so that for $t > \lambda$

$$\frac{e^t f(z,t)}{[1 + f(z,t)]^2} = \frac{e^\lambda f(z,\lambda)}{[1 + f(z,\lambda)]^2}.$$

We recall that, as $t \to \infty$ for fixed z, $e^t f(z,t) \in \mathfrak{S}$ and so remains bounded, so that $f(z,t) \to 0$. Thus

$$g(z,t) = e^t f(z,t) \to g(z) = \frac{e^\lambda f(z,\lambda)}{[1 + f(z,\lambda)]^2} \tag{7.34}$$

and $g(z) \in \mathfrak{S}$, $g(z)$ has real coefficients by (7.33) and $g(z)$ yields equality in Theorem 7.4, since (7.32) holds. Also $g(z)$ is given by (7.33) and (7.34).

Examples

7.1   Use Lemma 1.1 to show that if $f(z) \in \mathfrak{S}$ then $\left|a_3 - a_2^2\right| \le 1$, with equality if and only if $f(z) = f_\theta(z) = z/\left(1 - ze^{i\theta}\right)^2$.

(The function

$$g(z) = \frac{1}{f(z)} = \frac{1}{z} + \sum_{n=0}^{\infty} b_n z^n$$

is univalent in $0 < |z| < 1$ and $b_1 = a_2^2 - a_3$.)

7.2   Show that, if $\alpha < 0$ or $\alpha \ge 1$ and $f(z) \in \mathfrak{S}$, then

$$\left|a_3 - \alpha a_2^2\right| \le |4\alpha - 3|$$

with equality only for $f(z) = f_\theta(z)$.

(If $\alpha > 1$, note that $\left|\alpha a_2^2 - a_3\right| \le \left|a_2^2 - a_3\right| + (\alpha - 1)\left|a_2^2\right|$.)

7.3   If $f(z) = z + \sum_2^\infty a_n z^n \in \mathfrak{S}$ and

$$\left(\frac{f(z)}{z}\right)^\lambda = 1 + \sum_2^\infty a_n(\lambda)z^{n-1},$$

show that $a_2(\lambda) = \lambda a_2$ and $a_3(\lambda) = \lambda a_3 + \frac{1}{2}\lambda(\lambda - 1)a_2^2$. Deduce the sharp bounds $|a_2(\lambda)| \le 2\lambda$, $-\infty < \lambda < +\infty$ and

$$|a_3(\lambda)| \le |\lambda|\left(1 + 2e^{2(\lambda - 1)/(\lambda + 1)}\right), \quad -1 < \lambda < 1.$$

Find also the sharp bound for $a_3(\lambda)$ when $|\lambda| \geq 1$. (Hayman and Hummel [1986].)

7.4   If k is a positive integer and

$$g(z) = z + \sum_{n=1}^{\infty} b_{nk+1} z^{nk+1} \in \mathfrak{S}$$

we recall from the end of Section 5.8 that $g(z) = f(z^k)^{\frac{1}{k}}$, where $f(z) \in \mathfrak{S}$ and conversely. Deduce the sharp bounds

$$|b_{k+1}| \leq \frac{2}{k} \quad \text{and} \quad |b_{2k+1}| \leq \frac{1}{2}\left\{1 + 2e^{2(1-k)/(1+k)}\right\}.$$

([Fekete and Szegő 1933]. If $k = 2$ we obtain $|b_3| \leq 1$, $|b_5| \leq \frac{1}{2} + e^{-2/3} = 1.013\ldots$)

7.9 Coefficients of the inverse functions

Suppose again that $w = f(z) \in \mathfrak{S}$ and let the inverse function

$$z = \phi(w) = f^{-1}(w)$$

be given by

$$\phi(w) = w + \sum_{m=2}^{\infty} b_m w^m$$

near $w = 0$. By means of Theorem 7.1 Löwner obtained the exact bounds of all the coefficients b_m. His result is

Theorem 7.5   *If $w = f(z) \in \mathfrak{S}$ and $z = f^{-1}(w) = w + \sum_{m=2}^{\infty} b_m w^m$ then*

$$|b_m| \leq \frac{1.3.5.\ldots.(2m-1)2^m}{(m+1)!} \quad (m \geq 2).$$

Equality holds when

$$f(z) = \frac{z}{(1+z)^2}, \quad f^{-1}(w) = \left[1 - 2w - (1-4w)^{\frac{1}{2}}\right]/(2w).$$

It is again sufficient to consider instead of $f(z)$ the functions $e^{t_0}f(z,t_0)$ of Theorem 7.1. For these functions can be used to approximate a general $f(z)$ in $\mathfrak{S}$, so that their coefficients approximate those of $f(z)$, and the coefficients of the inverse functions $f^{-1}(w)$, which are polynomials in the coefficients of $f(z)$, can be similarly approximated.

Let then $f(z,t)$ be the function of Theorem 7.1 and let $z = \phi_t(w)$ be the inverse function so that

$$\phi_t[f(z,t)] = z. \qquad (7.35)$$

We write

$$\phi_t(w) = e^t \left[w + \sum_{m=2}^{\infty} b_m(t) w^m \right] \quad (0 \le t \le t_0), \qquad (7.36)$$

and $\beta = e^{-t_0}$. Thus

$$\phi(w) = \phi_{t_0}(\beta w) = w + \sum_{m=2}^{\infty} b_m w^m$$

is inverse to $\beta^{-1} f(z,t_0)$ in $\mathfrak{S}$, and so we need only prove our inequalities for the coefficients b_m of this function $\phi(w)$, where

$$b_m = \beta^{m-1} b_m(t_0).$$

We note that the equations (7.1) and (7.35) lead again to the analogue of (7.23), namely,

$$\frac{\partial \phi_t(w)}{\partial t} = \frac{\partial \phi_t(w)}{\partial w} w \frac{1 + \kappa(t)w}{1 - \kappa(t)w}.$$

Substituting from (7.36) we obtain just as in (7.24)

$$b_m'(t) + b_m(t) = m b_m(t) + 2 \sum_{r=1}^{m-1} r b_r(t) [\kappa(t)]^{m-r} \quad (0 \le t \le t_0, \ m \ge 2),$$

with the boundary conditions

$$b_1(t) \equiv 1 \ (0 \le t \le t_0), \quad b_m(0) = 0, \quad b_m(t_0) = \beta^{-m+1} b_m \ (m \ge 2).$$

These yield the inductive relation

$$b_m(t) = 2 e^{(m-1)t} \int_0^t \left\{ \sum_{r=1}^{m-1} r b_r(\tau) [\kappa(\tau)]^{m-r} \right\} e^{-(m-1)\tau} d\tau \quad (m > 1).$$

It is now clear that, if $t_0 > 0$, $m > 1$, $|b_m(t_0)|$ attains its maximum possible value if $\kappa(t) \equiv 1$. In this case all the $b_m(t)$ are real and positive. It is also evident that the corresponding value of $b_m = e^{-(m-1)t_0} b_m(t_0)$ increases with increasing t_0 and so we obtain the upper bound for variable t_0 in the limit as $t_0 \to \infty$.

We now take $\kappa(t) \equiv 1$ in the differential equation (7.1) and obtain on integration

$$\frac{f(z, t_0)}{1 + [f(z, t_0)]^2} = \frac{\beta z}{(1 + z)^2},$$

i.e.

$$\frac{w}{(1 + w)^2} = \frac{\beta \phi_{t_0}(w)}{[1 + \phi_{t_0}(w)]^2}.$$

Writing $\phi(w) = \phi_{t_0}(\beta w)$ we deduce

$$\frac{\phi(w)}{[1 + \phi(w)]^2} = \frac{w}{(1 + \beta w)^2}.$$

Thus $\beta \to 0$ $(t_0 \to \infty)$ and $\phi(w) \to \psi(w)$, where

$$\frac{\psi(w)}{[1 + \psi(w)]^2} = w, \quad \text{i.e.} \quad \psi(w) = \frac{1 - 2w - \sqrt{(1 - 4w)}}{2w} = \sum_{m=1}^{\infty} b_m w^m,$$

and

$$b_m = \frac{1.3.\ldots.(2m - 1)2^m}{(m + 1)!},$$

as required. The inverse function $w = \psi^{-1}(z) = z(1 + z)^{-2} \in \mathfrak{S}$, and so $\psi(w)$ has the largest coefficients in our class and Theorem 7.5 is proved.

7.10 The argument[†] of $f(z)/z$ While the elementary methods of Chapter 1 are adequate to obtain the bounds for $|f(z)|$, $|f'(z)|$, etc., when $f(z) \in \mathfrak{S}$, the bounds for $\arg(f(z)/z)$, $\arg f'(z)$, etc., lie deeper. Here the function $z(1 - z)^{-2}$ is no longer extremal. Following Grunsky [1932], we prove

Theorem 7.6   *Suppose that $f(z) \in \mathfrak{S}$. Then we have*

$$-\log \frac{1 + |z|}{1 - |z|} \le \arg \frac{f(z)}{z} \le \log \frac{1 + |z|}{1 - |z|}. \tag{7.37}$$

Both these inequalities are sharp for any fixed z in $|z| < 1$.

It is again sufficient to consider the functions $e^{t_0} f(z, t_0)$ of Theorem 7.1, and hence $f(z, t_0)$, since $\arg e^{t_0} = 0$. We write $f = f(z, t)$, $\kappa = \kappa(t)$ for short. Then (7.1) gives

$$\frac{\partial}{\partial t} \log f = -\frac{1 + \kappa f}{1 - \kappa f} = \frac{-(1 + \kappa f)(1 - \bar{\kappa}\bar{f})}{|1 - \kappa f|^2}.$$

[†] If $\phi(z) \neq 0$ in $|z| < 1$ and $\phi(0) > 0$ we define $\arg \phi(z)$ as the imaginary part of $\log \phi(0) + \int_0^z \phi'(\zeta)/\phi(\zeta) \, d\zeta$, where the integral is taken along a straight line segment.

Taking real and imaginary parts we deduce

$$\frac{\partial}{\partial t}\log|f| = -\frac{1-|f|^2}{|1-\kappa f|^2}, \tag{7.38}$$

$$\frac{\partial}{\partial t}\arg f = -\frac{2\Im(\kappa f)}{|1-\kappa f|^2}. \tag{7.39}$$

The equation (7.38) shows that $|f|$ decreases strictly with increasing t. Also (7.38) and (7.39) together give

$$d_t \arg f \le \frac{2|f|}{1-|f|^2} d_t \log\frac{1}{|f|}. \tag{7.40}$$

We integrate this from $t=0$ to t_0 and note that $f(z,0)=z$. Thus

$$\arg\frac{f(z,t_0)}{z} \le \log\frac{(1+|z|)(1-|f(z,t_0)|)}{(1-|z|)(1+|f(z,t_0)|)} \le \log\frac{1+|z|}{1-|z|}.$$

This yields the upper bound in (7.37). The lower bound follows similarly. It is worth noting that the methods of Chapter 1 only lead to

$$\left|\arg\frac{f(z)}{z}\right| \le 2\log\frac{1+|z|}{1-|z|}.$$

We next show that the right–hand inequality of (7.37) is sharp for a fixed value z_0 of z in $|z|<1$. To do this we have to find $\kappa(t)$ in an assigned range $0 \le t \le t_0$ so that the solution $f = f(z_0,t)$ of the differential equation

$$\frac{df}{dt} = -f\frac{1+\kappa f}{1-\kappa f} \quad (0 \le t \le t_0),$$

with the initial condition $f(z_0,0)=z_0$ satisfies

$$\Im(\kappa f) = -|f|.$$

For in this case we shall have equality in (7.40). Also (7.38) gives in this case

$$\frac{\partial}{\partial t}\log|f| = -\frac{1-|f|^2}{1+|f|^2},$$

so that $f = f(z_0,t)$ satisfies

$$\frac{|f|}{1-|f|^2} = e^{-t}\frac{|z_0|}{1-|z_0|^2}.$$

We define $|f| = |f(z_0,t)|$ by this equation, then $\arg f(z_0,t)$ by

$$d_t \arg f = -\frac{2d_t|f|}{1-|f|^2},$$

so that

$$\arg \frac{f(z_0, t)}{z_0} = \log \left(\frac{1 + |z_0|}{1 - |z_0|} \right) - \log \frac{1 + |f|}{1 - |f|},$$

and finally $\kappa(t)$ by

$$\kappa(t) = \frac{-i|f(z_0, t)|}{f(z_0, t)}.$$

With these definitions of $\kappa(t)$, $f(z_0, t)$ (7.38) and (7.39) are satisfied and $\arg[f(z_0, t)/z_0]$ can be chosen as close as we please to

$$\log[(1 + |z_0|)/(1 - |z_0|)],$$

since $f(z_0, t) \to 0$ $(t \to \infty)$. Thus the solution $e^{t_0} f(z_0, t_0)$ of the equation (7.1) corresponding to this value of $\kappa(t)$ belongs to $\mathfrak{S}$ and approaches the upper bound in (7.37) arbitrarily closely, so that this bound is sharp. By considering $\overline{f(z_0, t)}$ instead of $f(z_0, t)$ we can show that the lower bound is sharp. This completes the proof of Theorem 7.6.

7.11 Radii of convexity and starshapedness The function $f(z)$ maps

$$|z| = r$$

onto a convex curve $\gamma(r)$, if the tangent to $\gamma(r)$ at the point $f(re^{i\theta})$ turns continuously in an anticlockwise direction as θ increases. The condition for this is that $\arg[ire^{i\theta} f'(re^{i\theta})]$ increases with increasing θ, so that

$$\frac{\partial}{\partial \theta} \arg f'(re^{i\theta}) + 1 \geq 0 \quad (0 \leq \theta \leq 2\pi),$$

i.e.

$$\Im \left\{ \frac{\partial}{\partial \theta} \log f'(re^{i\theta}) \right\} = \Re \left\{ re^{i\theta} \frac{f''(re^{i\theta})}{f'(re^{i\theta})} \right\} \geq -1 \quad (0 \leq \theta \leq 2\pi).$$

This gives

$$\Re \left\{ z \frac{f''(z)}{f'(z)} \right\} \geq -1 \quad (|z| = r).$$

The inequality (1.6) of Chapter 1 gives for $f(z) \in \mathfrak{S}$

$$\Re \left\{ z \frac{f''(z)}{f'(z)} \right\} \geq \frac{r(2r - 4)}{1 - r^2} \quad (|z| = r),$$

so that our condition is satisfied if $2r^2 - 4r \geq r^2 - 1$, i.e.

$$r^2 - 4r + 1 \geq 0, \quad 0 \leq r \leq 2 - \sqrt{3}.$$

Thus $f(z)$ in $\mathfrak{S}$ maps $|z| = r$ onto a convex curve for $0 \le r \le 2 - \sqrt{3}$. On the other hand if $f(z) = z(1 - z)^{-2}$, then

$$z\frac{f''(z)}{f'(z)} = \frac{2z^2 + 4z}{1 - z^2},$$

and this is real and less than -1 for $-1 < r < \sqrt{3} - 2$. Thus this function does not map $|z| = r$ onto a convex curve for $r > 2 - \sqrt{3}$. The quantity $r_c = 2 - \sqrt{3}$ is called the *radius of convexity*.[†]

We may ask similarly for the radius of the largest circle $|z| = r$ such that the image $\gamma(r)$ of $|z| = r$ by $f(z)$ always bounds a starshaped domain with respect to $w = 0$. The condition for this was seen in Chapter 1, (1.15) to be that

$$\Re\left\{z\frac{f'(z)}{f(z)}\right\} \ge 0, \quad \text{i.e.} \quad \left|\arg\left\{z\frac{f'(z)}{f(z)}\right\}\right| \le \frac{\pi}{2} \quad (|z| = r).$$

If we write

$$\phi(z) = \frac{f\left(\dfrac{z_0 + z}{1 + \bar{z}_0 z}\right) - f(z_0)}{(1 - |z_0|^2)f'(z_0)}, \tag{7.41}$$

then $\phi(z) \in \mathfrak{S}$ if $f(z) \in \mathfrak{S}$. On applying the inequality (7.37) to $\phi(z)$ at $z = -z_0$, we obtain

$$\left|\arg\left(\frac{f(z_0)}{z_0 f'(z_0)}\right)\right| \le \log\frac{1 + |z_0|}{1 - |z_0|}, \tag{7.42}$$

and this is sharp.

Thus the *radius of starshapedness*, r_s, being the radius of the largest circle whose interior is always mapped onto a starshaped domain with respect to $w = 0$ by $f(z) \in \mathfrak{S}$, is given by[‡]

$$\frac{\pi}{2} = \log\frac{1 + r_s}{1 - r_s}, \quad r_s = \tanh\frac{\pi}{4} = 0.65\ldots \tag{7.43}$$

Examples

7.5   If $f(z) \in \mathfrak{S}$ and $\phi(z)$ is defined by (7.41) show that

$$\frac{-z_0\phi'(-z_0)}{\phi(-z_0)} = \frac{z_0}{f(z_0)(1 - |z_0|)^2}.$$

Deduce that (7.42) is sharp and verify (7.43).

[†] Gronwall [1916]
[‡] Grunsky [1933]

7.12 The argument of $f'(z)$ As a final application we prove the following result of Golusin [1936].

Theorem 7.7   *Suppose that $f(z) \in \mathfrak{S}$. Then we have the sharp inequalities*

$$\begin{cases} |\arg f'(z)| \leq 4\sin^{-1}|z| & \left(|z| \leq \tfrac{1}{\sqrt{2}}\right) \\ |\arg f'(z)| \leq \pi + \log\dfrac{|z|^2}{1-|z|^2} & \left(\tfrac{1}{\sqrt{2}} < |z| < 1\right). \end{cases}$$

We may again confine ourselves to the functions $f(z,t)$ of Theorem 7.1. We start with equation (7.1)

$$\frac{\partial}{\partial t}f(z,t) = -f\frac{1+\kappa f}{1-\kappa f} = f + \frac{2}{\kappa} - \frac{2}{\kappa(1-\kappa f)},$$

and differentiate both sides with respect to z. This leads to

$$\frac{\partial}{\partial t}f'(z,t) = f'(z,t)\left[1 - \frac{2}{(1-\kappa f)^2}\right],$$

i.e.

$$\frac{\partial}{\partial t}\log f'(z,t) = 1 - \frac{2}{(1-\kappa f)^2}.$$

Taking imaginary parts we deduce

$$\frac{\partial}{\partial t}\arg f'(z,t) = \frac{2\Im(1-\kappa f)^2}{|1-\kappa f|^4}.$$

We eliminate t between this and (7.38) and deduce

$$d_t\arg f' = \frac{2\Im(1-\kappa f)^2}{|1-\kappa f|^2}\cdot\frac{-d_t|f|}{|f|(1-|f|^2)},$$

as t increases.

Now since $|\kappa(t)| = 1$

$$\frac{|\Im(1-\kappa f)^2|}{|1-\kappa f|^2} = |\sin[2\arg(1-\kappa f)]|$$

$$\leq \begin{cases} 2|f|\sqrt{(1-|f|^2)}, & \left(|f| \leq \tfrac{1}{\sqrt{2}}\right), \\ 1, & \left(|f| \geq \tfrac{1}{\sqrt{2}}\right). \end{cases} \tag{7.44}$$

Recalling that $d_t|f| < 0$, we deduce that

$$d_t\arg f' \leq \begin{cases} \dfrac{-4d_t|f|}{\sqrt{(1-|f|^2)}} & \left(|f| \leq \tfrac{1}{\sqrt{2}}\right), \\ \dfrac{-2d_t|f|}{|f|(1-|f|^2)} & \left(|f| > \tfrac{1}{\sqrt{2}}\right). \end{cases}$$

Integrating from $t = 0$ to t_0, we obtain for $|z| \leq \frac{1}{\sqrt{2}}$

$$|\arg f'| \leq \int_{|f|}^{|z|} \frac{4dx}{\sqrt{(1-x^2)}} \leq 4\sin^{-1}|z|,$$

and for $|z| > \frac{1}{\sqrt{2}}$

$$|\arg f'| \leq \int_{|f|}^{\frac{1}{\sqrt{2}}} \frac{4dx}{\sqrt{(1-x^2)}} + \int_{\frac{1}{\sqrt{2}}}^{|z|} \frac{2dx}{x(1-x^2)} \leq \pi + \log\left(\frac{|z|^2}{1-|z|^2}\right).$$

This proves the inequalities of Theorem 7.7. To show that they are precise we must find $\kappa(t)$, such that if $f(z_0, t)$ is the solution of (7.1) with assigned initial value $f(z_0, 0) = z_0$, then equality holds in the inequalities (7.44). The resulting equation enables us to calculate $\kappa f(z_0, t)$ in terms of $|f(z_0, t)|$, hence $|f(z_0, t)|$ in terms of t by means of (7.38) and then $\arg f(z_0, t)$ in terms of t by means of (7.39). We can then choose $\kappa(t)$, so that equality holds in (7.44). When this is done for $0 \leq t \leq t_0$, then (7.38) and (7.39) and equality in (7.44) will hold simultaneously and we shall have

$$\arg f'(z_0, t_0) = \int_{|f(z_0,t_0)|}^{|z_0|} \frac{4dx}{\sqrt{(1-x^2)}} \quad \left(|z_0| < \frac{1}{\sqrt{2}}\right),$$

$$\arg f'(z_0, t_0) = \int_{|f(z_0,t_0)|}^{\frac{1}{\sqrt{2}}} \frac{4dx}{\sqrt{(1-x^2)}} + \log\frac{|z_0|^2}{1-|z_0|^2} \quad \left(|z_0| > \frac{1}{\sqrt{2}}\right),$$

if $|f(z_0, t_0)| < \frac{1}{\sqrt{2}}$, and so for all large t_0. Thus the upper bounds of Theorem 7.7 may be approached as closely as we please and so are sharp.

7.13 Conclusion The foregoing theorems represent some of the principal successes achieved by Löwner and his successors by means of Theorem 7.1. In the next chapter we shall use the technique to prove de Branges' Theorem and some of its consequences. At this point Schiffer's variational method[†] and Jenkins' theory of modules[‡] should be mentioned. These methods can be used to prove the results of this chapter and some others and in particular to give more information about the extremal functions. However Löwner Theory has so far proved to be an essential ingredient of the proofs of de Branges' Theorem.

[†] Schiffer [1943], see also Duren [1983, p. 318 *et seq.*]
[‡] Jenkins [1958].

8

De Branges' Theorem

8.0 Introduction In this chapter we prove de Branges' Theorem [1985], conjectured by Bieberbach [1916] that, if

$$f(z) = z + \sum_{2}^{\infty} a_n z^n \in \mathfrak{S}, \qquad (8.1)$$

we have $|a_n| \leq n$, $n = 2, 3, \ldots$ with equality only for the Koebe functions $f(z) = z(1 - ze^{i\theta})^{-2}$. This result had previously been proved for $n = 4$ by Garabedian and Schiffer [1955], for $n = 5$ by Pederson and Schiffer [1972] and for $n = 6$ by Pederson [1968] and Ozawa [1969]. (For a more detailed history see Duren [1983, p. 69].)

De Branges proved his theorem by first establishing a conjecture of Milin [1971], which Milin had shown to imply Bieberbach's conjecture. Suppose that

$$\log \frac{f(z)}{z} = \sum_{k=1}^{\infty} c_k z^k, \qquad (8.2)$$

then Milin conjectured that

$$\sum_{k=1}^{n} \left(\frac{4}{k} - k|c_k|^2 \right) (n - k + 1) \geq 0, \quad n = 1, 2, \ldots \qquad (8.3)$$

The proof of de Branges has been simplified successively by Milin [1984] and Emelyanov, by Fitzgerald and Pommerenke [1985] and Weinstein [1991]. Inevitably these simpler proofs however miss the operator theory basis of de Branges' subtle ideas. All the proofs rely on Löwner Theory, and a positivity result for the coefficients of certain special functions. The earlier proofs used an inequality of Askey and Gasper [1976] concerning Jacobi polynomials. Weinstein's proof, which we shall follow here, uses instead the addition formula for Legendre polynomials which

230

goes back to Legendre himself and seems simpler to establish. However Wilf [1993] has now shown rather surprisingly that the two results are equivalent. We start off by proving Legendre's formula, then prove Milin's conjecture, Milin's inequalities and de Branges' Theorem, which in fact generalises to give sharp bounds for the coefficients of $(f(z)/z)^\lambda$, when $\lambda \geq 1$. The analogous result fails for $0 < \lambda < 1$. A number of further generalisations and consequences, including proofs of conjectures, of Robertson [1936] and Rogosinski [1943], will be given at the end of the chapter. De Branges' Theorem, both the result itself and the subtlety of its proof, represents a milestone of twentieth century analysis.

8.1 Legendre polynomials   The Legendre polynomials $P_n(z)$ are defined by the expansion

$$\frac{1}{\left(1 - 2zh + h^2\right)^{\frac{1}{2}}} = 1 + \sum_1^\infty h^n P_n(z), \tag{8.4}$$

valid when $|h|$ is sufficiently small depending on z. We follow the account given in Whittaker and Watson [1946, Chapter 15] and start with the following simple lemma, that is easily proved by means of the calculus of residues.

Lemma 8.1   *Suppose that A, B, C are complex numbers, such that $A \neq 0$, $\Delta = B^2 - 4AC \neq 0$, that Γ is a circle containing in its interior $t_1 = (-B + \sqrt{\Delta})/(2A)$ but not $t_2 = (-B - \sqrt{\Delta})/(2A)$, and that Γ is described in the anticlockwise sense. Then*

$$\int_\Gamma \frac{dt}{At^2 + Bt + C} = \frac{2\pi i}{\sqrt{\Delta}}. \tag{8.5}$$

We shall also need the associated Legendre functions

$$P_n^k(z) = \left(z^2 - 1\right)^{\frac{1}{2}k} \left(\frac{d}{dz}\right)^k P_n(z), \quad 0 \leq k \leq n.$$

We suppose for the time being that $1 < z < \infty$, so that $(z^2 - 1)^{\frac{1}{2}}$ is real and positive. We choose for Γ the circle

$$t = z + \left(z^2 - 1\right)^{\frac{1}{2}} e^{i\theta}, \quad -\pi \leq \theta \leq \pi \tag{8.6}$$

which has centre z and encloses $z = 1$ but not $z = -1$. Then if h is sufficiently small Lemma 8.1 yields

$$\frac{1}{(1 - 2zh + h^2)^{\frac{1}{2}}} = \frac{1}{\pi i} \int_\Gamma \frac{dt}{2(t - z) + h(1 - t^2)}$$

$$= \frac{1}{\pi i} \int_\Gamma \frac{dt}{t^2 - 1 - 2h(t - z)}.$$

We expand in powers of h and equate coefficients in the result and in (8.4). This yields

$$P_n(z) = \frac{1}{2^{n+1}\pi i} \int_\Gamma \frac{(t^2 - 1)^n \, dt}{(t - z)^{n+1}} = \frac{2^n}{\pi i} \int_\Gamma \frac{(t - z)^n dt}{(t^2 - 1)^{n+1}}.$$

Differentiating under the integral signs we obtain

$$P_n^k(z) = \frac{(z^2 - 1)^{\frac{1}{2}k} (n + 1) \cdots (n + k)}{2^{n+1}\pi i} \int_\Gamma \frac{(t^2 - 1)^n \, dt}{(t - z)^{n+k+1}}$$

$$= \frac{2^n (z^2 - 1)^{\frac{1}{2}k} n(n - 1) \cdots (n - k + 1)(-1)^k}{\pi i} \int_\Gamma \frac{(t - z)^{n-k} dt}{(t^2 - 1)^{n+1}}.$$

Finally we substitute for t from (8.6) and note that $P_n^k(z)$ is real, when $z > 1$. Thus

$$P_n^k(z) = \frac{(n + 1) \cdots (n + k)}{2\pi} \int_{-\pi}^{\pi} \left\{ z + (z^2 - 1)^{\frac{1}{2}} \cos \theta \right\}^n \cos k\theta d\theta \qquad (8.7)$$

and

$$P_n^k(z) = \frac{n(n - 1) \cdots (n - k + 1)(-1)^k}{2\pi}$$

$$\times \int_{-\pi}^{\pi} \left(z + (z^2 - 1)^{\frac{1}{2}} \cos \theta \right)^{-n-1} \cos k\theta d\theta. \qquad (8.8)$$

We can now prove Legendre's addition theorem.

Lemma 8.2 *Suppose that* x, y, ω *are complex numbers and that* $Z = xy - (x^2 - 1)^{\frac{1}{2}}(y^2 - 1)^{\frac{1}{2}} \cos \omega$. *Then*

$$P_n(Z) = P_n(x)P_n(y) + 2\sum_{k=1}^{n}(-1)^k \frac{(n - k)!}{(n + k)!} P_n^k(x)P_n^k(y) \cos k\omega.$$

For the proof we shall assume that x, y, ω are real, $x > 1$, $y > 1$, and that the positive square roots are taken in the definitions of Z and $P_n^k(x)$, $P_n^k(y)$. The general result, with a suitable definition of the signs of square roots, is then obtained by analytic continuation.

We assume that h is small, and that ϕ is real. Then

$$\sum_{n=0}^{\infty} h^n \frac{\left\{x + (x^2 - 1)^{\frac{1}{2}} \cos(\omega - \phi)\right\}^n}{\left\{y + (y^2 - 1)^{\frac{1}{2}} \cos \phi\right\}^{n+1}}$$

$$= \frac{1}{y + (y^2 - 1)^{\frac{1}{2}} \cos \phi - h \left(x + (x^2 - 1)^{\frac{1}{2}} \cos(\omega - \phi)\right)}. \qquad (8.9)$$

We integrate both sides of this equation w.r.t. ϕ from $-\pi$ to π. The integral of the right-hand side takes the form

$$I = \int_{-\pi}^{\pi} \frac{d\phi}{a + b \cos \phi + c \sin \phi},$$

where

$$a = y - hx, \quad b = (y^2 - 1)^{\frac{1}{2}} - h (x^2 - 1)^{\frac{1}{2}} \cos \omega, \quad c = -h (x^2 - 1)^{\frac{1}{2}} \sin \omega.$$

To evaluate I we write $t = e^{i\phi}$, $d\phi = dt/(it)$, $\cos \phi = \frac{1}{2}(t + t^{-1})$, $\sin \phi = (t - t^{-1})/(2i)$. This yields

$$I = \int_{|t|=1} \frac{dt}{At^2 + Bt + C} = \frac{2\pi i}{\sqrt{\Delta}},$$

by Lemma 8.1, where

$$A = \tfrac{1}{2}(c + ib), \quad B = ai, \quad C = \tfrac{1}{2}(-c + ib).$$

(Since $|A| = |C|$, the Lemma is applicable.)
Also

$$
\begin{aligned}
\Delta &= B^2 - 4AC = b^2 + c^2 - a^2 \\
&= (y^2 - 1) + h^2 (x^2 - 1) - 2h (x^2 - 1)^{\frac{1}{2}} (y^2 - 1)^{\frac{1}{2}} \cos \omega - (y - hx)^2 \\
&= -\left\{1 + h^2 - 2hxy + 2h (x^2 - 1)^{\frac{1}{2}} (y^2 - 1)^{\frac{1}{2}} \cos \omega\right\} \\
&= -\left\{1 + h^2 - 2hZ\right\}.
\end{aligned}
$$

Thus

$$I = \frac{\pm 2\pi}{\sqrt{(1 + h^2 - 2hZ)}}.$$

To check the sign, we put $h = 0$ and obtain

$$I = \int_0^{2\pi} \frac{d\phi}{y + (y^2 - 1)^{\frac{1}{2}} \cos \phi} = 2\pi,$$

by (8.8) with $k = n = 0$. Thus for small h we have

$$\frac{I}{2\pi} = (1 + h^2 - 2hZ)^{-\frac{1}{2}} = \sum_{n=0}^{\infty} h^n P_n(Z)$$

by (8.4). On comparing coefficients with the integral of the left-hand side of (8.9) we obtain

$$P_n(Z) = \frac{1}{2\pi} \int_{-\pi}^{\pi} \frac{\left\{ x + (x^2 - 1)^{\frac{1}{2}} \cos(\omega - \phi) \right\}^n d\phi}{\left\{ y^2 + (y^2 - 1)^{\frac{1}{2}} \cos \phi \right\}^{n+1}}. \tag{8.10}$$

The right-hand side is a polynomial of degree n in $\cos \omega$ and $\sin \omega$ and is an even function of ω, as we see on making the substitution $\phi = -\psi$. Thus

$$P_n(Z) = \frac{1}{2} A_0 + \sum_{k=1}^{n} A_k \cos k\omega.$$

It remains to evaluate the coefficients A_k. We have

$$
\begin{aligned}
A_k &= \frac{1}{\pi} \int_{-\pi}^{\pi} P_n(Z) \cos(k\omega) d\omega \\
&= \frac{1}{2\pi^2} \int_{-\pi}^{\pi} \cos(k\omega) d\omega \int_{-\pi}^{\pi} \frac{\left\{ x + (x^2 - 1)^{\frac{1}{2}} \cos(\omega - \phi) \right\}^n d\phi}{\left\{ y + (y^2 - 1)^{\frac{1}{2}} \cos \phi \right\}^{n+1}}.
\end{aligned}
$$

We integrate first w.r.t. ω, setting $\omega = \phi + \theta$ and using (8.7). We integrate from $\omega = \phi - \pi$ to $\phi + \pi$, i.e. from $\theta = -\pi$ to π. This is legitimate since the integrand is periodic in ω. We also write $\cos k\omega = \cos k\theta \cos k\phi - \sin k\theta \sin k\phi$ and note that the integral containing $\sin(k\phi)$ vanishes since the integrand is odd. Using (8.7), and for the remaining integral in ϕ (8.8), we obtain

$$
\begin{aligned}
A_k &= \frac{2 P_n^k(x) P_n^k(y)(-1)^k}{(n+k)(n+k-1) \cdots n(n-1) \cdots (n-k+1)} \\
&= \frac{2(n-k)!(-1)^k}{(n+k)!} P_n^k(x) P_n^k(y).
\end{aligned}
$$

This proves Lemma 8.2.

8.1.1 We complete the section by proving the positivity result which is needed for Weinstein's proof.

(In a recent preprint by Ekhad and Zeilberger [1993] the authors produce a short direct argument for this.)

Lemma 8.3 *Suppose that $t \geq 0$, and that $w = w_t(z)$ is defined by*

$$\frac{e^t w}{(1-w)^2} = \frac{z}{(1-z)^2}, \quad w_t(0) = 0. \tag{8.11}$$

Then we have for $|z| < 1$

$$\frac{e^t w^{k+1}}{1-w^2} = \sum_{n=k}^{\infty} \Lambda_k^n(t) z^{n+1}, \tag{8.12}$$

where $\Lambda_k^n(t) \geq 0$, for $t \geq 0$ and $0 \leq k \leq n < \infty$.

We write $x = y = (1 - e^{-t})^{\frac{1}{2}}$ in Lemma 8.2 and note that

$$Z = 1 - e^{-t} + e^{-t} \cos \omega.$$

We choose $(x^2 - 1)^{\frac{1}{2}} = (y^2 - 1)^{\frac{1}{2}} = ie^{-\frac{1}{2}t}$.

$$P_n^k(x) = P_n^k(y) = \left(ie^{-\frac{1}{2}t}\right)^k \left(\frac{d}{dx}\right)^k P_n(x) = i^k p, \quad \text{where } p \text{ is real.}$$

Thus

$$(-1)^k P_n^k(x) P_n^{(k)}(y) = (-1)^{2k} p^2 \geq 0.$$

Now Lemma 8.2 yields

$$P_n(Z) = \sum_{k=0}^{n} A_{k,n}(t) \cos k\omega,$$

where $A_{k,n}(t) \geq 0$. Next

$$\frac{z}{1 - 2zZ + z^2} = z \left\{ \sum_{0}^{\infty} P_n(Z) z^n \right\}^2$$

$$= z \left\{ \sum_{m=0}^{\infty} z^m \sum_{k=0}^{m} A_{k,m} \cos k\omega \right\} \left\{ \sum_{n=0}^{\infty} z^n \sum_{l=0}^{n} A_{l,n} \cos l\omega \right\}.$$

We recall that $\cos k\omega \cos l\omega = \frac{1}{2}\{\cos(k-l)\omega + \cos(k+l)\omega\}$. Thus we obtain

$$\frac{z}{1 - 2zZ + z^2} = \sum_{n=0}^{\infty} z^{n+1} \sum_{k=0}^{n} B_{k,n} \cos k\omega, \tag{8.13}$$

where $B_{k,n}(t) \geq 0$. Again by (8.11)

$$
\begin{aligned}
\frac{z}{1 - 2zZ + z^2} &= \frac{1}{z + \frac{1}{z} - 2Z} \\
&= \frac{1}{2 + e^{-t}\left(\frac{1}{w} + w - 2\right) - 2Z} = \frac{e^t w}{1 + w^2 - 2w\cos\omega} \\
&= \frac{e^t w}{1 - w^2}\frac{1 - w^2}{1 + w^2 - 2w\cos\omega} \\
&= \frac{e^t w}{1 - w^2}\left\{1 + 2\sum_{k=1}^{\infty} w^k \cos(k\omega)\right\} \\
&= \sum_{k=0}^{\infty} \Lambda_0^n(t)z^{n+1} + 2\sum_{k=1}^{\infty}\sum_{n=k}^{\infty} \Lambda_k^n(t)z^{n+1}\cos k\omega.
\end{aligned}
$$

Equating coefficients first of z^{n+1} and then of $z^{n+1}\cos k\omega$ in this and (8.13) we obtain that

$$
\Lambda_0^n = B_{o,n} \quad \text{and} \quad \Lambda_k^n = \tfrac{1}{2}B_{k,n}, \quad k \geq 1
$$

and this proves Lemma 8.3. We remark that $w_t(z)$ maps $|z| < 1$ onto $|w| < 1$ cut from -1 to $-\tau$ along the real axis, where

$$
\tau = 2e^t - 1 - 2\sqrt{(e^{2t} - e^t)}.
$$

Thus $w_t(z)$ is regular and $|w_t(z)| \leq |z|$ for $|z| < 1$ and so the series (8.12) are valid for $|z| < 1$.

8.2 Proof of Milin's conjecture: preliminary results
In this section we prove

Theorem 8.1 *If $f(z) \in \mathfrak{S}$ and*

$$
\log\frac{f(z)}{z} = \sum_{k=1}^{\infty} c_k z^k, \tag{8.2}
$$

then we have for $n = 1, 2, \ldots$

$$
\sum_{k=1}^{n}\left(\frac{4}{k} - k|c_k|^2\right)(n - k + 1) \geq 0. \tag{8.3}
$$

Theorem 8.1 is the key step in de Branges' proof. As we shall see the result also leads to a number of conclusions which go beyond Bieberbach's original conjecture. Weinstein's proof actually shows that equality holds in (8.3) only for the Koebe functions $f_\theta(z) = z/\left(1 - ze^{i\theta}\right)^2$. He uses a refinement of Löwner Theory which applies to every function in $\mathfrak{S}$ and not just to a dense subclass. For this we refer the reader to Weinstein [1991] and Löwner [1923]. To prove Theorem 8.1 it is sufficient by Theorem 7.1 to consider the functions $f(z)$ in the subclass $\mathfrak{S}_1$ of that theorem. Thus

$$f(z) = \lim_{t \to \infty} e^t f(z, t),$$

where $f(z, t)$ satisfies Löwner's equation (7.1), $\kappa(t)$ is continuous and $|\kappa(t)| = 1$.

We define $g_t(\zeta)$ as in Theorem 7.2 by

$$g_t\{f(z, t)\} = f(z). \tag{8.14}$$

Thus $g_0(z) = f(z)$. Also for large t

$$g_t(z) = \frac{e^t z}{\left(1 - ze^{i\phi}\right)^2}.$$

We now write

$$h(z, t) = \log \frac{g_t(z)}{ze^t} = \sum_{k=1}^{\infty} c_k(t)z^k. \tag{8.15}$$

Thus

$$c_k(0) = c_k, \quad \text{and} \quad c_k(t) = \frac{2}{k}e^{ik\phi}, \quad t \geq t_0. \tag{8.16}$$

We deduce from (7.1), (8.14), just as in Section 7.8, that (7.23) holds, i.e., if $g = g_t(\zeta)$, $\zeta = f(z, t)$ we obtain

$$\frac{\partial g}{\partial t} = \frac{\partial g}{\partial \zeta}\zeta\frac{1 + \kappa\zeta}{1 - \kappa\zeta}$$

or writing z instead of ζ, $g = g_t(z)$ we have

$$\frac{\partial g}{\partial t} = \frac{\partial g}{\partial z}z\frac{1 + \kappa z}{1 - \kappa z}. \tag{8.17}$$

Using (8.15) we deduce

$$\frac{\partial}{\partial t}h(z, t) = \frac{\partial g/\partial t}{g} - 1$$

$$\frac{\partial}{\partial z}h(z, t) = \frac{\partial g/\partial z}{g} - \frac{1}{z}.$$

Thus (8.17) yields

$$1 + \frac{\partial h}{\partial t} = \left(\frac{1}{z} + \frac{\partial h}{\partial z}\right)z\frac{1 + \kappa z}{1 - \kappa z}$$

i.e.

$$\frac{\partial h}{\partial t} = \left(1 + z\frac{\partial h}{\partial z}\right)\left(\frac{1 + \kappa z}{1 - \kappa z}\right) - 1.$$

We substitute (8.15) in this and equate coefficients in the resulting series. This yields

$$c_k'(t) = kc_k + 2\kappa^k + 2\sum_{r=1}^{k-1} rc_r\kappa^{k-r}. \tag{8.18}$$

We need some crude bounds for $c_k(t), c_k'(t)$.

Lemma 8.4 *We have for $0 \le t < \infty$ and $1 \le k < \infty$,*

$$|c_k(t)| < 9, \quad |c_k'(t)| < 11k^2.$$

We deduce from (8.15) that

$$z\frac{\partial g/\partial z}{g} = 1 + \sum_{1}^{\infty} kc_k z^k.$$

Also $e^{-t}g_t(z) \in \mathfrak{S}$. Thus (1.4) and Cauchy's inequality yield

$$k|c_k| \le \frac{1}{r^k}\sup_{|z|=r}\left\{r\left|\frac{\partial g/\partial z}{g}\right|\right\} \le \frac{1+r}{r^k(1-r)}.$$

Choosing $r = k/(k+1)$, we obtain

$$k|c_k| < (2k+1)\left(1 + \frac{1}{k}\right)^k < 3ke < 9k.$$

Now (8.18) gives

$$|c_k'(t)| \le 2 + 9k + 2\sum_{r=1}^{k-1} 9r = 9k^2 + 2 < 11k^2.$$

This proves Lemma 8.4.

8.2.1 Completion of proof Following Weinstein we prove (8.3) by showing that

$$I(z) = \sum_{n=1}^{\infty}\left\{\sum_{k=1}^{n}\left(\frac{4}{k} - k|c_k(0)|^2\right)(n-k+1)\right\}z^{n+1}$$

$$= \sum_{n=1}^{\infty} z^{n+1}\int_0^{\infty} g_n(t)dt, \tag{8.19}$$

where

$$g_n(t) \geq 0 \quad \text{for} \quad t \geq 0 \quad \text{and} \quad n = 1, 2, \ldots$$

To prove (8.19) we fix z, such that $|z| < 1$, and define $w = w_t(z)$ by (8.11), i.e.

$$\frac{z}{(1-z)^2} = \frac{e^t w}{(1-w)^2}. \tag{8.20}$$

We recall that by (8.20) $|z| < 1$ corresponds to $|w| < 1$ cut along a segment of the negative real axis. So Schwarz's Lemma yields $|w_t(z)| \leq |z|$ for $0 \leq t < \infty$. Also

$$
\begin{aligned}
I(z) &= \sum_{n=1}^{\infty} \sum_{k=1}^{n} \left(\frac{4}{k} - k|c_k(0)|^2 \right) (n-k+1) z^{n+1} \\
&= \frac{z}{(1-z)^2} \sum_{k=1}^{\infty} \left(\frac{4}{k} - k|c_k(0)|^2 \right) z^k \\
&= \int_0^{t_0} -\frac{z}{(1-z)^2} \frac{d}{dt} \left\{ \sum_{k=1}^{\infty} \left(\frac{4}{k} - k|c_k(t)|^2 \right) w^k \right\} dt. \tag{8.21}
\end{aligned}
$$

For if

$$\psi(t) = \sum_{k=1}^{\infty} \left(\frac{4}{k} - k|c_k(t)|^2 \right) w_t(z)^k,$$

we have by (8.16)

$$\psi(t) = 0, \quad t \geq t_0, \quad \text{and} \quad \psi(0) = \sum_{k=1}^{\infty} \left(\frac{4}{k} - k|c_k(0)|^2 \right) z^k.$$

We differentiate the series in (8.21) term by term and integrate the result term by term. To justify this we need to show that the differentiated series converges uniformly in $[0, t_0]$. We write $x' = \partial x / \partial t$. Then (8.20) yields

$$w' = \frac{-(1-w)w}{(1+w)}.$$

Also by Lemma 8.4

$$\left| \frac{\partial}{\partial t} |c_k(t)|^2 \right| = \left| c_k(t) \overline{c_k'(t)} + \overline{c_k(t)} c_k'(t) \right| < 198k^2.$$

Thus

$$\frac{d}{dt} \left\{ \left(\frac{4}{k} - k|c_k(t)|^2 \right) w^k \right\} = -\frac{1-w}{1+w} \left(4 - k^2|c_k|^2 \right) w^k - kw^k \frac{\partial}{\partial t} \left(c_k(t) \overline{c_k(t)} \right).$$

Since by (8.20) $|w| \leq |z| < 1$, the right-hand side is bounded by $Ak^3|z|^k$, where A is a constant depending only on z. This implies the required uniform convergence. Using also (8.20) and (8.21) we obtain

$$I(z) = \int_0^\infty \frac{e^t w}{1-w^2} \left\{ \frac{1+w}{1-w} \sum_{k=1}^\infty k \left(c_k(t)\overline{c_k(t)} \right)' w^k + \sum_{k=1}^\infty \left(4 - k^2|c_k|^2 \right) w^k \right\} dt.$$

$$(8.22)$$

Next we write $z_1 = r_1 e^{i\theta}$, where $|z| < r_1 < 1$ and define

$$G(\theta) = \frac{1}{g(z_1,t)} \frac{\partial g(z_1,t)}{\partial t} = 1 + \frac{\partial h(z_1,t)}{\partial t} = 1 + \sum_{m=1}^\infty c'_m(t) r_1^m e^{im\theta} \quad (8.23)$$

by (8.15). Thus by Lemma 8.4, and since $r_1 < 1$ and $|w| < 1$, we have

$$\sum_{k=1}^\infty k \overline{c_k(t)} w^k \frac{1}{2\pi} \int_0^{2\pi} G(\theta)\overline{z}_1^k d\theta = \sum_{k=1}^\infty k c'_k(t)\overline{c_k(t)} r_1^{2k} w^k.$$

Hence

$$\frac{1+w}{1-w} \sum_{k=1}^\infty k c'_k(t)\overline{c_k(t)}(r_1^2 w)^k = \left\{ 1 + 2\sum_{m=1}^\infty w^m \right\} \left\{ \sum_{k=1}^\infty k c'_k(t)\overline{c_k(t)}(r_1^2 w)^k \right\}$$

$$= \sum_{k=1}^\infty \frac{1}{2\pi} \int_0^{2\pi} G(\theta) \left\{ 1 + 2\sum_{m=1}^\infty w^m \right\} \sum_{k=1}^\infty k \overline{c_k(t)} w^k r_1^k e^{-ik\theta} d\theta$$

$$= \sum_{l=1}^\infty w^l \frac{1}{2\pi} \int_0^{2\pi} G(\theta) \overline{\left\{ 2\sum_{m=1}^l m c_m z_1^m - l c_l z_1^l \right\}} d\theta$$

$$= \sum_{l=1}^\infty w^l \left\{ \frac{1}{2\pi} \int_0^{2\pi} G(\theta) \overline{\left\{ 2\left(1 + \sum_{m=1}^l m c_m z_1^m \right) - l c_l z_1^l \right\}} d\theta - 2 \right\}. \quad (8.24)$$

Similarly

$$\frac{1+w}{1-w} \sum_{k=1}^\infty k c_k(t)\overline{c'_k(t)}(r_1^2 w)^k$$

$$= \sum_{l=1}^\infty w^l \left\{ \frac{1}{2\pi} \int_0^{2\pi} \overline{G(\theta)} \left\{ 2\left(1 + \sum_{m=1}^l m c_m z_1^m \right) - l c_l z_1^l \right\} d\theta - 2 \right\}. \quad (8.25)$$

Next we deduce from (8.15) that

$$1 + z_1 \frac{\partial h(z_1,t)}{\partial z_1} = z_1 \frac{\partial g(z_1,t)/\partial z}{g(z_1,t)} = 1 + \sum_{m=1}^\infty m c_m(t) z_1^m.$$

In particular

$$\left\{1 + \sum_{m=1}^{l} mc_m(t)z_1^m\right\} \bigg/ \left\{\frac{z_1\partial g(z_1,t)/\partial z_1}{g(z_1,t)}\right\} = 1 + P_{l+1}(z_1),$$

where $P_{l+1}(z_1)$ is a power series having a zero of order at least $(l+1)$ at $z_1 = 0$. Thus we may multiply the integrand in the coefficient of w^l in (8.24) by $1 + P_{l+1}(z_1)$ without affecting the value of the integral. Substituting for $G(\theta)$ from (8.23) and using (8.17) we obtain from (8.24)

$$\frac{1+w}{1-w}\sum_{k=1}^{\infty}\left\{k\overline{c_k(t)}c_k'(t)(r_1^2 w)^k\right\} + 2\sum_{k=1}^{\infty}w^k$$

$$= \sum_{l=1}^{\infty}w^l\frac{1}{2\pi}\int_0^{2\pi}\frac{1+\kappa(t)z_1}{1-\kappa(t)z_1}$$

$$\times\left\{1 + \sum_{n=1}^{l}mc_m z_1^m\right\}\overline{\left\{2\left(1 + \sum_{m=1}^{l}mc_m z_1^m\right) - lz_1^l\right\}}d\theta,$$

$$= \sum_{l=1}^{\infty}w^l\left\{\frac{1}{2\pi}\int_0^{2\pi}\frac{1+\kappa z_1}{1-\kappa z_1}2\left|\left(1 + \sum_{m=1}^{l}mc_m z_1^m\right) - \frac{1}{2}lc_l z_1^l\right|^2 d\theta\right.$$

$$\left. + \frac{1}{2}l^2|c_l|^2 r_1^{2l}\right\}.$$

We rearrange (8.25) similarly and let r_1 tend to 1. The series all converge uniformly and absolutely for fixed z, variable t and $|z| < r_1 < 1$. We also note that

$$\Re\frac{1+\kappa z_1}{1-\kappa z_1} = u(z_1,t)$$

where $u \geq 0$. Thus (8.22) yields finally

$$I(z) = \int_0^{\infty}\frac{e^t w dt}{1-w^2}\sum_{l=1}^{\infty}w^l$$

$$\times \lim_{r_1\to 1}\frac{1}{2\pi}\int_0^{2\pi}u(z_1,t)\left|2\left(1 + \sum_{m=1}^{l}mc_m z_1^m\right) - lc_l z_1^l\right|^2 d\theta$$

$$= \int_0^{\infty}\frac{e^t w dt}{1-w^2}\sum_{l=1}^{\infty}A_l(t)w^l,$$

where $A_l(t) \geq 0$. On combining this with Lemma 8.3 we obtain

$$I(z) = \int_0^{\infty}dt\sum_{l=1}^{\infty}A_l(t)\sum_{n=l}^{\infty}\Lambda_l^n(t)z^{n+1} = \sum_{n=1}^{\infty}z^{n+1}\int_0^{\infty}g_n(t)dt,$$

where

$$g_n(t) = \sum_{l=1}^{n} A_l(t)\Lambda_l^n(t) \geq 0.$$

This proves (8.19) and so Theorem 8.1.

8.2.2 An extension Following de Branges [1985] we deduce the following slight generalisation of Theorem 8.1 which will be useful in the sequel.

Theorem 8.2 *Suppose that $\sigma_1, \sigma_2, \ldots, \sigma_{N+1}$ is a sequence such that*

$$\sigma_1 \geq \sigma_2 \geq \cdots \geq \sigma_{N+1} = 0 \tag{8.26}$$

and

$$\sigma_k - \sigma_{k+1} \geq \sigma_{k+1} - \sigma_{k+2}, \quad 1 \leq k \leq N-1. \tag{8.27}$$

Then

$$\sum_{1}^{N} \left(k|c_k|^2 - \frac{4}{k} \right) \sigma_k \leq 0. \tag{8.28}$$

We choose nonnegative numbers α_1 to α_N, multiply (8.3) by $-\alpha_n$, and add for $n = 1$ to N. We deduce that

$$\sum_{k=1}^{N} \left(k|c_k|^2 - \frac{4}{k} \right) \sigma_k \leq 0,$$

where

$$\sigma_k = \sum_{n=k}^{N} \alpha_n(n - k + 1). \tag{8.29}$$

It remains to show that if the σ_k satisfy (8.26) and (8.27) we can solve this system of equations for nonnegative α_n. In fact (8.29) yields

$$\alpha_N = \sigma_N$$

and if $k < N$, and α_{k+1} to α_N have been chosen we obtain from (8.29)

$$\sigma_k - 2\sigma_{k+1} + \sigma_{k+2} = \alpha_k.$$

Thus $\alpha_n \geq 0$ by (8.26) and (8.27). Hence (8.28), i.e. Theorem 8.2 is proved.

8.3 The Milin–Lebedev inequalities   Milin and Lebedev (see Milin [1971])
proved some subtle inequalities involving coefficients of power series.
Milin showed that by means of these inequalities Bieberbach's conjecture
could be deduced from Theorem 8.1. We proceed to develop these results.
We suppose that

$$\omega(z) = \sum_{k=1}^{\infty} A_k z^k \qquad (8.30)$$

is a power series with a positive radius of convergence, though in fact all
the results remain valid for formal power series. We write

$$\phi(z) = \exp\{\omega(z)\} = \sum_{k=0}^{\infty} D_k z^k. \qquad (8.31)$$

We also define the binomial coefficients $d_k(\lambda)$ by

$$(1-z)^{-\lambda} = \exp\left\{\lambda \sum_{1}^{\infty} \frac{z^k}{k}\right\} = \sum_{0}^{\infty} d_k(\lambda) z^k; \quad d_k(\lambda) = \frac{\Gamma(\lambda+k)}{\Gamma(\lambda)\Gamma(k+1)}. \qquad (8.32)$$

Here $\Gamma(x)$ is the gamma function. Then we have (Milin [1971, Lemma
2.2, p. 44])

Theorem 8.3   *With the above notation we have for $n = 1, 2, \ldots$ and $\lambda > 0$*

$$\sum_{k=0}^{n} \frac{|D_k|^2}{d_k(\lambda)} \le d_n(\lambda+1) \exp\left\{\frac{1}{d_n(\lambda+1)} \sum_{k=1}^{n} \left(\frac{k^2|A_k|^2 - \lambda^2}{k\lambda}\right) d_{n-k}(\lambda+1)\right\}. \qquad (8.33)$$

Equality holds if and only if

$$A_k = \frac{\lambda}{k}\eta^k, \quad \text{for} \quad k = 1, 2, \ldots, n, \quad \text{where} \quad |\eta| = 1. \qquad (8.34)$$

Theorem 8.2 says roughly that if $|A_k| \le \lambda/k$ in a suitable average sense,
then $|D_k| \le d_k(\lambda)$ also in some average sense. We follow Milin's argument
and note that

$$z\phi'(z) = \phi(z)z\omega'(z).$$

Equating coefficients of z^k we obtain

$$kD_k = \sum_{v=1}^{k} v A_v D_{k-v}.$$

We write $a_v = v|A_v|$ and deduce from Schwarz's inequality

$$|D_k|^2 \le \frac{1}{k^2} \sum_{v=1}^{k} a_v^2 d_{k-v}(\lambda) \sum_{v=0}^{k-1} \frac{|D_v|^2}{d_v(\lambda)}. \tag{8.35}$$

We apply this with $k = n$ and deduce

$$\sum_{k=0}^{n} \frac{|D_k|^2}{d_k(\lambda)} \le \sum_{k=0}^{n-1} \frac{|D_k|^2}{d_k(\lambda)} + \frac{1}{n^2 d_n(\lambda)} \sum_{k=1}^{n} d_{n-k}(\lambda) a_k^2 \sum_{k=0}^{n-1} \frac{|D_k|^2}{d_k(\lambda)}$$

$$= \frac{n+\lambda}{n} \sum_{k=0}^{n-1} \frac{|D_k|^2}{d_k(\lambda)} \left[\frac{n}{n+\lambda} + \frac{\sum_{k=1}^{n} d_{n-k}(\lambda) a_k^2}{n(n+\lambda) d_n(\lambda)} \right].$$

We denote the term in square brackets by x, and use (8.32) and the fact that $x \le \exp(x-1)$, $-\infty < x < \infty$, with equality if and only if $x = 1$. We obtain

$$\sum_{k=0}^{n} \frac{|D_k|^2}{d_k(\lambda)} \le \frac{d_n(\lambda+1)}{d_{n-1}(\lambda+1)} \sum_{k=0}^{n-1} \frac{|D_k|^2}{d_k(\lambda)} \exp\left\{ \frac{\sum_{k=1}^{n} d_{n-k}(\lambda) a_k^2}{n(n+\lambda) d_n(\lambda)} - \frac{\lambda}{n+\lambda} \right\}. \tag{8.36}$$

We now write

$$t_n = \frac{1}{d_n(\lambda+1)} \sum_{k=1}^{n} \frac{a_k^2 - \lambda^2}{k\lambda} d_{n-k}(\lambda+1), \tag{8.37}$$

and deduce, using (8.32) with the convention $d_{-1}(\lambda) = 0$, that

$$t_n - t_{n-1} = \sum_{k=1}^{n} \frac{(a_k^2 - \lambda^2) d_{n-k}(\lambda)}{n(n+\lambda) d_n(\lambda)}. \tag{8.38}$$

In fact by (8.32)

$$\frac{d_{n-k}(\lambda)}{d_n(\lambda)} = \frac{\Gamma(n-k+\lambda)}{\Gamma(n-k+1)} \frac{\Gamma(n+1)}{\Gamma(n+\lambda)}.$$

Thus

$$\frac{d_{n-k}(\lambda+1)}{d_n(\lambda+1)} - \frac{d_{n-1-k}(\lambda+1)}{d_{n-1}(\lambda+1)}$$

$$= \frac{\Gamma(n-k+\lambda)\Gamma(n)}{\Gamma(n-k)\Gamma(n+\lambda)} \left\{ \frac{(n-k+\lambda)n}{(n-k)(n+\lambda)} - 1 \right\}$$

$$= \frac{\lambda k}{n(n+\lambda)} \frac{d_{n-k}(\lambda)}{d_n(\lambda)}.$$

Now (8.37) yields (8.38). Further

$$\frac{\lambda^2}{n(n+\lambda)} \sum_{k=1}^{n} \frac{d_{n-k}(\lambda)}{d_n(\lambda)} = \frac{\lambda^2}{n(n+\lambda)d_n(\lambda)} \sum_{p=0}^{n-1} d_p(\lambda)$$

$$= \frac{\lambda^2 d_{n-1}(\lambda+1)}{n(n+\lambda)d_n(\lambda)} = \frac{\lambda}{n+\lambda}. \qquad (8.39)$$

The identity $\sum_{p=0}^{n-1} d_p(\lambda) = d_{n-1}(\lambda+1)$ follows from

$$(1-z)^{-\lambda}(1-z)^{-1} = (1-z)^{-(\lambda+1)}.$$

Using (8.38) and (8.39) we can write (8.36) in the form

$$\frac{1}{d_n(\lambda+1)} \sum_{k=0}^{n} \frac{|D_k|^2}{d_k(\lambda)} \exp(-t_n) \leq \frac{1}{d_{n-1}(\lambda+1)} \sum_{k=0}^{n-1} \frac{|D_k|^2}{d_k(\lambda)} \exp\{-t_{n-1}\}.$$

$$\leq \cdots \leq \frac{1}{d_0(\lambda+1)} \frac{|D_0|^2}{d_0(\lambda)} \exp(-t_0) = 1. \qquad (8.40)$$

This yields (8.33).

If equality holds in (8.33), then equality must hold in (8.40) and this is only possible if equality holds in (8.36) when n is replaced by $n, (n-1), \ldots, 1$. Since $x < e^{x-1}$, for $x \neq 1$, this implies

$$t_n = t_{n-1} = \cdots = t_0 = 0.$$

Since $t_1 = 0$, we have $|A_1|^2 - \lambda^2 = 0$, and now induction on k yields

$$k^2 |A_k|^2 = \lambda^2, \quad \text{i.e. } kA_k = \lambda e^{i\theta_k}, \ k = 1 \text{ to } n.$$

This yields (8.34) for $n = 1$. Suppose next that $n \geq 2$. Then equality must hold in (8.35) for $k = 2, \ldots, n$. Suppose first that $\theta_1 = 0$, i.e. $A_1 > 0$. Then, since $D_0 = 1$, $D_1 = A_1$ by (8.30) and (8.31), we deduce, when $k = 2$, that

$$0 = \arg(A_1 D_1) = \arg(2A_2 D_0),$$

so that $A_2 > 0$. By induction on n we deduce that $A_k > 0$ and $D_k > 0$, for $k = 1$ to n. If $\theta_1 \neq 0$, we apply the above argument with $\omega_1(z) = \omega\left(ze^{-i\theta_1}\right)$ and $\phi(z) = \phi\left(ze^{-i\theta_1}\right)$ instead of $\omega(z)$, $\phi(z)$. Then $|A_k|$, $|D_k|$ are unchanged so that equality still holds in (8.33) but now $\omega_1'(0) > 0$ so that the first n coefficients of $\omega_1(z)$ are positive. Thus

$$A_k = |A_k| e^{ik\theta_1} = \frac{\lambda}{k} e^{ik\theta_1}, \quad k = 1, 2, \ldots, n.$$

This yields (8.34) with $\eta = e^{i\theta_1}$ and the proof of Theorem 8.3 is complete.

8.3.1 We can also obtain a corresponding bound for D_k. This is Milin's second inequality [1971, Theorem 2.4, p. 50].

Theorem 8.4 *With the hypotheses of Theorem 8.3 we have*

$$|D_n| \le d_n(\lambda) \exp \left\{ \frac{1}{2d_n(\lambda)} \sum_{k=1}^{n} d_{n-k}(\lambda) \left(\frac{k^2 |A_k|^2 - \lambda^2}{k\lambda} \right) \right\}. \qquad (8.41)$$

Again equality holds if and only if (8.34) holds.

To prove Theorem 8.4 we deduce from (8.35) and (8.33) for $n = 1, 2, \ldots$

$$|D_n|^2 \le \frac{1}{n^2} \sum_{v=1}^{n} a_v^2 d_{n-v}(\lambda) \sum_{v=0}^{n-1} \frac{|D_v|^2}{d_v(\lambda)}$$

$$\le \frac{d_{n-1}(\lambda+1)}{n^2}$$

$$\times \sum_{v=1}^{n} a_v^2 d_{n-v}(\lambda) \exp \left\{ \frac{1}{d_{n-1}(\lambda+1)} \sum_{v=1}^{n-1} \left(\frac{a_v^2 - \lambda^2}{v\lambda} \right) d_{n-v-1}(\lambda+1) \right\}.$$

$$(8.42)$$

We have, since $x \le e^{x-1}$,

$$\sum_{v=1}^{n} |a_v|^2 d_{n-v}(\lambda)$$

$$= \lambda^2 d_{n-1}(\lambda+1) \left\{ \frac{1}{\lambda^2 d_{n-1}(\lambda+1)} \sum_{v=1}^{n} a_v^2 d_{n-v}(\lambda) \right\}$$

$$\le \lambda^2 d_{n-1}(\lambda+1) \exp \left\{ \frac{1}{\lambda^2 d_{n-1}(\lambda+1)} \sum_{v=1}^{n} a_v^2 d_{n-v}(\lambda) - 1 \right\}$$

$$= \lambda^2 d_{n-1}(\lambda+1) \exp \left\{ \frac{1}{\lambda^2 d_{n-1}(\lambda+1)} \sum_{v=1}^{n} (a_v^2 - \lambda^2) d_{n-v}(\lambda) \right\}$$

by (8.39). On combining this with (8.42) and using (8.32) we have

$$|D_n| \le \frac{\lambda d_{n-1}(\lambda+1)}{n}$$

$$\times \exp \frac{1}{2d_{n-1}(\lambda+1)} \sum_{v=1}^{n} (a_v^2 - \lambda^2) \left(\frac{d_{n-v-1}(\lambda+1)}{v\lambda} + \frac{d_{n-v}(\lambda)}{\lambda^2} \right)$$

$$= d_n(\lambda) \exp \left\{ \frac{1}{2d_n(\lambda)} \sum_{v=1}^{n} \left(\frac{a_v^2 - \lambda^2}{\lambda v} \right) d_{n-v}(\lambda) \right\}.$$

This proves (8.41). Equality holds only if equality holds in (8.42). This yields equality in (8.33) with $n - 1$ instead of n, and so (8.34) holds with $n-1$ instead of n. Also equality holds in (8.35) with $k = n$ and this shows that (8.34) holds also for n. This completes the proof of Theorem 8.4.

8.4 Proof of de Branges' Theorem We can now prove Bieberbach's conjecture and a little more.

Theorem 8.5 *Suppose that $f(z) \in \mathfrak{S}$ and that*

$$\left(\frac{f(z)}{z} \right)^\lambda = \sum_{n=1}^{\infty} a_n(\lambda) z^{n-1}. \tag{8.43}$$

Then, if $\lambda \geq 1$, and $n \geq 2$, we have

$$|a_n(\lambda)| \leq d_{n-1}(2\lambda) = \frac{\Gamma(n - 1 + 2\lambda)}{\Gamma(n)\Gamma(2\lambda)}. \tag{8.44}$$

Equality holds if and only if $f(z)$ is a Koebe function

$$k_\theta(z) = \frac{z}{\left(1 - z e^{i\theta} \right)^2}, \quad \text{where } \theta \text{ is real.} \tag{8.45}$$

If $\lambda = 1$, $a_n(\lambda) = a_n$, we obtain Bieberbach's conjecture. We also recall from Example 7.3 that (8.44) fails for $-1 < \lambda < 1$ and $n = 3$. In fact the sharp bound for $|a_3(\lambda)|$ in this case is

$$|\lambda| \left\{ 1 + 2e^{2(\lambda-1)/(\lambda+1)} \right\} > \left| \frac{\Gamma(2 + 2\lambda)}{2\Gamma(2\lambda)} \right| = |\lambda||2\lambda + 1| = |d_2(2\lambda)|.$$

To prove (8.44) we deduce from Theorem 8.1 that (8.3) holds. Thus

$$\left(\frac{f(z)}{z} \right)^\lambda = \exp \left\{ \sum_{k=1}^{\infty} \lambda c_k z^k \right\},$$

where

$$\sum_{k=1}^{n} \left(k|c_k|^2 - \frac{4}{k} \right) (n - k + 1) \leq 0, \quad n = 1, 2, \ldots$$

We apply Theorems 8.2, 8.4 with $\mu = 2\lambda$ instead of λ, and $A_k = \lambda c_k$, $D_k = a_{k+1}(\lambda)$. In order to apply Theorem 8.2, we need to check that, if

$\sigma_k = d_{N-k}(\mu)$, $k = 1, 2, \ldots, N+1$, $\mu \geq 2$, (8.26) and (8.27) are satisfied. By our convention

$$\sigma_{N+1} = d_{-1}(\mu) = 0.$$

Also $\sigma_k > 0$ for $1 \leq k \leq N$, and by (8.32)

$$\frac{\sigma_{k+1}}{\sigma_k} = \frac{d_{N-k-1}(\mu)}{d_{N-k}(\mu)} = \frac{N-k}{\mu + N - k - 1} < 1, \quad \text{since} \quad \mu > 1.$$

Thus (8.26) holds. Next

$$\sigma_{k+1} - \sigma_k = \left(\frac{N-k}{\mu + N - k - 1} - 1 \right) \sigma_k, \quad \sigma_k - \sigma_{k-1} = \left(1 - \frac{\mu + N - k}{1 + N - k} \right) \sigma_k.$$

Thus

$$
\begin{aligned}
\sigma_{k+1} - 2\sigma_k + \sigma_{k-1} &= \sigma_k \left\{ \frac{1-\mu}{\mu + N - k - 1} - \frac{1-\mu}{1 + N - k} \right\} \\
&= \frac{(\mu - 1)(\mu - 2)\sigma_k}{(N - k + \mu - 1)(N - k + 1)} \geq 0
\end{aligned}
$$

if $1 \leq k \leq N$. Hence (8.26) and (8.27) are satisfied and we deduce that

$$\sum_{k=1}^{N} \left(k|c_k|^2 - \frac{4}{k} \right) d_{N-k}(\mu) = \frac{2}{\lambda} \sum_{k=1}^{N} \left(\frac{k^2|A_k|^2 - 4\lambda^2}{2k\lambda} \right) d_{N-k}(2\lambda) \leq 0.$$

Now (8.41), with 2λ instead of λ, yields

$$|a_{N+1}(\lambda)| = |D_n| \leq d_N(2\lambda), \quad \text{if} \quad N \geq 1.$$

This is (8.44). If $N \geq 1$, equality is only possible if (8.34) holds. In particular we have

$$|A_1| = 2\lambda, \quad |c_1| = |a_2(1)| = 2.$$

Now it follows from Theorem 1.1 that $f(z)$ is given by (8.45).

8.5 Some further results   We note some consequences of Theorem 8.3, which go beyond Bieberbach's conjecture.

Theorem 8.6   *If* $f_2(z) = \sum_{n=1}^{\infty} a_{2n-1} z^{2n-1} \in \mathfrak{S}$, *then*

$$|a_1|^2 + |a_3|^2 + \cdots + |a_{2N-1}|^2 \leq N + 1. \tag{8.46}$$

Equality holds for $N \geq 2$ *if and only if* $f(z) = z/(1 - z^2 e^{i\theta})$, *for a real* θ.

The result was conjectured by Robertson [1936] and proved by de Branges. We recall from the end of Chapter 5 that, since $f_2(z)$ is odd and univalent, $f(z) = \{f_2(z^{\frac{1}{2}})\}^2$ is univalent. Thus

$$1 + a_3 z + a_5 z^2 + \cdots = \frac{f_2\left(z^{\frac{1}{2}}\right)}{z^{\frac{1}{2}}} = \left(\frac{f(z)}{z}\right)^{\frac{1}{2}}.$$

Hence with the notation of (8.43) we can write (8.46) in the form

$$\sum_{k=1}^{N} |a_n(\tfrac{1}{2})|^2 \le N. \tag{8.47}$$

To prove (8.47) we apply Theorem 8.3 with $A_k = \frac{1}{2}c_k$ and $\lambda = 1$. Then it follows from Theorem 8.1 that

$$\sum_{k=1}^{n} \left(\frac{k^2|A_k|^2 - \lambda^2}{k}\right) d_{n-k}(\lambda + 1) = \sum_{k=1}^{n} \left(\frac{k|c_k|^2}{4} - \frac{1}{k}\right)(n - k + 1) \le 0.$$

Now (8.33) yields

$$\sum_{k=0}^{n} |D_k|^2 \le n + 1.$$

Equality holds for $n \ge 1$ only if $|a_3| = 1$. In this case

$$f(z) = z + 2a_3 z^2 + \cdots$$

and since $f(z) \in \mathfrak{S}$, $f(z) = z\left(1 - ze^{i\theta}\right)^{-2}$ by Theorem 1.1. Thus

$$f_2(z) = f\left(z^2\right)^{\frac{1}{2}} = \frac{z}{\left(1 - z^2 e^{i\theta}\right)}.$$

This proves Theorem 8.6.

Theorem 8.6 implies a conjecture by Rogosinski [1943], as was shown by Robertson [1970].

Theorem 8.7 *Suppose that* $\omega_1(z)$, $\omega_2(z)$ *are regular in* $|z| < 1$ *and satisfy* $|\omega_j(z)| < 1$ *there, and further* $\omega_2(0) = 0$. *Suppose that* $f(z) \in \mathfrak{S}$ *and that*

$$F(z) = \omega_1(z)f\{\omega_2(z)\} = \sum_{n=1}^{\infty} A_n z^n.$$

Then $|A_n| \le n$, $n = 1, 2, \ldots$ *Equality holds if and only if* $F(z) = e^{i\lambda}k_\theta(z)$, *where* $k_\theta(z)$ *is the Koebe function* (8.45).

If $\omega_1(z) \equiv 1$, then $F(z)$ is said to be subordinate to $f(z)$. This is the case for instance if $f(z)$ maps $|z| < 1$ onto a simply connected domain D and if $F(z)$ is any function regular in $|z| < 1$ and with values in D. For these functions $F(z)$ the result was conjectured by Rogosinski [1943]. To prove Theorem 8.7 we follow the exposition of Duren [1983, p. 196]. We write

$$H(z) = f\left(z^2\right)^{\frac{1}{2}} = z + c_3 z^3 + \cdots$$

and proceed to show that

$$|A_n| \leq \sum_{k=1}^{n} |c_{2k-1}|^2 \leq n \qquad (8.48)$$

by Theorem 8.6. We define

$$\phi(z) = \frac{H\left(\sqrt{z}\right)}{\sqrt{z}} = 1 + c_3 z + c_5 z^2 + \cdots$$

Then

$$\phi(z)^2 = \frac{f(z)}{z} \quad \text{and}$$

$$F(z) = \omega_1(z) f\{\omega_2(z)\} = \omega_1(z)\omega_2(z) \left\{1 + c_3\omega_2(z) + c_5\omega(z)^2 + \cdots\right\}^2.$$

We denote by $s_n(z) = \sum_{k=1}^{n} c_{2k-1} z^{k-1}$ the nth partial sum of $\phi(z)$. Since $\omega_2(0) = 0$ we can modify the Cauchy representation of the coefficients A_n of $F(z)$ to

$$A_n = \frac{1}{2\pi i} \int_{|z|=r} \frac{\omega_1(z)\omega_2(z) \{s_n[\omega_2(z)]\}^2 \, dz}{z^{n+1}}. \qquad (8.49)$$

It now follows from Littlewood's subordination theorem (see e.g. Hayman and Kennedy [1976, p. 76]) that

$$\int_{|z|=r} |s_n[\omega_2(z)]|^2 \, |dz| \leq \int_{|z|=r} |s_n(z)|^2 |dz|. \qquad (8.50)$$

Thus (8.49) yields, since $|\omega_1(z)\omega_2(z)| \leq 1$,

$$|A_n| \leq \frac{1}{2\pi r^n} \int_{|z|=r} |s_n[\omega(z)]^2| \, |dz| \leq \frac{1}{2\pi r^2} \int_{|z|=r} |s_n(z)|^2 |dz|$$

$$= \frac{1}{2\pi r^n} \sum_{k=1}^{n} |c_{2k-1}|^2 r^{2k-1}. \qquad (8.51)$$

Letting r tend to 1 we obtain (8.48) and Theorem 8.7 is proved, apart from the cases of equality.

To deal with these we note that if $n = 1$

$$A_1 = \omega_1(0)\omega_2'(0)$$

so that $|A_1| = 1$ only if $\omega_1(z) \equiv e^{i\mu}$ and $\omega_2(0) \equiv ze^{i\lambda}$, by Schwarz's Lemma. If $n > 1$ and $|A_n| = n$, then (8.48) shows that $\sum_{k=1}^{n} |c_{2k-1}|^2 = n$, and so

$$H(z) = \frac{z}{1 - z^2 e^{i\theta}}, \quad f(z) = H(z^{\frac{1}{2}})^2 = \frac{z}{\left(1 - ze^{i\theta}\right)^2}$$

by Theorem 8.6. We absorb the constant $e^{i\theta}$ into ω_2 and so assume that

$$H(z) = \frac{z}{\left(1 - z^2\right)},$$

$$s_n(z) = \sum_{k=1}^{n} z^{k-1}.$$

We next write

$$s_n[\omega_2(z)] = \sum_{k=0}^{\infty} B_k z^k$$

and prove that

$$B_k = 0, \quad k > n. \tag{8.52}$$

In fact if we replace $s_n[\omega_2(z)]$ in (8.49) by

$$\sigma_n(z) = \sum_{k=0}^{n} B_k z^k$$

then the integral is unaffected. We deduce that

$$|A_n| \le \frac{1}{2\pi} \int_{|z|=1} |\sigma_n(z)|^2 |dz| = \sum_{k=0}^{n} |B_n|^2.$$

If (8.52) is false we deduce, letting r tend to 1 in (8.51) that

$$|A_n| < \sum_{k=0}^{\infty} |B_n|^2 \le \int_{|z|=1} |s_n(z)|^2 dz = n,$$

and this contradicts our assumption.

It follows from (8.52) that $\omega_2(z)$ is algebraic and so remains continuous in $|z| \le 1$. Also letting r tend to 1 in (8.49), (8.50) we deduce that

$$n = |A_n| \le \frac{1}{2\pi} \int_{|z|=1} |\omega_2(z)||s_n[\omega(z)]|^2 |dz| < \int_{|z|=1} |s_n(z)|^2 |dz| = n$$

unless

$$|\omega_2(z)| = 1 \quad \text{on} \quad |z| = 1, \tag{8.53}$$

so that (8.53) must hold. Thus $\omega_2(z)$ can be continued into the closed plane by reflection [C. A. pp. 172–3]. Either $\omega_2(z) = \eta_2 z^p$, where p is a positive integer and η_2 is a constant, $|\eta_2| = 1$; or else $\omega_2(z)$ has in $0 < |z| < 1$ at least one zero at $z = z_0$ of order q say. In the latter case $\omega_2(z)$ has a pole of order of q at $1/\bar{z}_0$ and

$$s_n\{\omega_2(z)\} = 1 + \omega_2 + \omega_2^2 + \cdots + \omega_2^{n-1}$$

has a pole of order qn at $1/\bar{z}_0$ and this contradicts (8.52). Thus $\omega_2(z) = \eta_2 z^p$ and we again obtain a contradiction from (8.52) unless $p = 1$. So $\omega_2(z) = \eta_2 z$. Thus

$$F(z) = \eta_2 \omega_1(z) \frac{z}{(1 - \eta_2 z)^2}.$$

We replace $F(z)$ by $F(z/\eta_2)$ and so assume that $\eta_2 = 1$. We also assume now $A_n = n$, since otherwise we may consider $e^{-i\lambda} F(z)$ instead of $F(z)$. Then, since $\omega_2(z) \equiv z$ we obtain from (8.49)

$$n = \frac{1}{2\pi i} \int_{|z|=r} \frac{\omega_1(z) s_n(z)^2 dz}{z^{n+1}}.$$

Since $|\omega_1(z)| < 1$, $\omega_1(z)$ has radial limits $\omega_1(e^{i\theta})$ for almost all θ on $|z| = 1$. (see e.g. Hayman [1989, Theorem 7.44, p. 515]). Thus, letting r tend to 1, we obtain

$$n = \frac{1}{2\pi} \int_{-\pi}^{\pi} \omega_1\left(e^{i\theta}\right) s_n\left(e^{i\theta}\right)^2 e^{-in\theta} d\theta.$$

Again by (8.49) and (8.50) with $A_n = n$, we have

$$n \leq \frac{1}{2\pi} \int_0^{2\pi} \left| s_n\left(e^{i\theta}\right) \right|^2 d\theta = n$$

and

$$n = \frac{1}{2\pi} \int_{-\pi}^{\pi} s_n\left(e^{i\theta}\right)^2 e^{-in\theta} d\theta.$$

Since $\left| \omega_1\left(e^{i\theta}\right) \right| \leq 1$, we deduce that for almost all θ

$$\omega_1\left(e^{i\theta}\right) s_n\left(e^{i\theta}\right)^2 e^{-in\theta} = s_n\left(e^{i\theta}\right)^2 e^{-in\theta} > 0$$

so that

$$\omega_1\left(e^{i\theta}\right) = 1.$$

Using Cauchy's integral formula

$$\omega_1(z) = \frac{1}{2\pi i} \int_{|\zeta|=r} \frac{\omega_1(\zeta)d\zeta}{\zeta - z}$$

and letting r tend to 1, we deduce that $\omega_1(z) \equiv 1$ in $|z| < 1$. This completes the case of equality in Theorem 8.7.

8.5.1 *Examples* We suppose that $f(z) \in \mathfrak{S}$ and define $k_\theta(z)$ by (8.45).

Examples

8.1   Why doesn't Theorem 8.5 follow from Theorem 8.2, when $\lambda = \frac{1}{2}$?

8.2   Prove that if $|z| < 1$ and p is a positive integer we have

$$\left|f^{(p)}(z)\right| \leq \frac{p!(p+r)}{(1-r)^{p+2}}, \quad |z| = r.$$

8.3   If λ is an even integer prove that

$$I_\lambda\left(r, f^{(p)}\right) \leq I_\lambda\left(r, k_\theta^{(p)}\right) ;$$

(use the fact that if $g(z) = \sum_0^\infty b_n z^n$ then

$$I_2(r, f) = \sum_0^\infty |b_n|^2 r^{2n}).$$

8.4   Prove that if $p = 1, 2, 3, \ldots$

$$\int_{|z| \leq r} \left|f^{(p)}\left(re^{i\phi}\right)\right|^2 r dr d\phi \leq \int_{|z| \leq r} \left|k_\theta^{(p)}\left(re^{i\phi}\right)\right|^2 r dr d\phi.$$

8.5   If $\lambda \geq 2$, prove that

$$I_\lambda(r, f) \leq I_\lambda(r, k_\theta).$$

Baernstein [1975] has shown that this result remains true for all positive λ and even more general means (see e.g. Hayman [1989, Theorem 9.8, p. 674]).

8.6   With the hypotheses of Theorem 8.7 prove that

$$I_2(r, F) \leq I_2(r, k_\theta), \quad 0 < r < 1.$$

8.7 If $S_\lambda(r,f) = r\frac{d}{dr}I_\lambda(r,f)$ prove that, for $\lambda \geq 2$

$$S_\lambda(r,f) \leq S_\lambda(r,k_\theta).$$

(Consider

$$\frac{1}{2\pi}\int_0^{2\pi}\left|\phi'\left(re^{i\theta}\right)\right|^2 d\theta,$$

where $\phi(z) = f(z)^{\frac{1}{2}\lambda}$ and apply Theorems 3.1 and 8.5.)

8.8 If $\lambda \geq \frac{1}{2}$ and $d_k(\lambda)$, $a_k(\lambda)$ are defined by (8.32), (8.43) respectively, prove that

$$\sum_{k=0}^{N}\frac{|a_{k+1}(\lambda)|^2}{d_k(2\lambda)} \leq d_N(2\lambda+1).$$

Bibliography

Books and papers listed here are referred to in the text by the authors and year of publication. When two papers are quoted by the same author in the same year, they are distinguished by a, b. The figures in square brackets denote the pages in the text where the work in question is referred to.

Ahlfors, L. V., Untersuchungen zur Theorie der konformen Abbildungen und der ganzen Funktionen, *Acta Soc. Sci. Fenn.* Nova Ser. A **1** no. 9. (1930) [28, 29].

Ahlfors, L. V., An extension of Schwarz's Lemma, *Trans. Amer. Math. Soc.* **43** (1938), 359–64 [136].

Ahlfors, L. V., *Conformal invariants: Topics in geometric function theory.* McGraw–Hill, New York 1973 [29].

Ahlfors, L. V., *Complex analysis*, 3rd edn. McGraw-Hill, New York (1979). This book is referred to in the text as C. A. [x, 3, 7, 11, 34, 44, 103, 104, 108, 110, 116, 129, 147, 198, 204, 252].

Ahlfors, L. V. & Grunsky, H., Über die Blochsche Konstante, *Math. Zeit.* **42** (1937), 671–73 [136].

Askey, R. & Gasper, G. Positive Jacobi sums, II, *Amer. J. Math.* **98** (1976), 709–37 [230].

Baernstein, A., Integral means, univalent functions and circular symmetrization. *Acta Math.* **133** (1975), 139–169 [11, 140, 254].

Baernstein, A., Coefficients of univalent functions with restricted maximum modulus. *Complex Variables* **5** (1986), 225–36 [67, 73, 79, 83, 96].

Bieberbach, L., Über die Koeffizienten derjenigen Potenzreihen, welche eine schlichte Abbildung des Einheitskreises vermitteln. *S. B. preuss. Akad. Wiss.* **138** (1916), 940–55 [1, 4, 230].

Biernacki, M., Sur les fonctions multivalentes d'ordre *p. C. R. Acad. Sci., Paris,* **203** (1936), 449–51 [67, 78].

Biernacki, M., Sur les fonctions en moyenne multivalentes. *Bull. Sci. Math.* (2) **70** (1946), 51–76 [144].

Bloch, A., Les théorèmes de M. Valiron sur les fonctions entières et la théorie de l'uniformisation. *Ann. Fac. Sci. Univ. Toulouse* (3) **17** (1926), 1–22 [136].

255

Bohr, H., Über einen Satz von Edmund Landau. *Scr. Bibl. Univ. Hierosolym* **1** no. 2 (1923) [135].

Bonk, M. On Bloch's constant. *Proc. Amer. Math. Soc.* **110** (1990), 889–94 [103, 136].

Burkill, J. C. *The Lebesgue Integral*, (Cambridge, 1951). This book is referred to in the text as B. [31, 32, 105].

Carleson, L. & Jones, P. W. On coefficient problems for univalent functions and conformal dimension. *Duke. Math J.* **66** (1992), 169–206 [82].

Cartwright, M. L., Some inequalities in the theory of functions. *Math. Ann.* **111** (1935), 98–118 [28, 29, 38].

Clunie, J. & Pommerenke, Ch., On the coefficients of univalent functions. *Michigan Math. J.* **14** (1967), 71–8 [78].

de Branges, L. A., Proof of the Bieberbach conjecture. *Acta Math.* **154** (1985), 137–52 [xi, 4, 230, 242].

Dieudonné, J., Sur les fonctions univalentes. *C. R. Acad. Sci., Paris,* **192** (1931), 1148–50 [13].

Dubinin, U. N., Symmetrization in the geometric theory of functions (Russian) Uspechi Mat Nauk **48** (1993) [133].

Duren, P. L., *Univalent functions*. Springer, New York (1983) [xi, 16, 161, 198, 217, 220, 229, 230, 250].

Dvoretzky, A., Bounds for the coefficients of univalent functions. *Proc. Amer. Math. Soc.* **1** (1950), 629–35 [94].

Eke, B. G., Remarks on Ahlfors' distortion theorem. *J. Analyse Math.* **19** (1967a), 97–134 [49].

Eke B. G., The asymptotic behaviour of areally mean valent functions. *J. Analyse Math.* **20** (1967b), 147–212 [29, 49, 65, 150, 187].

Ekhad, S. B. & Zeilberger, D. A., A high-school algebra, wallet-sized proof, of the Bieberbach conjecture [after L. Weinstein]. Preprint 1993 [235].

El Hosh, M. M., On successive coefficients of close–to–convex functions. *Proc. Roy. Soc. Edinburgh*, Sect. A **96** (1984), 47–9 [166].

Fekete, M & Szegö, G., Eine Bemerkung über ungerade schlichte Funktionen. *J. London Math. Soc.* **8** (1933), 85–9 [161, 217, 222].

Fitzgerald, C. H., Exponentiation of certain quadratic inequalities for schlicht functions. *Bull. Amer. Math. Soc.* **78** (1972), 209–10 [16].

Fitzgerald, C. & Pommerenke, Ch., The de Branges Theorem on univalent functions. *Trans. Amer. Math. Soc.* **290** (1985), 683–90 [230].

Flett, T. M., Note on a function theoretic identity. *J. London Math. Soc.* **29** (1954), 115–18 [78].

Garabedian, P. R. & Royden, H. A. L., The one–quarter theorem for mean univalent functions. *Ann. Math.* (2) **59** (1954) 316–24 [149].

Garabedian, P. R. & Schiffer, M., A proof of the Bieberbach conjecture for the fourth coefficient. *J. Rat. Mech. Anal.* **4** (1955), 427–65 [230].

Golusin, G. M., On distortion theorems of schlicht conformal mappings (Russian, German summary) *Rec. Math., Moscou* (2) **1** (1936), 127–135 [228].

Golusin, G. M., *Interior problems of the theory of schlicht functions*, translated by T. C. Doyle, A. C. Schaeffer and D. C. Spencer. Office of Naval Research, Washington (1947) [ix].

Goodman, A. W., On some determinants related to *p*–valent functions. *Trans. Amer. Math. Soc.* **63** (1948), 175–92 [xi, 162].

Goodman, A. W., An invitation to the study of univalent and multivalent functions. *Internat. J. Math. and Math. Sci.* **2** (1979), 163–86 [163].

Goodman, A. W. & Robertson, M. S., A class of multivalent functions. *Trans. Amer. Math. Soc.* **70** (1951), 127–36 [163].

Grinspan, A. Z., Improved bounds for the difference of the moduli of adjacent coefficients of univalent functions, in some questions in the Modern Theory of Functions (*Sib. Inst. Mat. Novosiborsk*, 1976, 41–5 (in Russian) [186].

Gronwall, T. H., Sur la déformation dans la représentation conforme. *C. R. Acad. Sci., Paris* **162** (1916), 249–52 [4, 227].

Grunsky, H., Neue Abschätzungen zur konformen Abbildung ein–und mehrfach zusammenhängender Bereiche. *Schr. Inst. angew. Math. Univ. Berlin* **1** (1932), 95–140 [224].

Grunsky, H., Zwei Bemerkungen zur konformen Abbildung. *Jber. dtsch. Math. Ver.* **43** (1933) 1. Abt. 140–3 [227].

Hamilton, D. H., The successive coefficients of univalent functions. *J. London Math. Soc.* (2) **25** (1982), 122–138 [187].

Hamilton, D. H., A sharp form of the Ahlfors' distortion theorem with applications. *Trans. Amer. Math. Soc.* **282** (1984), 799–806 [187].

Hardy, G. H., The mean value of the modulus of an analytic function. *Proc. London Math. Soc.* (2) **14** (1915), 269–77 [67].

Hayman, W. K., *Symmetrization in the theory of functions.* Tech. Rep. no. 11. Navy Contract N. 6–ori–106 Task Order 5, (Stanford University (1950). [127, 129, 145, 149].

Hayman, W. K., Some applications of the transfinite diameter to the theory of functions. *J. Anal. Math.* **1** (1951), 155–79 [103, 124, 128, 129, 141, 145].

Hayman, W. K., Functions with values in a given domain. *Proc. Amer. Math. Soc.* **3** (1952), 428–32 [94, 143].

Hayman, W. K., The asymptotic behaviour of p–valent functions. *Proc. London Math. Soc.* (3) **5** (1955), 257–84 [15, 65, 145, 150, 162].

Hayman, W. K., On successive coefficients of univalent function. *J. London Math Soc.* **38** (1963), 228–43 [166, 176, 186].

Hayman, W. K. Mean p–valent functions with gaps. *Colloq. Math.* **16** (1967), 1–21 [101].

Hayman, W. K., Mean p–valent functions with mini–gaps. *Math. Nachr.* **39** (1969), 313–24 [101].

Hayman, W. K., *Subharmonic functions. Vol. II.* Academic Press, London (1989) [11, 140, 143, 252, 254].

Hayman, W. K. & Hummel, J. A., Coefficients of powers of univalent functions. *Complex Variables* **7** (1986), 51–70 [222].

Hayman, W. K. & Kennedy, P. B., On the growth of multivalent functions. *J. London Math. Soc.* **33** (1958), 331–41 [46].

Hayman, W. K. & Kennedy, P. B., *Subharmonic functions. Vol. I.* Academic Press, London (1976), [250].

Hayman, W. K. & Nicholls, P. J., On the minimum modulus of functions with given coefficients. *Bull. London Math. Soc.* **5** (1973), 295–301 [33].

Heins, M., Entire functions with bounded minimum modulus subharmonic function analogues. *Ann. of Math.* (2) **49** (1948), 200–13 [49].

Heins, M., On a class of conformal metrics. *Nagoya Math. J.* **21** (1962), 1–60 [136].

Hempel, J., The Poincaré metric on the twice punctured plane and the theorems of Landau and Schottky. *J. London Math. Soc.* (2) **20** (1979), 435–45 [140, 143].

Jenkins, J. A., Some uniqueness results in the theory of symmetrization. *Ann. Math.* (2) **61** (1955), 106–15 [126, 128].

Jenkins, J. A., *Univalent functions and conformal mappings.* Springer, Berlin (1958) [x, 162, 163, 198, 229].

Jenkins, J. A. & Oikawa, K., On results of Ahlfors and Hayman. *Ill. J. of Math.* **15** (1971), 664–71 [33, 38].

Kaplan, W., Close-to-convex schlicht functions. *Michigan Math. J.* **1** (1953), 169–85 [14].

Koebe, P., Über die Uniformisierung der algebraischen Kurven. II. *Math. Ann.* **69** (1910), 1–81 [1].

Lai, W. T., The exact value of Hayman's constant in Landau's Theorem. *Scientia Sinica* **22** (1979), 129–34 [140, 143].

Landau, E., Über eine Verallgemeinerung des Picardschen Satzes. *S. B. preuss. Akad. Wiss.* (1904), 1118–33 [140].

Landau, E., Zum Koebeschen Verzerrungssatz. *R. C. Circ. mat. Palermo,* **46** (1922), 347–48 [33, 132].

Landau, E., Über die Blochsche Konstante und zwei verwandte Weltkonstanten. *Math. Z.* **30** (1929), 608–34 [217].

Lelong-Ferrand, J., *Représentation conforme et transformations à intégrale de Dirichlet bornée* (Paris, 1955) [29].

Leung, Y., Successive coefficients of starlike univalent functions. *Bull. London Math. Soc.* **10** (1978), 193–6 [188].

Levin, V. I., Ein Beitrag zum Koeffizientenproblem der schlichten Funktionen. *Math. Z.* **38** (1934) [96].

Littlewood, J. E., On inequalities in the theory of functions. *Proc. London Math. Soc.* (2) **23** (1925), 481–519 [9, 67, 96].

Littlewood, J. E., On the coefficients of schlicht functions. *Quart. J. Math.* **9** (1938), 14–20 [96].

Littlewood, J. E. & Paley, R. E. A. C., A proof that an odd schlicht function has bounded coefficients. *J. London Math Soc.* **7** (1932), 167–9 [67, 96].

Livingston, A. E., *p*–valent close-to-convex functions. *Trans. Amer. Math. Soc.* **115** (1965), 161–79 [163].

Livingston, A. E., The coefficients of multivalent close-to-convex functions. *Proc. Amer. Math Soc.* **21** (1969), 545–52 [163].

Löwner, K., Untersuchungen über die Verzerrung bei konformen Abbildungen des Einheitskreises $|z| < 1$, die durch Funktionen mit nicht verschwindender Ableitung geliefert werden. *Leipzig Ber.* **69** (1917), 89–106 [11].

Löwner, K., Untersuchungen über schlichte konforme Abbildungen des Einheitskreises. I. *Math. Ann.* **89** (1923), 103–21 [4, 197, 215, 237].

Lucas, K. W., A two-point modulus bound for areally mean *p*-valent functions. *J. London Math. Soc.* **43** (1968), 487–94 [176].

Lucas, K. W. On successive coefficients of areally mean *p*-valent functions. *J. London Math. Soc.* **44** (1969), 631–42 [166, 185].

MacGregor, T. H., An inequality concerning analytic functions with a positive real part. *Canad. J. Math.* **21** (1969), 1172–7 [189].

Milin, I. M., Hayman's regularity theorem for the coefficients of univalent functions. *Dokl. Akad. Nauk. S. S. R.* **192** (1970), 738–41 (in Russian) [15].

Milin, I. M., Univalent functions and orthonormal systems, *Izdat. Nauka, Moscow* 1971, (Russian) [190, 230, 243, 246].

Milin, I. M., L. de Branges' proof of the Bieberbach conjecture. *Lomi preprint* 1984, (Russian) [230].

Montel, P., *Leçons sur les fonctions univalentes ou multivalentes*. (Paris, 1933) [ix].

Netanyahu, E., The minimal distance of the image boundary from the origin and the second coefficient of a univalent function in $|z| < 1$. *Arch. Rational Mech. Anal.* **32** (1969), 100–12 [142].

Nevanlinna, R., Über die konforme Abbildung von Sterngebieten. *Öfvers. Finska Vet. Soc. Förh.* **53** (A) (1921), Nr. 6 [14].

Ozawa, M., On the Bieberbach conjecture for the sixth coefficient. *Kodai Math. Sem. Rep.* **21** (1969), 97–128 [230].

Pedersen, R. N., A proof of the Bieberbach conjecture for the sixth coefficient. *Arch. Rational Mech. Anal.* **31** (1968), 331–51 [230].

Pedersen, R. N. & Schiffer M., A proof of the Bieberbach conjecture for the fifth coefficient. *Arch. Rational Mech. Anal.* **45** (1972), 161–93 [230].

Pólya, G., Sur la symétrisation circulaire. *C. R. Acad. Sci., Paris*, **230** (1950), 25–7 [113].

Pólya, G. & Szegö, G., *Isoperimetric inequalities in mathematical physics*. (Princeton, 1951) [103, 112, 119, 125].

Pommerenke, Ch., Über die Mittelwerte und Koeffizienten multivalenter Funktionen. *Math. Ann.* **145** (1961/62), 285–96 [73].

Pommerenke, Ch., Relations between the coefficients of a univalent function. *Inventiones Math.* **3** (1967), 1–15 [96].

Pommerenke, Ch., *Univalent functions*. Göttingen, Vandenhoek and Ruprecht (1975) [xi, 16, 83, 96, 143].

Pommerenke, Ch., On the integral means of the derivative of a univalent function. *J. London math. Soc.* (2) **32** (1985a), 254–8 [79].

Pommerenke, Ch., On the integral means of the derivative of a univalent function II. *Bull. London Math. Soc.* **17** (1985b), 565–70 [81].

Reade, M. O., On close–to–convex univalent functions. *Michigan Math. J.* **3** (1955), 59–62 [14].

Robertson, M. S., A remark on the odd schlicht functions. *Bull. Amer. Math. Soc.* **42** (1936), 366–70 [231, 249].

Robertson, M. S., Multivalent functions of order *p*. *Bull. Amer. Math. Soc.* **44** (1938), 282–5 [96].

Robertson, M. S., Quasi-subordination and coefficient conjectures. *Bull. Amer. Math. Soc.* **76** (1970), 1–9 [249].

Rogosinski, W. W., Über positive harmonische Sinusentwicklungen. *Jber. dtsch. Math. Ver.* **40** (1931), 2. Abt. 33–5 [13].

Rogosinski, W. W., On the coefficients of subordinate functions. *Proc. London Math. Soc.* **48** (1943), 48–82 [231, 249, 250].

Schaeffer, A. C. & Spencer, D. C., The coefficients of schlicht functions. *Duke Math. J.* **10** (1943), 611–35 [161].

Schaeffer, A. C. & Spencer, D. C., *Coefficient regions for schlicht functions*. (New York, 1950) [x].

Schiffer, M., Variation of the Green function and theory of the *p*–valued functions. *Amer. J. Math.* **65** (1943), 341–60 [229].

Schottky, F., Über den Picardschen Satz und die Borelschen Ungleichungen. *S. B. preuss. Akad. Wiss.* (1904), 1244–63 [140].

Shen, D., On successive coefficients of univalent functions. *Proc. Roy. Soc. Edinburgh* **120A** (1992), 153–64 [167, 191, 192].

Spencer, D. C., Note on some function-theoretic identities. *J. London Math. Soc.* **15** (1940a), 84–6 [67].

260   *Bibliography*

Spencer, D. C., On finitely mean valent functions II. *Trans. Amer. Math. Soc.* **48** (1940b), 418–35 [28, 29, 37, 38, 42, 67, 96].

Spencer, D. C., On finitely mean valent functions. *Proc. London Math. Soc.* (2) **47** (1941a), 201–11 [37, 67, 69, 78, 96].

Spencer, D. C., On mean one-valent functions. *Ann. Math.* (2) **42** (1941b), 614–33 [149, 163].

Stein, P., On a theorem of M. Riesz. *J. London Math. Soc.* **8** (1933), 242–7 [67].

Szász, O., Über Funktionen die den Einheitskreis schlicht abbilden. *Jber. dtsch. Math. Ver.* **42** (1932), 1. Abt. 73–5 [13].

Szegö, G., Zur Theorie der schlichten Abbildungen. *Math. Ann.* **100** (1928), 188–211 [4].

Titchmarsh, F. C., *The theory of functions, 2nd ed.* (Oxford, 1939), [x, 21, 26, 30, 81, 151, 213].

Weinstein, L., The Bieberbach conjecture. International Mathematics Research Notices, 1991 No. 5, *Duke Math J.* **64** (1991), 61–4 [230, 237].

Weitsman, A., Symmetrization and the Poincaré metric. *Ann. Math.* **124** (1986), 159–69 [140].

Whittaker, E. T. & Watson, G. N., *A course of modern analysis.* 4th ed. Cambridge University Press (1946) [231].

Wilf, H. S., A footnote on two proofs of the Bieberbach–de Branges Theorem. *Bull. London Math. Soc.* 1993 [231].

Index

This index contains page references to topics of major importance and to terminology: usually when the topic or symbol is first used, or is used in a new sense, or features in a major theorem.

261